Hans Bisswanger

Practical Enzymology

Related Titles

Bommarius, A. S., Riebel-Bommarius, B. R.

Biocatalysis

A Textbook

2011

ISBN: 978-3-527-31868-1

Fessner, W.-D., Anthonsen, T. (eds.)

Modern Biocatalysis

Stereoselective and Environmentally Friendly Reactions

2009

ISBN: 978-3-527-32071-4

Bisswanger, H.

Enzyme Kinetics

Principles and Methods

2008

ISBN: 978-3-527-31957-2

Aehle, W. (ed.)

Enzymes in Industry

Production and Applications

2007

ISBN: 978-3-527-31689-2

Hans Bisswanger

Practical Enzymology

Second, Completely Revised Edition

The Author

Prof. Dr. Hans Bisswanger
Interfakultäres Institut für Biochemie
Hoppe-Seyler-Str. 4
72076 Tübingen
Germany

Cover
Laborbild mit freundlicher
Genehmigung von Fotolia/Sven Hoppe
© Sven Hoppe-Fotolia.com

Library of Congress Card No.: applied for

British Library Cataloguing-in-Publication Data
A catalogue record for this book is available from the British Library.

Bibliographic information published by the Deutsche Nationalbibliothek
The Deutsche Nationalbibliothek lists this publication in the Deutsche Nationalbibliografie; detailed bibliographic data are available on the Internet at <http://dnb.d-nb.de>.

Wiley-Blackwell is an imprint of John Wiley & Sons, formed by the merger of Wiley's global Scientific, Technical, and Medical business with Blackwell Publishing.

Composition Laserwords Private Ltd., Chennai, India
Printing and Binding betz-druck GmbH, Darmstadt
Cover Design Grafik-Design Schulz, Fußgönheim

Printed in Federal Republic of Germany
Printed on acid-free paper

ISBN: 978-3-527-32076-9

Contents

Preface to the Second Edition

The principal concept of the first edition: general aspects of enzymes and presentation of special enzyme assays and related tests such as protein determination and enzyme immobilization, as well as instrumental aspects, remain conserved in the second edition. Additional enzyme assays and tests, for example, for the determination of glycoproteins and inorganic phosphate, have been included, considering the principles of broad interest and diversity of methods. Features of the enzymes, which are of importance for the assay conditions, such as cofactor requirement, molecular mass, state of aggregation, kinetic constants, and pH optimum, are indicated, but it is not intended to present all known data. Actually, an overwhelming quantity of data accumulated within the last years, owing to the violent progress in gene technology, and features of the same enzyme species can differ extremely, depending on the organisms from where it originates. For practical treatment of an enzyme assay, such a variety of data is more disturbing than helpful; therefore, only the features of one or few representative enzyme species, preferentially from human or mammalian origin, are mentioned. Particular attention has been drawn to frequent pitfalls and error detection for the methods described.

Enzymes applied for molecular biology and gene technology, such as restriction enzymes and polymerases, are not taken into consideration. They are extensively described in text books and manuals of the respective field, but the general rules for handling and for the assay conditions are also valid for these enzymes. This holds also for RNA enzymes (ribozymes), antibody enzymes (antizymes), and artificial enzymes derived, for example, from cyclodextrans or crown ethers, and for enzymes modified by site-specific mutations.

The layout has been improved, for example, by introduction of colors; the structure of text is clearer; and the essential points of the sections are summarized in separated boxes. A companion web site (*www.wiley-vch.de/home/enzymology*) provides animations for all figures together with supplementary material, for deeper understanding of the partially abstract matter.

Especially emphasized are the valuable contributions of Klaus Möschel and Rainer Figura to the chapter of immobilized enzymes.

Tübingen, January 2011 *Hans Bisswanger*

Note to the Reader

Animations should assist the comprehension of the text. They are principally self-explaining, but knowledge of the corresponding book chapter and the respective figure legend is presupposed. Most animations are subdivided into sequential steps, which are initiated by the cursor button (→) or a mouse click. A green arrow at the bottom of each figure gives the signal for pressing, during the animation run it disappears. 'X' indicates the end of the animation before passing to the next figure.

Abbreviations[1)]

A, B, C	specific binding ligands
{A]	ligand concentration
$\{A]_0$	total ligand concentration
A	absorption
ABTS	2,2′-azino-bis-3-ethylbenzothiazoline-6-sulfonic acid
ADH	alcohol dehydrogenase
ANS	anilinonaphthalene sulfonate
BAPNA	*N*′-benzoyl-L-arginine-*p*-nitroanilide
BCA	bicinchoninic acid
BSA	bovine serum albumine
CDI	carbonyldiimidazol
CoA	coenzyme A
CPG	controlled pore glass
d	density
DMSO	dimethylsulfoxide
DPIP	2,6-dichlorophenolindophenol
DTE	dithioerythritol
DTNB	5,5′-dithio-bis(2-nitrobenzoic acid), Ellman's reagent
DTT	dithiothreitol, cleland's reagent
E, [E]	enzyme, enzyme concentration
$\{E]_0$	total enzyme concentration
ε_{nm}	absorption ("extinction") coefficient at the wavelength indicated
ε_r	dielectric constant
EDTA	ethylenediaminetetraacetic acid
EIA	enzyme immunoassays
ELISA	enzyme-linked immunoadsorbent assays
FAD	flavine adenine dinucleotide
FMN	flavine mononucleotide
GOD	glucose oxidase

1) Only repeatedly used abbreviations, special abbreviations are defined at the respective section.

h	Planck's constant
HK	hexokinase
I	light intensity
IU	International enzyme unit (μmol min^{-1})
k	rate constant
kat	Katal (mol s^{-1})
k_{cat}	catalytical constant
K_d	dissociation constant
K_m	Michaelis constant
LDH	lactate dehydrogenase
MDH	malate dehydrogenase
M_r	relative molecule mass
n	number of subunits
NAD	nicotinamide adenine dinucleotide[2)]
NADH	nicotinamide adenine dinucleotide[2)]
NADP	nicotinamide adenine dinucleotide phosphate[2)]
NADPH	nicotinamide adenine dinucleotide phosphate[2)]
ONPG	*o*-nitrophenyl *β*-D-galactopyranoside
ORD	optical rotatory dispersion
P, Q, R	products
PAGE	polyacrylamide gel electrophoresis
PBS	phosphate buffered saline
PEG	polyethylene glycol
PLP	pyridoxal 5-phosphate
PMSF	phenylmethylsulfonyl fluoride
POD	peroxidase
R	Gas constant (8.3145 J mol^{-1} K^{-1})
RIA	radioimmunoassay
RN	recommended name
rpm	rotations per min
RT	room temperature
S	substrate
SA	specific enzyme activity
SDS	sodium dodecyl sulfate
SN	systematic name
ThDP	thiamine diphosphate
TCA	trichloroacetic acid
TRIS	tris(hydroxymethyl)aminomethane
v	reaction velocity
V	maximum reaction velocity
v_i	initial reaction velocity

[2)] For simplicity the charge ($NAD(P)^+$), for the reduced form the free proton ($NAD(P)H + H^+$) is omitted.

1
Introduction

Enzymes are the most important catalysts and regulators indispensably involved in each process in living organisms. Any investigation of the cell metabolism requires a thorough understanding of enzyme action. Enzymes are very sensitive markers for correct function and, consequently, also for dysfunction of the metabolism, serving as indicators both for health and manifestation of diseases. Accordingly, they are used as invaluable tools in medical diagnostics. Beyond that, enzymes are applied in many technical operations. They play an essential role in the environmental processes in the microbial world in waters, rivers, lakes, and soil, and are important for filter plants as well as for fermentation procedures in dairies and breweries.

According to current estimates, about 25 000 enzymes are expected to exist in the living world, where more than 3000 are described in detail, and some hundreds are commercially available. Enzymes are extremely efficient catalysts, enhancing the turnover rates of spontaneous reactions by factors between 10^8 and 10^{10}, sometimes even up to 10^{12} (Menger, 1993). Orotidine-5′-phosphate decarboxylase is a striking example: the spontaneous reaction proceeds with a half-life of 78 million years and the enzyme increases the velocity by a factor of 10^{17} (Radzicka and Wolfenden, 1995). Triosephosphate isomerase accelerates the enolization of dihydroxyacetone phosphate by more than 10^9 (Alberty and Knowles, 1976).

Even reactions spontaneously proceeding with a considerable rate, such as the formation of water from hydrogen and oxygen in the respiratory chain, are subject to enzyme catalysis: each reaction step in metabolism is controlled by a special enzyme. Thus the role of enzymes in the metabolism is broader than to act only as biocatalysts. The peculiarity of catalysis is not only restricted to the acceleration of spontaneous reactions, but it also allows controlling reactions. Spontaneous reactions, after initiating, run off to the end and cannot be stopped. Catalyzed reactions, in contrast, proceed only in the presence of the catalyst; its activity and amount determines the reaction rate. Consequently, tuning the activity of an enzyme from the outside by activating or inhibiting mediates an exact control of the velocity. In the living cell, a strictly coordinate network of regulation exists, comprising enzymes whose activity is controlled by the concentration levels of metabolites, hormones, and transmitter substances. The precise interaction of all these components is a prerequisite of life.

Practical Enzymology, Second Edition. Hans Bisswanger.

The protein nature of enzymes is excellently suited for this dual function as catalyst and regulator; it supplies functional groups of amino acids to form specific binding sites and catalytic centers, and it provides flexibility to promote formation and stabilization of transition states and to induce conformational changes for modulation of the catalytic efficiency. The 20 proteinogenic amino acids with their hydrophilic, hydrophobic, acidic, and basic side chains permit most enzymes to realize both functions such as specific binding of substrates and regulator molecules and catalytic conversion. More difficult catalytic mechanisms cannot be brought forth only by the amino acid side chains; rather, nonproteinogenic compounds are included, which can either be dissociable as *coenzymes*,[1] or nondissociable as *prosthetic groups*. Dissociable coenzymes are NAD(P), thiamine diphosphate, or coenzyme A, while FAD, cytochromes, porphyrins, pyridoxamine, lipoic acid, biotin, and tetrahydrofolic acid function as nondissociable, partly covalently bound prosthetic groups. Often also metal ions are required, both for catalysis and for stability of the enzyme, Mg^{2+} serves to neutralize the phosphate groups in compounds such as ADP, ATP, and thiamine diphosphate and mediates their binding to the enzyme. Iron (in cytochromes), cobalt (in the corrin ring system), copper (e.g., in cytochrome oxidase and tyrosinase), zinc (in carboanhydase and alcohol dehydrogenase), molybdenum (in nitrogenase), manganese (in arginase and xylose/glucose isomerase), and selenium ion (in glutathion peroxidase) support the enzyme reactions.

The protein nature enables enzymes to adapt their specificity to any desired ligand by mutations. This feature is applied in biotechnology using *site-directed mutagenesis* to modify the specificity and function of enzymes. By the method of *molecular modeling* (protein design) distinct modifications are simulated and thereafter the respective mutations are executed. An example is hydroxyisocaproate dehydrogenase, an enzyme catalyzing the reductive conversion of α-oxo acids to chiral hydroxycarbonic acids as hydroxyanalogs of amino acids. Its preferred substrate is α-oxocaproic acid. α-Isocaproic acid, an analogous compound, is accepted only with reduced efficiency. By site-directed mutagenesis the catalytic efficiency (k_{cat}/K_m) for this compound has been increased by four orders of magnitude, as compared to the physiological substrate (Feil, Lerch, and Schomburg, 1994).

Owing to their protein nature, enzymes are very sensitive to environmental influences such as pH, ionic strength, and temperature and, consequently, to attain optimum activity, stringent conditions must be established. In the physiological milieu of the living cell, these conditions are maintained as far as possible, although with respect to temperature, this cannot be permanently guaranteed (with the exception of warm-blooded vertebrates). However, enzymes are remarkably

1) The terms *coenzyme* and *cosubstrate* are not always clearly differentiated. Coenzymes, in contradistinction to cosubstrates, are supposed to support the catalytic mechanism and should not be converted. For example, pyridoxal phosphate in transamination reactions accepts an amino residue becoming pyridoxamine phosphate, but in the second step of the reaction the amino group is transferred to an α-oxoacid and the coenzyme regains its original form at the end of the reaction. NAD(P), on the other hand, is reduced in a dehydrogenase reaction and must be reoxidized by a separate enzyme reaction, therefore it is more a cosubstrate than a coenzyme.

able to adapt to extreme conditions. Although proteins are regarded as being very temperature sensitive, distinct microorganisms such as *Thermus, Thermotoga,* and *Thermoplasma,* including their complete enzymatic inventory, persist in temperatures up to 100 °C. It must be assumed that during evolution the ancient organisms have had to bear much higher temperatures. The ancient precursors of the present enzymes must have all been thermophilic, but obviously they lost this feature with the decrease in environmental temperature. This can also explain the fact that proteins, instead of the more stable nucleic acids, are preferred by nature as biocatalysts, although some catalytic activities are retained in RNA.

As an introduction to the practical work with enzymes, at least some fundamental theoretical rules must be discussed. They will be addressed in the first part, followed by a description of the general features of enzymes, which must be considered when dealing with them. This is followed by a presentation of the most important techniques. This general part should enable the reader to work with enzymes; for instance, to develop an assay for a newly isolated enzyme without further need to consult the literature. The following special part presents detailed descriptions of enzyme assays and related methods such as protein determination. A multitude of assays corresponding to the immense number of different enzymes exists, which cannot all be considered within the scope of a laboratory manual; rather, only a selection can be presented. Criteria for the selection are not only the frequency of application, but also the broad variety of enzyme types and methods. Certainly, such a selection cannot satisfy all expectations and the choice will sometimes appear rather arbitrary. For further information the reader is referred to standard books and databases of enzymology (see References section below). Procedures for immobilization of enzymes and special aspects of analysis of immobilized enzymes, principles of enzyme reactors, and enzyme electrodes are presented in separate sections of the book.

References

Alberty, W.J. and Knowles, J.R. (1976) *Biochemistry*, **15**, 5631–5640.

Feil, I.K., Lerch, H.P., and Schomburg, D. (1994) *Eur. J. Biochem.*, **223**, 857–863.

Menger, F.M. (1993) *Acc. Chem. Res.*, **26**, 206–212.

Radzicka, A. and Wolfenden, R. (1995) *Science*, **267**, 90–93.

Standard Books, Series, and Databases

Advances in Enzymology and Related Areas of Molecular Biology, John Wiley & Sons, Inc., New York.

Bergmeyer, H.U. (ed.) (1983) *Methods of Enzymatic Analysis*, 3rd edn, Verlag Chemie, Weinheim.

Methods in Enzymology, Academic Press, San Diego.

Schomburg, D. (ed.) (2001f) *Springer Handbook of Enzymes*, Springer, Berlin.

www.expasy.org/enzymes.

www.brenda-enzymes.org.

2
General Aspects of Enzyme Analysis

2.1
Basic Requirements for Enzyme Assays

The task of enzymes as biocatalysts is to render feasible reactions, which cannot proceed in their absence. Therefore, the first requirement when dealing with a special enzyme is to study its reaction. In a simple generalization it can be stated that one compound (or more than one), designated as the *substrate*, gets converted into another compound, the *product*, with the aid of the enzyme. Thus, to identify an enzyme its reaction must be demonstrable or measurable, that is, a method must be developed for the quantitative detection of the reacting components. The prerequisite for this is a detectable *signal* for the reacting components. But a signal alone is not sufficient; rather, a clear distinction between the substrate and the product is necessary. Absorption ultraviolet and visible (UV/Vis) spectroscopy may be an illustrative example. It is an easy and convenient detection method and in fact, each biological substance shows absorption at least in the UV region. Therefore, this method is principally suited for the quantification of every substance and, thus, may be the method of choice for any enzyme assay. However, in most cases, the substrate and the product of the same reaction show similar absorption features. So, even if the compounds possess pronounced absorption spectra, they are not useful to detect the reaction. This is the case with sugars, such as glucose and fructose, which cannot be distinguished by absorption spectroscopy and so this method is not applicable here.

Hence, the first step is to find a clear signal for detection of the substrate and/or the product, and the second step is to uncover differences between both compounds. The compound showing the clearer difference signal will be used. Principally, this is irrespective of whether the substrate or the product will be detected, as it can be assumed, that the amount of substrate converted corresponds exactly to the amount of product formed. Observing the decay of the substrate or the formation of the product must give the same result, only changing the sign. However, if possible, product formation is preferred because of practical reasons. At the start of the reaction the product concentration and, consequently, its signal, is zero and any increase is a direct indication of the progressing reaction. Conversely, the concentration of the substrate and thus its signal is highest at

Practical Enzymology, Second Edition. Hans Bisswanger.

the beginning of the reaction. This can influence the detection method. Each method shows some scatter and usually higher signals cause stronger scatter, and small changes, for example in the case of slow reaction rates are difficult to detect. Some substrates are unstable and decay spontaneously appearing to be an enzyme-catalyzed reaction; an effect, which is also more pronounced at high concentrations.

Various methods are available to search for an appropriate difference signal for an enzyme assay. Any analytic method for identifying substances such as the substrate and the product can be considered and usually the method yielding the clearest difference signal will be chosen; however, other criteria must also be considered. A simple, but practical aspect, is the availability of an appropriate instrument. For enzyme analysis, frequent assays series must be performed and the appropriate instrument should be permanently accessible. Accordingly, it must be affordable and handling should be easy. Such demands limit the kind of methods that can be employed. One important aspect concerns the mode of registration. As discussed in detail later, a progressive reaction should be pursued continuously (**continuous assay**) as far as possible, while various methods allow only detection of single points of the reaction after defined time periods (**stopped assay**, Box 2.1). In fact, often a method enabling continuous registration is superior to a method allowing only stopped assays even if a weaker signal must be accepted.

Box 2.1: Fundamental Demands for Enzyme Assays

- *Product* formation (increasing reaction) or *substrate* consumption (decreasing reaction) can be detected, alternatively
- *Enzyme activity* is defined as the amount of product formed – substrate consumed – within a distinct time unit (second, minute, hour). During this time unit the reaction must proceed *strictly linearly* (zero order)
- *Continuous* monitoring of the reaction (e.g., spectral change, pH change)
- *Stopped assay* (if continuous monitoring is not possible): stop of the reaction after a defined time and subsequent analysis of the amount of product formed – substrate consumed – within the time unit

The most frequently used methods for enzyme assays are summarized in Box 2.2. All spectroscopic methods (absorption, fluorescence, circular dichroism (CD), optical rotatory dispersion (ORD), turbidimetry) and electrochemical techniques such as pH stat, can principally be performed as a continuous test (but only if the substrate or the product can be identified directly), while trapping and separation methods allow only stopped assays. Therefore, such methods will be used only if the other methods do not work. For routine assays, simple devices and instruments belonging to the standard equipment of an analytical

Box 2.2: How to Determine the Enzyme Activity?

$$E + A \underset{k_{-1}}{\overset{k_1}{\rightleftharpoons}} EA \xrightarrow{k_2} E + P$$

Substrate A and product P must differ in at least one detectable feature.

Feature	Detection method	Examples
Absorption spectrum	UV/Vis spectroscopy	NAD/NADH
Fluorescence spectrum	Fluorescence spectroscopy	Umbelliferone-coupled substrates
Formation of ATP or NAD(P)H	Luminometry	Kinases, dehydrogenases
Optical rotation	Polarimetry, ORD, CD	Sugars (glucose)
pH	pH stat	Cleavage of triglycerides
Gas release or consumption	O_2 and CO_2 electrodes, manometry	Decarboxylase reaction
Turbidity	Turbidimetry	Degradation of starch
Chemical reactivity	Trapping reactions, colorimetry	Peroxidase reaction with dianisidin
General features (size, polarity)	Separation methods, HPLC, FPLC	Aggregation, depolymerization (cellulose, starch)

or a biochemical laboratory are preferred. These criteria are fulfilled in the best manner by absorption (UV/Vis) spectroscopy, which is an easy method with various applications, for example, determinations of proteins, nucleic acids, and phosphate (cf. Box 2.15). Suitable apparatus are available at moderate prices and computer-controlled instruments with monitoring and calculation modes make the evaluation of reactions easy so that photometric enzyme assays are the first choice. Other spectroscopic methods, such as fluorescence, CD, and ORD are for special applications and thus usually not present as standard equipment in laboratories. Manipulation is more difficult, intense knowledge for appropriate operation is required, and high-quality instruments are rather expensive; all these are aspects not supporting their application. Nevertheless, these spectroscopic methods possess significant advantages. They are more selective and, especially for fluorescence, much more sensitive compared with absorption spectroscopy, a feature important especially for enzyme studies. If such an instrument is not available, the work must be done in specialized laboratories, which usually provide not only the appropriate apparatus, but also a thorough knowledge of the technique, indispensable to avoid inappropriate procedures and misinterpretations. These considerations hold also for other instruments such as the pH stat, a very useful device, if demanded by the type of assay, such as the digestion of lipids, but superfluous if not really required. The main instruments applied for enzyme tests are described in detail in Section 2.3.

2.2
What Must Be Observed for an Enzyme Assay?

As already mentioned, for an enzyme assay the progression of the reaction (**progress curve**) is decisive and should be carefully observed. For normal enzyme reactions this curve should obey a common pattern, that is, it should be a straight line reflecting linear progression of the reaction proportional to time. In reality, however, nonlinear behavior is often observed, which is frequently a smooth curvature, but sometimes even irregular deviations. Evaluation of such behavior requires some knowledge of the theoretical background; the essential rules are discussed in the following chapter.

2.2.1
Order of Reactions

The progression of a reaction is determined by its order. The simplest chemical reaction is the conversion of a substance A (in chemical terms: *educt*) into the product P, as the spontaneous decomposition of instable substances, for example, the radioactive decay:

$$\mathrm{A} \longrightarrow \mathrm{P}$$

The velocity v of this reaction depends on the initial concentration of A and is expressed as:

$$-\frac{\mathrm{d[A]}}{\mathrm{d}t} = \frac{\mathrm{d[P]}}{\mathrm{d}t} = k[\mathrm{A}] = v \tag{2.1}$$

t is the time, k the rate constant with the dimension of s^{-1}. It is obvious that the higher the amount of A, the faster the reaction. Because A decays during the reaction, the velocity declines permanently, and the reaction follows a curve, which is steepest at the start and decreases steadily (Figure 2.1a). A similar curve, only in a positive sense, is obtained, when the formation of P is observed. Mathematically, this curve is described by an exponential relationship ($[A]_0$ is the initial substrate concentration):

$$[\mathrm{A}] = [\mathrm{A}]_0 \mathrm{e}^{-kt} \tag{2.2}$$

This is the equation for a **first-order reaction**, because only one substrate is involved. Hence, an exponential curve is indicative of a first-order reaction. However, an exponential progression of a reaction is not easy to recognize unequivocally, because other reaction types (higher orders) show similar nonlinear curves. Although they follow no simple exponential relationship, in practice they are often difficult to discern from real exponential curves. In such cases of ambiguity it is a good principle to transform the nonlinear relationship into a linear form, where only dependencies obeying the original relationship will yield straight lines, while others show characteristic deviations. By transformation of the first-order equation (2.2)

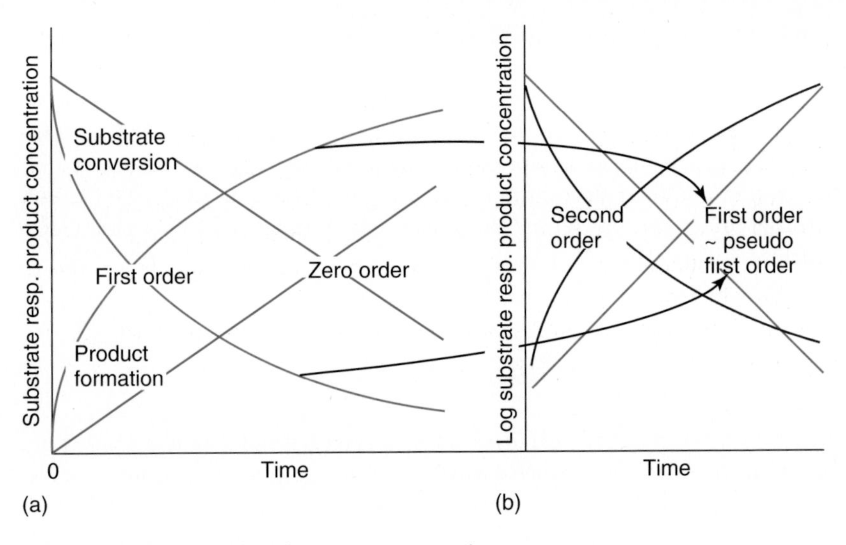

Figure 2.1 Progress curves of various reaction orders. (a) Direct plotting and (b) semilogarithmic plotting.

into a half logarithmic form

$$\ln[\mathrm{A}] = \ln[\mathrm{A}]_0 - k_1 t \quad (2.3)$$

the curves in Figure 2.1a become linear if a logarithmic ordinate scale is chosen (Figure 2.1b).

In nature, spontaneous decays are rather seldom.[1] More frequent are reactions initiated by collisions of two or more reactive substances. The number of substrates involved determines the reaction order,[2] for example, a reaction:

$$\mathrm{A} + \mathrm{B} \longrightarrow \mathrm{P} + \mathrm{Q}$$

is of **second order**.[3] The velocity of a second-order reaction depends on two variables. As shown in Figure 2.1a, nonlinear behavior is observed and no straight line results in the half logarithmic plot (Figure 2.1b). This feature allows the distinction of first and second orders (and similarly for higher orders, which are not dealt with here). When performing experiments, the dependency of the second-order reaction on two independent variables is impracticable. Under normal conditions both substrates may be present in comparable amounts, but this is not a necessary condition. If we assume that one component (e.g., B)

1) With the exception of radioactive decay, it can be assumed that spontaneous reactions already came to their end.
2) Substrates are written in alphabetic order A, B, C . . . , products as P, Q, R . . . .
3) The reaction order is only defined by the number of substrates; the number of products formed is only of significance if the reverse reaction is considered. Accordingly, the second order reaction can also be written as $\mathrm{A} + \mathrm{B} \rightarrow \mathrm{P}$ or $\mathrm{A} + \mathrm{B} \rightarrow \mathrm{P} + \mathrm{Q} + \mathrm{R}$.

is present in surplus in comparison with the other one (A), conversion of the lesser amount of A will not essentially change the higher amount of B, so that its concentration can be considered as constant. Under this condition, the reaction depends only on one, the minor component (A), and becomes similar to a first-order reaction, following an exponential time course, which now becomes linear in the half logarithmic plot (Figure 2.1b). As this reaction is only formally first order, but second order in reality, it is designated as **pseudo-first order**.

2.2.2
Significance of the Reaction Order for Enzyme Reactions

It has already been mentioned that enzyme reactions should, ideally, proceed in a linear manner, while we now see that the simplest chemical reaction, the first-order reaction, is already exponential. Are enzyme reactions simpler than simple? Linear progression can only be expected if the reaction rate is completely independent of the substrate concentration, so that the amount of product formed per time unit remains constant, irrespective of whether low or high substrate concentrations are present:

$$-\frac{d[A]}{dt} = \frac{d[P]}{dt} = k = v \tag{2.4}$$

$$[A] = [A]_0 - k_1 t \tag{2.5}$$

To explain this apparent contradiction let us turn to enzyme reactions. The simplest enzyme reaction is the conversion of one substrate catalyzed by the enzyme

$$A + E \underset{k_{-1}}{\overset{k_1}{\rightleftharpoons}} EA \xrightarrow{k_2} E + P \tag{2.6}$$

obviously, a second-order reaction.[4] However, there is an important difference compared with the second-order reaction described above: the enzyme takes part in the reaction, but does not become converted. It appears unchanged at the product site, according to its function as catalyst, and enters again into the reaction cycle from the substrate site. So the rate Eq. (2.1): $d[A]/dt = k[A]$ is not valid for the enzyme, rather it must be written $d[E]/dt = 0$, because the amount of the enzyme remains unchanged during the reaction. Also the expression $-d[A]/dt = k[A]$ for the substrate is not true. It cannot be written as a first-order reaction, because substrate can only be converted in the presence of the enzyme, and only that portion of the substrate actually bound to the enzyme reacts. Therefore, the reaction rate depends **not** on the substrate concentration, as for a first- or higher-order reaction, but only on the amount of enzyme. As the same

4) For each partial reaction, a rate constant k is defined with consecutive positive digits in the forward direction and negative digits in the backward direction.

enzyme molecule can repeatedly take part in the reaction, very low amounts of the enzyme compared with the substrate (**catalytic amounts**) are sufficient: $[E] \ll [A]$ (Box 2.3). Since the reaction depends only on – constant – enzyme concentration, and the amount of product formed per time unit is also constant, the reaction proceeds in a strictly linear manner (Figure 2.1a). Such a reaction is called a **zero-order** reaction. To answer the above question, the course of the enzyme reaction is more simple, but the reaction mechanism is more complicated compared with first-order reactions. This is an essential feature of catalytic reactions and it must be kept in mind that this condition holds only as long as the catalyst is clearly limiting. When during the reaction course the amount of substrate declines (or when the reaction is started with low amounts of substrate and/or high amounts of enzyme), this condition no longer prevails and the reaction course becomes nonlinear (first order). The linear zero-order range is called **steady state**. It can be regarded as a time-dependent equilibrium, existing only as long as the condition $E \ll A$ predominates, in contrast to a true time-independent equilibrium. Linearity of the progress curve is a clear indication for the presence of the steady state phase. As follows from the upper discussion, the duration of the steady-state phase depends on the relative amounts of both the substrate and the enzyme.

Box 2.3: How Much Enzyme Is Required for an Assay?

The velocity of enzyme-catalyzed reactions is strictly proportional to the enzyme amount:

$$v = k_{cat}[EA]$$

For substrate saturation:

$$V = k_{cat}[E]_0$$

Deviation from linear relationship is an indication for nonideal conditions.

The enzyme concentration $[E]_0$[5] *in the assay must adhere to the following rules*:

- concentration should be as low as possible, according to the steady-state theory [enzyme] ≪ [substrate]
- however, it must be sufficient to detect the initial velocity

The central relationship of enzymology, the **Michaelis–Menten equation**, is based on this steady-state assumption. As already mentioned, under steady-state

5) For exact calculation of the enzyme amount see Box 2.18.

conditions the enzyme concentration remains constant ($d[E]/dt = 0$) and, consequently, also the amount of substrate bound to the enzyme, the *Michaelis–Menten complex*, $d[EA]/dt = 0$. Therefore, the reaction rate v is determined solely by the concentration of EA. The derivation of the Michaelis–Menten equation is based on this assumption. For simplicity the one-substrate reaction (Eq. (2.6)) is taken. Separate equations are derived for the time-dependent change of each component:

$$\frac{d[A]}{dt} = -k_1[A][E] + k_{-1}[EA] \tag{2.7}$$

$$\frac{d[E]}{dt} = -k_1[A][E] + (k_{-1} + k_2)[EA] \tag{2.8}$$

$$\frac{d[EA]}{dt} = k_1[A][E] - (k_{-1} + k_2)[EA] \tag{2.9}$$

$$\frac{d[P]}{dt} = k_2[EA] = v \tag{2.10}$$

The overall reaction velocity v is defined as the rate of product formation (Eq. (2.10)). In addition to these four equations the mass conservation relationships

$$[A]_0 = [A] + [EA] \tag{2.11}$$

$$[E]_0 = [E] + [EA] \tag{2.12}$$

are considered. However, even these six relationships yield no simple solution. But recalling the above stated steady-state condition, Eqs. (8) and (9) can be simplified by $[E]/dt = [EA]/dt = 0$ and combined to yield the Michaelis–Menten equation

$$v = \frac{k_2[E]_0[A]}{\frac{k_{-1} + k_2}{k_1} + [A]} = \frac{V[A]}{K_m + [A]} \tag{2.13}$$

which shows the dependence of the reaction velocity v on the substrate concentration [A]. The equation is directly derived in the form of rate constants (left term), which is usually simplified and presented in the form of the right term, where $k_2[E]_0 = V$ is defined as a new constant, the **maximum velocity**. It is obtained from the rate constant k_2, the so-called **catalytic constant** (k_{cat}) for the conversion of the Michaelis–Menten complex to product (and enzyme), and the total enzyme amount $[E]_0$, which is assumed to remain constant during the reaction. The highest possible rate under the given conditions, the maximum velocity V, is attained when all enzyme molecules present in the assay ($[E]_0$) contribute at the same time to the reaction. The three rate constants of the denominator term are combined to one single constant, the **Michaelis constant** K_m. It should be noted, that the ratio of the two rate constants k_{-1}/k_1 represents the **dissociation constant** K_d, a thermodynamic equilibrium constant for the binding equilibrium between the substrate and the enzyme (first part of Eq. (2.6)). But in contrast to this constant, the Michaelis constant contains the catalytic constant k_2 as an additional term, which characterizes the chemical conversion of substrate to product. Since the chemical reaction is usually slower than the fast binding equilibrium, the value of the Michaelis constant is mainly dominated by the dissociation constant, with the contribution of the catalytic constant remaining small. Actually, the early derivations of this equation by Brown, Hill, Michaelis, and Menten considered only the dissociation constant, without regarding k_2. However, the modification introduced by the steady-state theory describes the real situation more.

The derivation of the Michaelis–Menten equation on the basis of constancy of the EA complex accentuates its strict limitation to the linear zero-order range. Nonlinear deviations are indications for nonvalidity of this relationship and it can now be understood that linear progress curves are a prerequisite for analyzing enzymes.

The Michaelis–Menten equation describes the dependency of the substrate concentration on the reaction velocity v. However, it has been stated above, that the zero-order range should be independent of the substrate concentration, depending only on the enzyme amount. How can this contradiction be understood? To explain this, the term **saturation** is introduced. If a very small amount of enzyme is added to a surplus of substrate to establish the condition $[E] \ll [A]$, it may intuitively be assumed, that the enzyme must be saturated, that is, the substrate occupies all available enzyme molecules. If this will be the case, the reaction indeed becomes completely independent of the substrate concentration. However, in reality, the enzyme will not be completely saturated; only a part of the enzyme molecules binds the substrate, while the other part remains unoccupied and does not contribute to the reaction. Only the part actually binding substrate, not the total enzyme amount, determines the reaction rate. To understand this, it must be considered that the degree of binding is determined by the binding affinity, expressed by the dissociation constant K_d. Its value expresses just the concentration of substrate required for half saturation of the enzyme; low substrate amounts cause a lower degree of saturation and more substrate amounts a higher degree. For example, consider a K_d value of 10^{-5} M. Substrate is added in this concentration to a 10^{-9} M enzyme solution. As the substrate concentration is the same as the K_d value, the enzyme is only half saturated – in spite of a 10 000-fold surplus of substrate. Variation of the enzyme concentration will not influence the degree of saturation, but will change the reaction velocity by the same factor, because the amount of enzyme molecules occupied with substrate changes to the same extent. On the other hand, change of the substrate concentration at constant enzyme amount alters the degree of saturation corresponding to the Michaelis–Menten law and the velocity changes accordingly. This demonstrates the mutual dependence of the velocity on both the enzyme and the substrate concentration, the first one being strictly linear and the second one being dependent on the Michaelis–Menten law. For our example, increase of the enzyme concentration by a factor of 10 increases the velocity 10-fold, while a 10-fold increase of substrate raises the degree of saturation, and consequently the reaction velocity, from 50 to 90.9%, which is less than twofold!

2.2.3 Determination of the Velocity of Enzyme Reactions

In the following section, the basic principles of the Michaelis–Menten equation are discussed. The dependency of the velocity of an enzyme reaction on the substrate concentration is described, but before going into details one must understand how velocity is to be determined. For this, let us recall the statement, that an enzyme-catalyzed reaction should proceed in a linear manner, with the **progress curve** forming a straight line. The slope of this line, expressed by the amount of product formed and substrate converted within a defined time unit, respectively, comprises the actual value of the velocity v (Figure 2.3).

2.2.3.1 Enzyme Units

Since the velocity is directly proportional to the enzyme concentration, it can be taken as a measure for the amount of active enzyme. For an exact quantification a clear definition of the dimension of the velocity is necessary. In early literature, various definitions with respect to the concentration and the time unit, dependent on the respective enzyme assay can be found. Therefore, in 1971 the Enzyme Commission of the International Union of Biochemistry (IUB) standardized one **International Unit** (**IU**) as the enzyme activity that converts 1 **μmol substrate within 1 min** (Box 2.4). Only two years later, in 1973, the Commission of Biological Nomenclature adapted the enzyme units to the Système International d'Unités (SI) with the mole as the concentration and seconds as time units. The valid enzyme unit is now 1 **katal** (kat), the enzyme activity converting **1 mol substrate per second**. To compare both enzyme units: 1 kat = 60 000 000 IU, 1 IU = 0.000 000 016 67 kat. Owing to this unpractical dimension, katal has problems to replace the older IU. For example, for the assay of a dehydrogenase an enzyme amount producing an absorbance change of about 0.1 per min is required. This corresponds to 0.016 IU, which is 0.000 000 000 26 kat. Commercial enzymes are frequently sold in quantities between 100 and 10 000 IU, corresponding to 0.000 001 667 and 0.000 1667 kat. Another reason is that, in comparison to a minute, a second as a time unit is too short for rate determination. Actually, for accurate measurements the velocity must be observed for a longer time interval and the value obtained must be recalculated for getting katals. Certainly, this is not a crucial problem, but cumbersome. Therefore, in literature and also commercially IU is still preferred. It is advisable to deal with nanokatal (nkat), 1 nkat = 0.060 IU, 1 IU = 16.67 nkat. Different expressions for the enzyme activity are presented in Box 2.4.

Box 2.4: Expressions for the Enzyme Activity

Enzyme units	Measure of the enzyme activity (calculated from the maximum velocity)
Katal (kat)	Amount of enzyme converting 1 mol substrate per second (according to the SI system)
International units (IU)	Amount of enzyme converting 1 μmol substrate per minute 1 kat = 60 000 000 IU ⇌ 1 IU = 0.000 000 01667 kat 1 nkat = 0.016 IU ⇌ 1 IU = 16.667 nkat
Maximum velocity	$V = k_{cat}[E]_0$ (concentration per time unit, in katal or IU); dependent on the enzyme concentration in the assay
Turnover number	$k_{cat} = V/[E]_0$ (s^{-1}) first-order rate constant; maximum velocity, divided by the enzyme concentration, independent on the enzyme concentration in the assay
Catalytic constant	k_{cat} (s^{-1}); identical with the turnover number
Michaelis constant	K_m (M); measure for substrate affinity (dissociation constant, extended by the catalytic constant k_{cat}
Catalytic efficiency	k_{cat}/K_m (M^{-1} S^{-1}); specificity constant, higher values indicating higher specificity
Volume activity	Enzyme activity per volume unit (enzyme units per milliliter)
Specific activity	Enzyme activity per protein concentration (enzyme units per milligram protein)

Still some special units (e.g., Anson Units for proteases, cf. Section 3.3.3.18) are used, especially if it is difficult to determine exact product molarities. For microbial transformations, instead of enzyme activity a *Productivity Number*, PN = $n_{prod}/m_{dry} \times t$, is applied, where n_{prod} is the amount of product formed, m_{dry} the dry weight of the cells, and t the transformation time.

For exact determination of any enzyme unit the enzyme assay must be carried out under *standard conditions*. They are discussed in detail in the following sections. On the other hand, for investigation of enzyme features, such as determination of the Michaelis constant, temperature stability or inhibition mechanisms, distinct parameters, for example, the substrate concentration, must be varied and in this respect standard conditions can no longer be maintained. In such cases the values obtained from velocity determinations correspond no longer to defined enzyme units, but the actual enzyme velocity should still be indicated in the respective dimensions of moles/second or micromoles/minute.

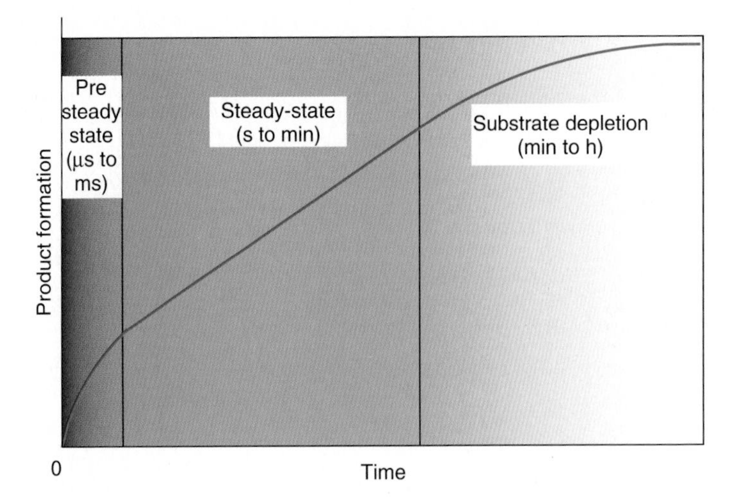

Figure 2.2 Schematic representation of the three phases of a progress curve.

2.2.3.2 Progress Curves

The progression of an enzyme reaction is schematically shown in Figure 2.2. Three phases can be discerned: initially a steep, nonlinear *pre-steady-state phase*, which is too short (~μs) to be detected in normal enzyme assays. Immediately thereafter the linear *steady-state phase* follows until the reaction ceases after the nonlinear *phase of substrate depletion*.

Linear progress curves as an indication for the prevalence of steady-state conditions and the validity of the Michaelis–Menten equation can best be expected at substrate saturation. Total saturation, however, exists only at infinite substrate concentration, while under real assay conditions saturation is not complete and decreases during the reaction course due to the substrate conversion. Therefore, strict linearity can only be expected at the start of the reaction, when the substrate concentration is nearly saturating. During the progression of the reaction, the velocity slows down and deviates from linearity (Figures 2.2 and 2.3). Therefore, the value of the velocity should not be estimated directly from the progress curve, rather a tangent to the initial linear range must be aligned as shown in Figure 2.3. Its slope, expressed as the concentration of substrate or product converted or formed per time unit, respectively gives the actual velocity (**tangent method**). If the substrate concentration is reduced within an experimental series, or if higher substrate concentrations cannot be realized (e.g., due to low solubility), the situation becomes more difficult, the linear range becomes shorter, and the deviation more pronounced (Figure 2.3, curves to the right). Even in such cases the velocity is obtained from tangents aligned to the initial range. If, however, the linear range becomes so short that it is hardly detectable, the prevalence of steady-state conditions and the validity of the Michaelis–Menten equation are no longer verified. As long as high accuracy is not demanded it may be assumed that steady-state conditions exist at least at the start of the reaction. This simplification,

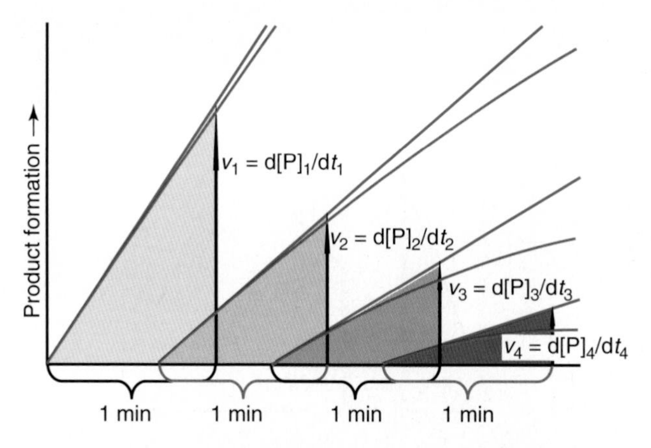

Figure 2.3 Determination of velocities of enzyme reactions from progress curves. The progress curves (red) from left to right are measured with decreasing substrate concentrations. Tangents (green) are aligned to the initial range; 1 min is taken as the time unit for the evaluation of the velocity v according to the definition of IU.

however, does not hold for exact determinations, for which some restrictions must be considered.

2.2.3.3 Difficulties in Determination of Initial Velocities

The most serious restriction in the determination of enzyme reaction velocities is the **dead time**. From the start of an enzyme reaction to the beginning of detection, that is, onset of registration, a certain time interval, at least some seconds, passes. This is no problem as long as the progress curve is linear (Figure 2.4a), but with nonlinear progress curves registration starts when the initial velocity has already passed. Aligning a tangent to the now apparent initial range causes a severe underestimation of the velocity (Figure 2.4b) and an attempt should be made to make the linear range visible. The reason for nonlinear progress curves is frequently a too high enzyme concentration. If an enzyme assay does not work immediately there is a tendency to add more and more enzyme. But too much enzyme converts the substrate instantly during the dead time, so that after start of recording no reaction will be observed at all. Such a situation is easily misinterpreted as lack of activity, not recognizing that the problem is in fact too much activity. As a rule, the enzyme amount should be as low as possible. The lower limit is given by the sensitivity of the method, when a very slow turnover cannot be detected within the scatter (Box 2.5). The higher the enzyme concentration the shorter the steady-state range. In Figure 2.5, a nonlinear progress curve due to a high enzyme concentration is shown. Dilution of the enzyme (e.g., 10-fold) reduces the velocity by just this factor extending the steady-state range by the same factor. If, after enzyme dilution, the velocity becomes too slow, the assay time can be prolonged. The same amount of product is produced within 10 min as in 1 min with a 10-fold

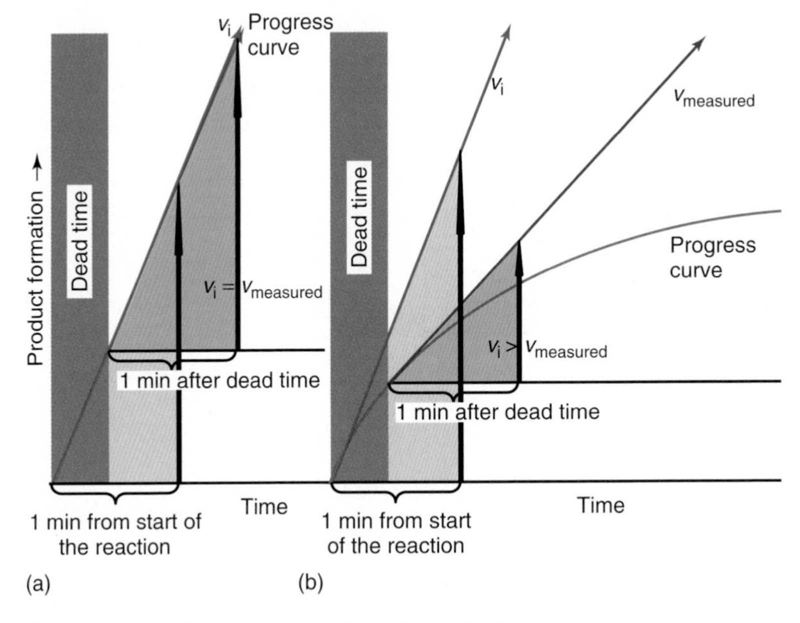

Figure 2.4 Disturbance of velocity determination by the dead time. Progress curve (red); tangent to real initial velocity (green); and tangent to progress curve after dead time (blue). (a) Linear progress curve and (b) nonlinear progress curve.

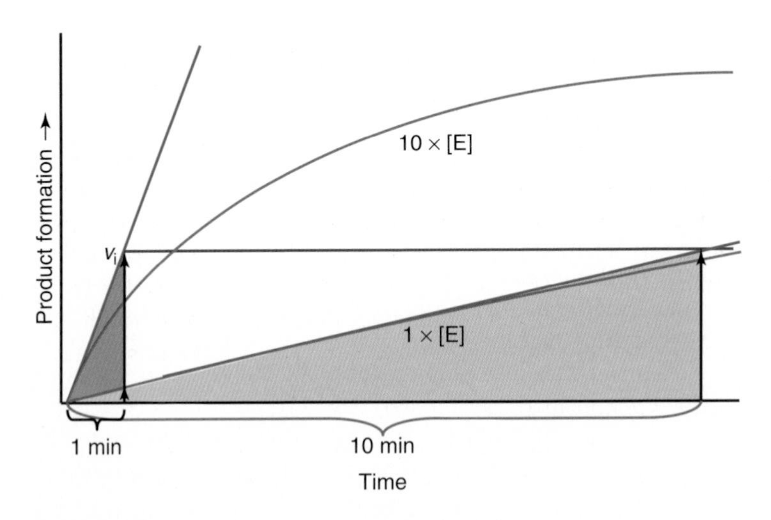

Figure 2.5 Linearity of progress curve depends on the enzyme amount: less enzyme (1 × [E]) yields slower velocities, but longer linearity than more enzyme (10 × [E]).

enzyme concentration. Principally there exists no general rule for the **assay time**. Usually short times (e.g., 1 min) are preferred, but with low enzyme activities, assay times of hours or even days can be chosen. Only the stability of the assay components and, in particular, of the enzyme must be established for such long time periods.

Box 2.5: Errors in Determination of Initial Velocities

Error source	Error elimination
Too much enzyme	Reduction of enzyme concentration Increase of detection sensitivity
Dead time longer than the linear steady-state range	Immediate recording after reaction start (rapid mixing system) Start of recording before reaction start Attenuation of reaction velocity (less enzyme, lower temperature)
Measure points outside the linear range (for stopped tests)	Shorter time intervals Reduction of reaction velocity
Scattering of progress curves or measure points	Intense mixing, clean devices Avoidance of dust, air bubbles, turbidity Shielding from external influence
Blank drift	Assay components of highest purity Suppression of oxidative processes (oxygen trapping) Blank controls

Particular care on linear progression must be observed with stopped assays. As long as the complete progress curve is continuously recorded, any deviation will be recognized. In stopped assays, however, there is only one measuring point and the velocity is calculated from the slope of a connecting line between the start of the reaction (usually the blank) and the measure point (Figure 2.6). Any deviation occurring during this time interval cannot be detected. Repeated measurements can only compensate for the common scatter, but not for extraordinary or systematic deviations. To reduce this problem, the time intervals should be chosen as short as possible and several measuring points after particular time intervals should be taken to confirm the expected linearity. In cases where this cannot be realized (because of scarcity of time or substances) at least a standard series with sufficient measuring point must be performed (Figure 2.7). Even if such a procedure demonstrates that the chosen time interval lies fairly within the linear range, it must be considered, that this is valid only for actual test conditions. Any change, such as temperature, pH, substrate or enzyme concentration, can shorten the linear range so that the chosen time interval is no longer appropriate. Possible error sources in determination of initial velocities are summarized in Box 2.5.

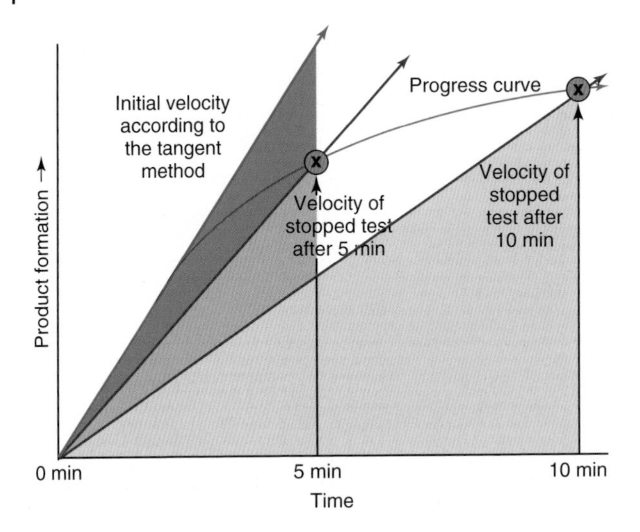

Figure 2.6 Stopped assay performed under nonideal conditions. The measurements ⊗ are carried out beyond the linear range so that the determined velocity is lower than the real initial velocity.

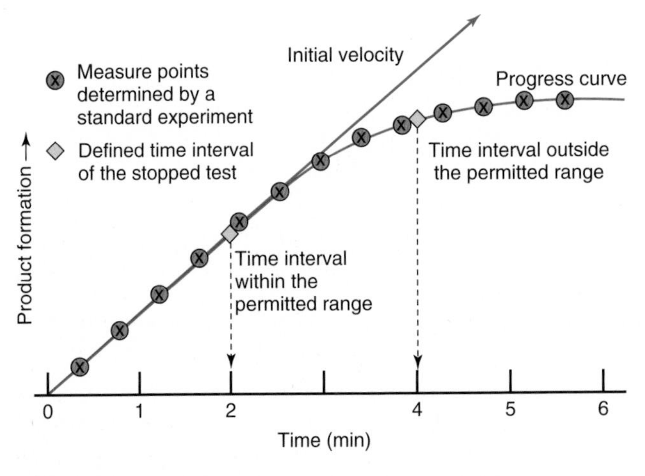

Figure 2.7 Stopped assay performed under ideal conditions. A series of measurements ⊗ allows the control of the linear range. The first measurement of the stopped test ◇ lies within this range, the second one is still outside.

An alternative procedure for evaluation of progress curves is the **integrated Michaelis–Menten equation**, which is obtained by integrating Eq. (2.14) with respect to time

$$v = -\frac{d[A]}{dt} = \frac{V[A]}{K_m + [A]} \tag{2.14}$$

$$K_m \ln \frac{[A]_0}{[A]} + [A]_0 - [A] = Vt \tag{2.15}$$

The integrated Michaelis–Menten equation (Eq. (2.15)) describes the complete progress curve, including the nonlinear range, so that there is obviously no necessity to obtain the linear initial range. According to this equation, the Michaelis constant and the maximum velocity can be derived from one single progress curve. By rearranging

$$\frac{[A]_0 - [A]}{t} = V - K_m \frac{\ln \frac{[A]_0}{[A]}}{t} \tag{2.16}$$

the curve is linearized and V and K_m obtained from the ordinate intercept and the slope, respectively. The particular advantage of this procedure is the fact that in computer-controlled instruments the data can be directly transformed according to this equation and the constants can be displayed immediately after the experiment. This method appears tempting, but it is only reliable if the respective progress curve obeys the Michaelis–Menten equation completely. Unfortunately, this is usually not the case, because the product formed influenced the reaction by inhibiting the enzyme and inducing the backward reaction. Therefore, for accurate measurements the determination of initial velocities from the linear part of progress curves is emphatically recommended.

2.2.3.4 A Short Discussion on Error

Errors severely affect scientific experiments and their avoidance would be a desirable goal. However, errors will always occur and the only realistic aim is to minimize them as far as possible. Here, potential sources of errors, their consequences, and possible methods to limit their perturbing influences on enzyme assays are discussed. At first, a diagnosis of the error should be undertaken to detect its origin. Generally, spontaneous and systematic errors can be differentiated. **Spontaneous errors** arise from inexact manipulation (e.g., pipetting). Any handling bears intrinsically a certain inaccuracy, which may be reduced by careful performance, but cannot be eliminated completely. Spontaneous errors cause randomly distributed deviations from the true value in the positive and negative directions, both to a comparable extent (Figure 2.8a). Such errors will be observed in any experiment, but as long as the deviations are not too large, they can be mastered by common error calculations and regression analysis, which are described in the relevant literature of statistics. In contrast, **systematic errors** are not randomly distributed, but rather cause deviations in a particular direction and, therefore, distort the results and can lead to wrong interpretations. Causes of such errors may be imperfect operation of the instrument (e.g., incorrectly adjusted apparatus, such as a defective pipette used for the same step always) or faulty manipulation by the operator (making the same mistake always in routine assays).

Before discussing both error types in detail, one important argument must be mentioned. Errors are regarded as artificial, and, thus, unintended deviations from

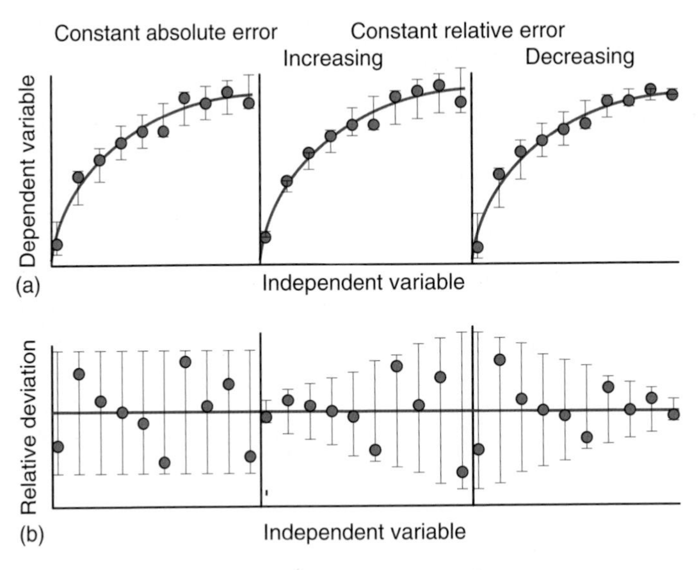

Figure 2.8 Error distributions. (a) Direct plotting of experimental data; (b) residual plots of the same data; error bars and deviations from the adapted curve are enlarged.

the real value. The aim is to approach the true value, for example, by calculating a medium value out of a series of repeated measurements, or to adapt a curve, such as a straight line or an exponential function, where it is generally understood, that nature will make no jumps, and physiological processes will always follow clear, smooth functions. Albeit, it must be kept in mind, that any such function, even based on statistical treatments, is an *interpretation* of experimental results. It is the task of the scientist to demonstrate convincingly the appropriateness of the fit to the data, while the experimental data are the only stringent result. So, **outliers**, which are values obviously not fitting into the system, are commonly assumed to originate from unusual strong artificial deviations. Because outliers distort the regression analysis dramatically so that the calculated function will no longer fit the majority of data, one feels authorized to suppress such strongly deviating values. This may, in many cases, be justified, but, principally, elimination of particular values out of a data series is an arbitrary selection revealing the preference of the scientist for a particular solution, neglecting any other interpretation. Therefore, it is a strict rule in science to present all original data together with the interpretation (i.e., the fitted function), indicating, if particular data points remained unconsidered for regression analysis, so that an independent observer can judge whether the applied function is an adequate description for the presented data.

Even for spontaneous errors different possibilities of expression exist. It may be supposed, that during an experimental series the extent of error, that is, the error limits to both directions, may be *constant* throughout (Figure 2.8a, left diagram). For such a series often one parameter is varied, such as the substrate concentration for enzyme kinetic determinations, and the question is, whether the error remains independent or whether the varied parameter influences the error.[6] Figure 2.8a shows different error developments. They can be easier visualized by **residual diagrams**, where the deviation of the data from the assumed function is enlarged and plotted around the x axis (Figure 2.8b). For equally distributed errors, the error bars should be the same throughout (Figure 2.8a,b, left diagrams), but the error can increase or decrease during the test series (Figure 2.8a,b, middle and right diagrams, respectively).

It must be considered that the dependent variable (velocity) increases with the substrate concentration and if the error limits remain constant throughout (Figure 2.8a left), they actually increase relative to the velocity in the lower concentration range. The behavior of the error limits depends on the error source. For example, scatter of the instrument will cause a constant error, while pipetting of different volumes influences the extent of the error, with very small volumes producing higher errors than larger ones. In this case the error decreases with increasing substrate concentration.

Pipetting is one of the severest error sources; careful pipetting improves considerably the accuracy of experiments. Various pipette systems are available, producing different error types. Automatic pipettes with variable volumes are mostly used. If they are of good quality they are very precise, but some general rules must be regarded (common manipulation is described by the producer's instructions and is not mentioned here). It is common usage to pour out the pipette by dipping the tip into the assay solution and to mix the solution thereafter with the tip. This is commodious and time saving, but not a very accurate (Figure 2.9, I). When extracting an aliquot from the stock solution some liquid will adhere to the outer surface of the tip (Figure 2.9, II) and get into the assay solution (Figure 2.9, II). Wiping the tip is not recommended, as tissue fibers can extract some liquid from inside the tip (Figure 2.10). When dipping into the assay solution for ejecting and mixing, the assay solution can penetrate into the tip and wash out the residual liquid, which remains inevitably in the tip after emptying (this remaining portion is already considered in the normal pipetting process, Figures 2.9, III; 2.10). Such an error is more severe when the pipetting volume is smaller.Very small volumes ($<5 \mu l$) cannot be pipetted with high accuracy, even if such volumes are within the range of the pipette. On the other hand small volumes have the advantage of avoiding significant dilution of the assay solution.[7] Volumes of 10–20 μl are recommended for pipetting, they are easy to manipulate and do not essentially change the common assay volumes of 1 or a few milliliters. The aliquot should not be directly added into the assay solution, but placed on the flat end of a (commercially available) plastic spatula (Figure 2.9, IV), which can also be used to mix the assay solution (Figure 2.9, V).

6) The parameter varied is designated as *independent*, and the result, for example, the reaction rate, as a *dependent* variable; it is, by definition, generally accepted that only the dependent and not the independent variable is subject to an error, an unfounded, but practical assumption.

7) For accurate tests all additions must be accounted in the final assay volume, but in preliminary tests they are often neglected.

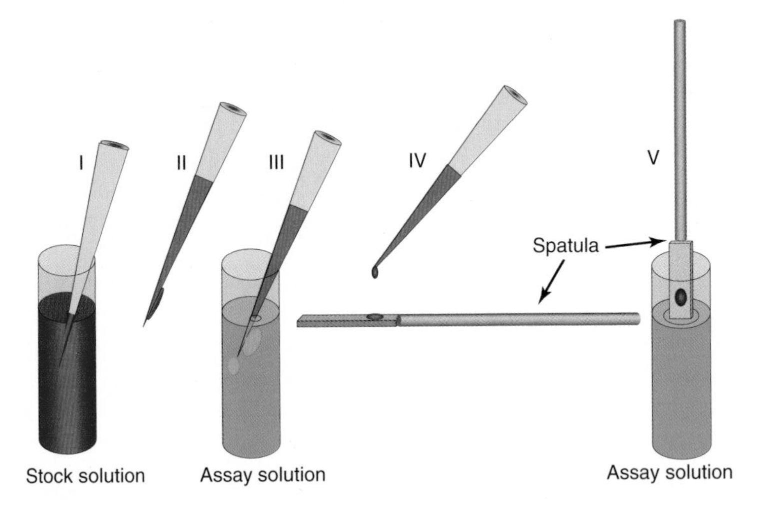

Figure 2.9 Pipetting and mixing in assay solutions. I–III, pipetting of a sample from the stock to the assay solution and direct mixing. A drop of the stock solution remains attached at the outside of the tip and gets into the assay solution. IV and V, placing the sample on a spatula, inserting of the sample, and mixing of the assay solution.

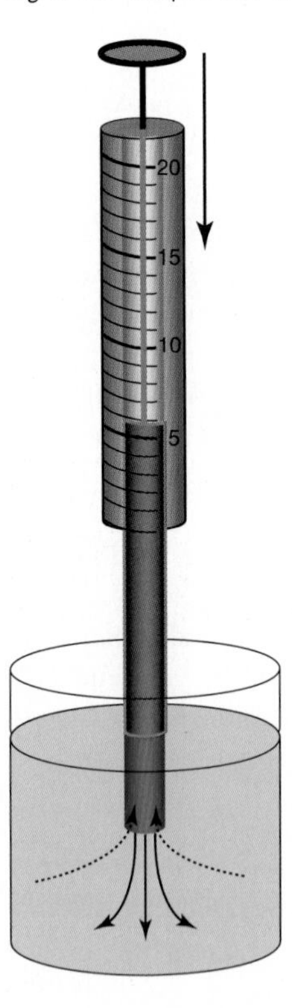

Figure 2.10 Plunging the pipette directly into the assay solution: some solution penetrates into the tip and more of the stock solution gets into the assay solution than intended. A microliter syringe is shown, the inner canal is enlarged.

2.2.3.5 Preparation of Dilution Series

If lower concentrations of the added sample are required, instead of reducing the volume, the stock solution should be diluted to avoid pipetting of very small volumes. When the concentration of substrate or cofactor must be varied during a test series, this is after realized by modifying the added volume of the stock solution correspondingly with an automatic pipette (Figure 2.11a). This, however, is the less exact procedure not only because of inaccuracy of repeated pipette adjusting and varying dilution of the assay solution, but, because of the dependency of the extent of the error on the pipetted volume. Therefore, the preparation of a dilution series with varying concentrations of the respective components is strongly recommended. Such series must be prepared very carefully. From such a series, an equal volume (e.g., 10 μl) is always given to the assay solution to yield the desired final assay concentration. For this the pipette volume must not be changed and the aliquot added to the assay has always the same volume and can easily be considered for calculation of the final assay volume. There are two modes to prepare a dilution series. The first is to prepare the dilutions in a stepwise manner (Figure 2.11b). A defined volume of the stock solution (e.g., 0.2 ml) is taken in a test tube and filled up to the final sample volume (e.g., 1 ml) with a buffer solution (resulting in a fivefold dilution of the stock solution). After mixing a similar aliquot (0.2 ml) is removed from this sample and taken in a second tube and also filled up to the final sample volume with buffer (resulting in a fivefold dilution of the previous solution). This procedure is repeated several times until the desired dilution range is covered. This is an easy procedure. However, any erroneous deviation in one step will be carried forward to all following steps and all errors during the procedure accumulate to the end. For the second mode of preparing a dilution series, different volumes of the stock solution are added to the test tubes and they are brought up to an equal final volume (e.g., 1 ml) with a buffer solution (Figure 2.11c). This procedure is more reliable; an error concerns only the respective sample and not all the following ones. A disadvantage is that for a broad dilution range also very small volumes are required. To avoid this intermediated stock dilutions, for example, for each order of magnitude, should carefully be prepared.

More accurate than automatic pipettes, especially, for very small volumes are microliter glass syringes (Figure 2.10). They are available for different volume ranges, but the criteria discussed above must be regarded for these pipettes also. They consist of a glass cylinder with a central hole, ending in a steel tubing. A stainless steel piston fitted to the hole serves to draw up the solution. Care must be taken to ensure that the pipette is completely filled with liquid without any air bubbles included.

Mixing is a further source of error. Besides the problems discussed above, inappropriate handling can cause two particular errors. Insufficient mixing, especially with thin tips, produces inhomogeneities in the assay solution with regions of lower and higher concentrations. The result of such an effect in a photometric assay is shown in Figure 2.12, where areas of higher and lower velocity float through the light path consecutively, simulating an oscillating behavior. During the progression of the reaction, the oscillation smoothes slowly approaching the actual reaction

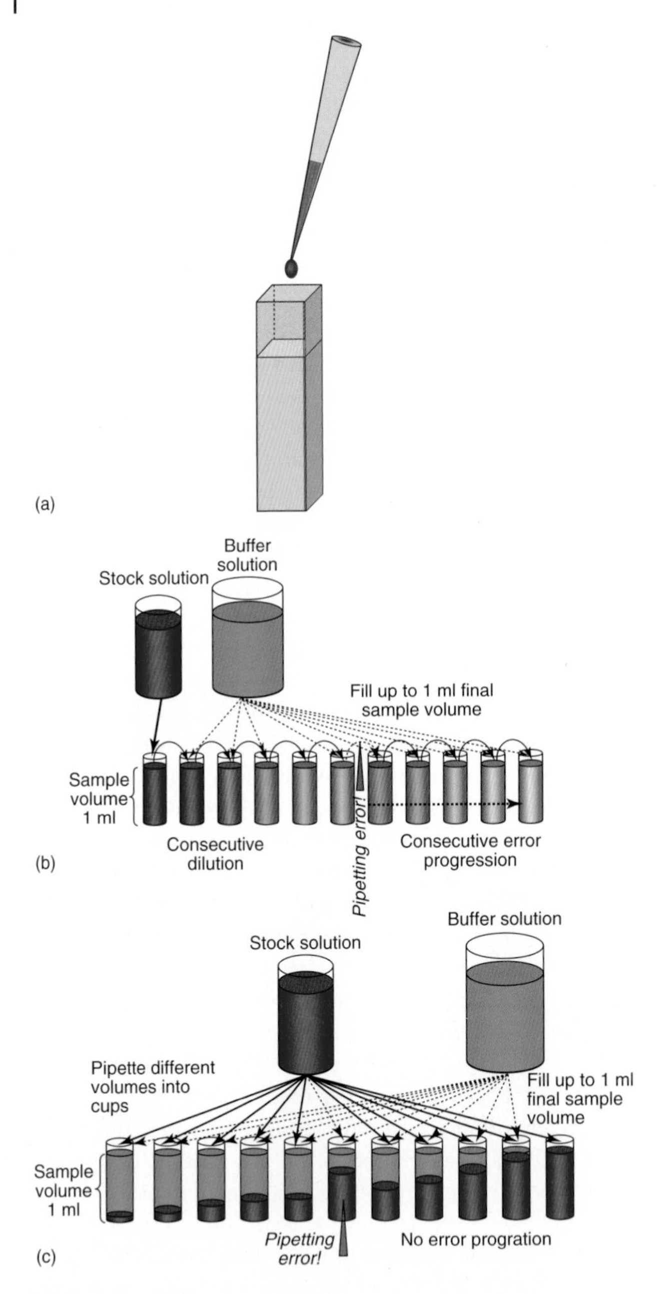

Figure 2.11 Different modes for preparation of a dilution series. (a) Direct pipetting into the assay solution has the lowest accuracy. (b) Stepwise dilution from one tube to the next, an error will be propagated into the following samples. (c) Dilution from the same stock solution, best method, no error propagation.

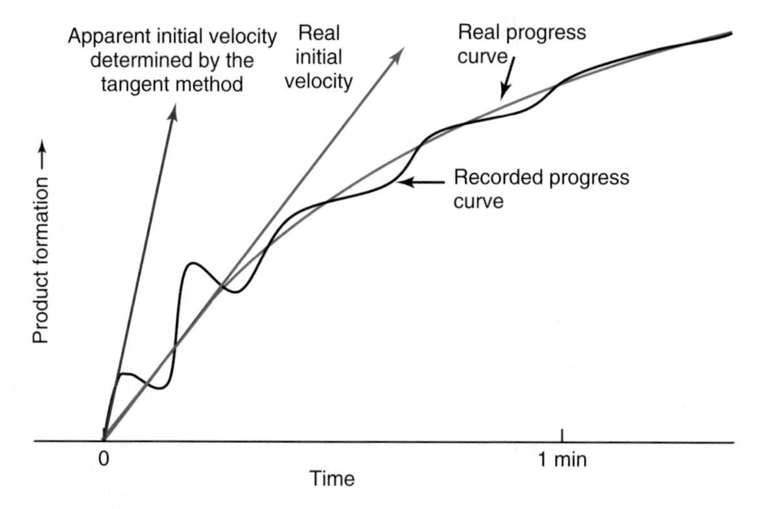

Figure 2.12 Artificial deviations from the progress curve, due to insufficient mixing. The initial velocity taken from the recorded curve deviates considerably from the real one.

velocity – the real initial rate, however, has already been passed. Too vigorous mixing, on the other hand, causes the enclosure of air bubbles, disturbing especially optical methods, and of oxygen, favoring oxidative processes. Various mixing devices are commercially available, such as small magnetic stirrers, which can be installed in optical instruments for direct mixing in the cuvette. Special apparatus, for example, the stopped-flow device, enable rapid mixing within parts of seconds.

2.2.3.6 Considerations about Statistical Treatments

It is particularly requested to perform repeating experiments, from which statistically established mean values can be calculated to avoid misinterpretations of the final result by one or few wrong measurements. So one should not be acquiesced by one single determination of an enzyme assay, rather several independent measurements should be performed. How much repeats are required? It may be asserted, the more determinations, the better the security, but the principle of parsimony dictates a limit both with respect to material (especially enzyme) and time, and accuracy does not essentially increase with a higher number of repetitions. For routine tests, three independent determinations can be regarded as sufficient, and five should be enough for higher accuracy. It is often advantageous to perform repeated assays not under completely identical conditions, but to modify one parameter, which causes strictly proportional changes. For enzyme assays the amount of enzyme (not of substrate!), and for protein tests that of protein may be changed. The expected linear dependency of the values is a further control for the method. Deviations from linearity, especially when steadily inclining in the same direction, are indications for errors. Any assay is reliable only within a particular range. Too low values disappear in the scatter, too high values fall outside the linearity of the method (due to depletion of assay components or the limited range of the Lambert–Beer law in optical assays). Linear dependency establishes that the assay has been performed completely within the reliable range of the method. A mean value can be calculated, regarding the different amounts of the parameter that is varied.

For a test series, such as the determination of the Michaelis constant, numerous (~10) measurements with increasing substrate concentrations will be required. For a series to be confirmed by three independent measurements, 30 assays must be carried out. If each single test needs 5 min, the whole series will last 2.5 h. This is not only the question of time, but also of the stability of the assays, especially of the enzyme. Most enzymes are not stable for longer durations in diluted solutions, as required for enzyme assays. Therefore, the test series must be performed within a short duration to establish equal conditions for all measurements. Interruptions for hours or over night must be avoided. If for example, an enzyme loses 10% of its activity per hour, the value of the last measurement of the test series will be 25% lower than that of the first one under otherwise identical conditions. Obviously, the improvement in statistical reliability by repeated measurements is obtained at the price of a systematic error. Constants derived from such an experiment will be seriously underestimated, although statistical treatment demonstrates more confidence. If the same series is performed with only single determinations, less than 1 h is needed with an activity difference of only 9% between the start and end of the experiment. In this case, it will be better to repeat the whole series three times and to calculate the constants independently for each series and derive the mean value for each constant. Such a procedure has the further advantage that any systematic deviation, caused either artificially or as a special feature of the system, can be easily identified by its independent repeated occurrence. Indeed systematic deviations are often overlooked and taken for usual scatter.

Even if an error source becomes obvious, it cannot always be avoided (such as scatter of instruments), but it can be observed and, as far as possible, reduced. Sometimes, however, unexpected deviations appear without any obvious reason. If such deviations occur by chance, it is often easier to repeat the experiment carefully, than to search for the reason for the disturbance (which may often be caused by impurities, such as dust or soiling, or wrong handling). Only if disturbances appear repeatedly in a similar manner, is a particular examination of the error source advisable.

2.2.4 Michaelis–Menten Equation

2.2.4.1 General Considerations

The Michaelis–Menten equation has already been derived in Section 2.2.2, and after the description of aspects of determination of the enzyme velocity, a prerequisite for treating the Michaelis–Menten equation, this fundamental relationship should be considered more closely. The conditions for a special enzyme assay, such as the concentration of substrate, are usually given in the protocol, and performing the

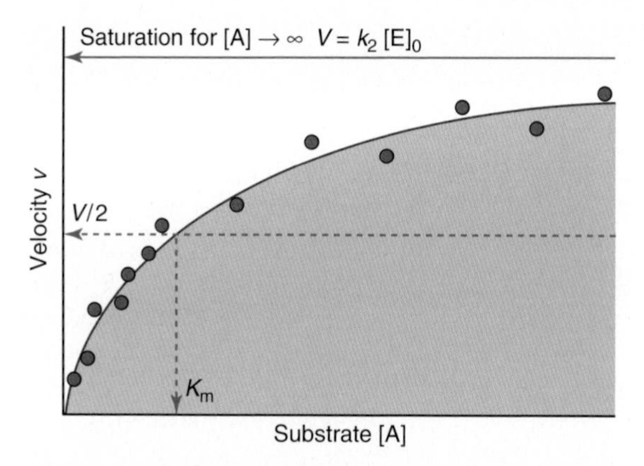

Figure 2.13 Michaelis–Menten presentation of enzyme velocities measured in dependence on varying initial substrate concentrations. The data points, the resulting hyperbolic function, and the determination of the kinetic constants are shown.

assay accordingly, one value for the velocity will be obtained from which enzyme units, as a measure of the enzyme activity, can be calculated as described above. In many cases, such as routine tests, this information is sufficient. Sometimes, however, a more detailed analysis of the enzyme is required. Such an analysis comprises at least the determination of the **kinetic constants** of the Michaelis–Menten equation, the Michaelis constant K_m, and the maximum velocity V; but it must be borne in mind, that for correct performance of even simple enzyme assays, the fundamental principles of the Michaelis–Menten equation must be considered. Therefore, after treating this equation in detail we turn back to the basic conditions of enzyme assays.

The two kinetic constants K_m and V can be obtained by determining the velocity of the enzyme reaction at various initial substrate concentrations. If plotted in the **Michaelis–Menten diagram**, with substrate concentration [A] as the x coordinate, and velocity v as the y coordinate, a hyperbolic function results, as shown in Figure 2.13. To obtain such a curve, not only must the Michaelis–Menten equation be obeyed, the substrate concentration must also cover the appropriate range. Actually, it can be assumed that most enzymes obey the equation, and even those that do not, show only weak deviations, which may easily be overlooked in a preliminary study, so that in a first approach this equation can be applied. The second point is more critical: how can we know the appropriate substrate range for a particular enzyme? For this the Michaelis constant must be known, that is, the result should be obvious before making the experiment! In fact, the problem is not so difficult. In any case it cannot be assumed that an accurate result is obtained already from the first experiment. Rather, some preliminary experiments must be undertaken, allowing a crude estimation, which serves in performing the final experiment (Box 2.6).

Box 2.6: How Much Substrate is Required for an Assay?

- *Saturating amounts* of all substrates and cofactors are required
- *Saturating amounts* means infinite concentration – according to theory
- *Saturation* can practically be related to the K_m value of the respective substrate: 50-fold $K_m \sim 98\%$ saturation is sufficient (10-fold $K_m \sim 91\%$ saturation is not enough)
- However, saturating concentrations can often not be attained, because of:
  1. substrate inhibition
  2. unspecific inhibition due to high ionic strength
  3. limited solubility
- *Example 1:* K_m of NADH is 1×10^{-5} M, 50-fold $K_m = 5 \times 10^{-4}$ M, yielding an absorption of 3.15 at 340 nm, this is far outside the range of the Lambert–Beer law!
- *Example 2:* K_m of benzoylarginine p-nitroanilide, a substrate for trypsin, is 1×10^{-3} M, solubility in water is only 1.3 mg/ml (3 mM) $\sim 2.3 \times K_m$. More concentrated solutions can be prepared in DMSO, but the solvent influences the enzyme activity
- *Conclusion:* Saturating conditions (50-fold K_m), if possible; if not, highest possible concentration and calculation of the real saturation according to the Michaelis–Menten law

If the substrate concentration is increased until $[A] \gg K_m$ (and finally $[A] \to \infty$) K_m can be neglected and $v = k_2[E]_0$ according to the denominator of Michaelis–Menten equation (Eq. (2.14)), that is, all enzyme molecules present in the assay will be involved in the reaction, yielding the highest possible reaction rate under these conditions, the **maximum velocity** V. Since, in reality, infinite substrate concentration is not possible, V must be regarded as a limiting value, to which the curve approaches, but which will never be attained. This point is often disregarded and V is taken directly from tests with an apparently high substrate concentration. Figure 2.14 shows several degrees of saturation in dependence on the substrate concentration, indicated in relationship to the K_m value. The true V can only be obtained by extrapolation to infinite substrate amounts, as is discussed below. For $v = V/2$, the substrate concentration becomes equal to the **Michaelis constant** K_m and its value can be directly obtained from the abscissa scale as shown in Figure 2.13. So the knowledge of V is necessary also for the determination of K_m. This is an unsatisfying situation, since any error of V will be transferred to K_m.

2.2.4.2 Linear Representations of the Michaelis–Menten Equation

The restrictions in the determination of the kinetic constants mentioned above can be overcome by transforming the Michaelis–Menten law into a linear relationship. Three equivalent transformations can be derived, the simplest being the reciprocal

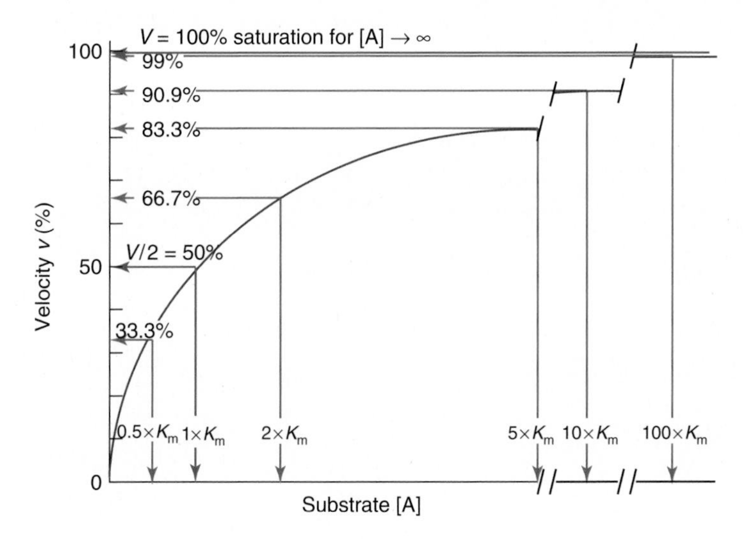

Figure 2.14 Relative saturation (in %) of an enzyme in dependence on the substrate concentration, indicated as a multiple of the K_m value.

form

$$\frac{1}{v} = \frac{1}{V} + \frac{K_m}{V[A]} \tag{2.17}$$

yielding a straight line in a plot of $1/[A]$ against $1/v$ (Figure 2.15a). It is known as the **Lineweaver–Burk diagram.**[8] Upon extrapolating to the left, the line intersects the ordinate at $1/V$ and the abscissa at $1/K_m$. Both constants can be obtained independent of each another. A serious disadvantage of this plot is the distortion of the error limits, which are extended from left to right (to lower substrate concentrations!), so that linear regression analysis is not applicable. In this respect, the **Hanes diagram** is superior. It is obtained by multiplying Eq. 2.17 with [A]

$$\frac{[A]}{v} = \frac{[A]}{V} + \frac{K_m}{V} \tag{2.18}$$

Plotting of $[A]/v$ against [A] results in a straight line with a positive slope and the ordinate and abscissa intercepts K_m/V and K_m (Figure 2.15b). This and the following diagram have the disadvantage, that both variables are not separated. This also holds for the constants determined from the ordinate intercept, and $1/V$ can be obtained separately from the slope.

Multiplication with Vv of Eq. (2.17) and rearrangement leads to the third transformation

$$v = V - \frac{vK_m}{[A]} \tag{2.19}$$

8) It was actually Woolf who mentioned this and the following two transformations first.

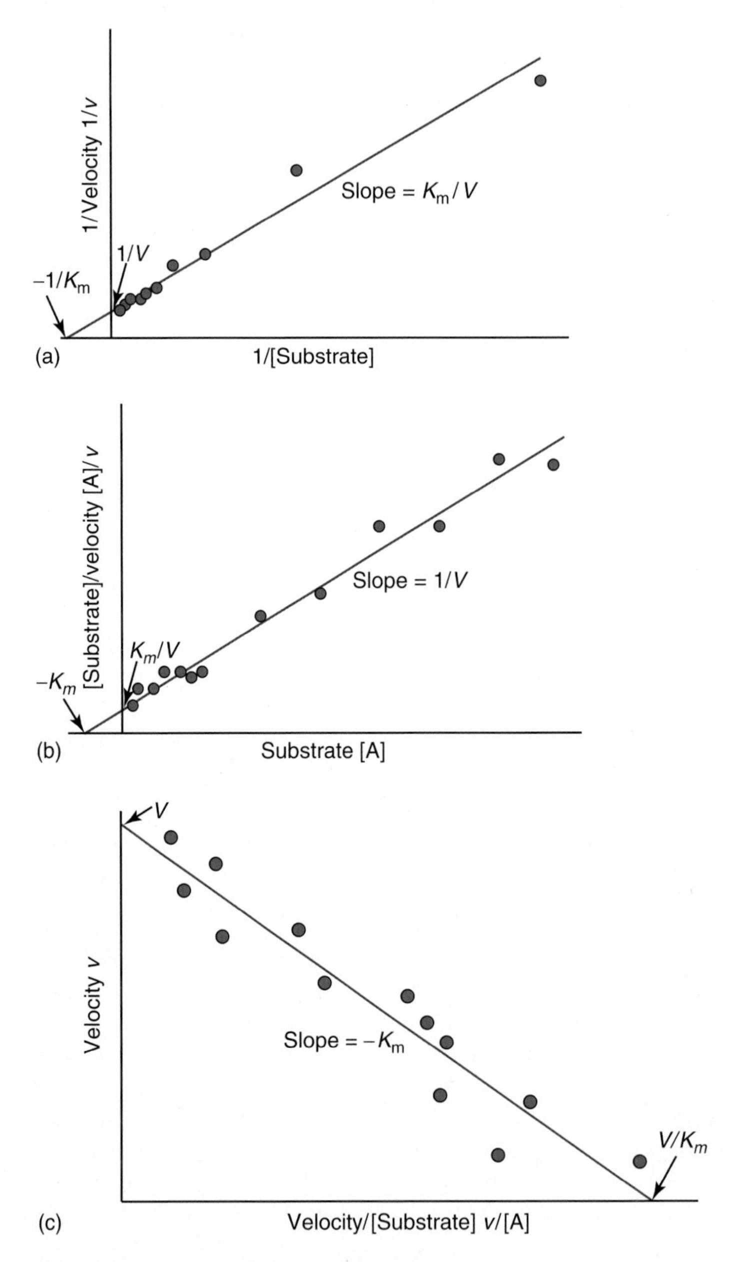

Figure 2.15 Diagrams for linear presentation of the Michaelis–Menten data. Data points, resulting lines, and determination of kinetic constants are shown. (a) Lineweaver-Burk plot; (b) Hanes plot; and (c) Eadie–Hofstee plot.

For this **Eadie–Hofstee diagram** $v/[A]$ is plotted against v. The slope of the straight line is $-K_m$, the ordinate intercept V, and the abscissa intercept V/K_m (Figure 2.15c).

Owing to the various advantages and disadvantages it is advisable to plot the data in all three diagrams. Linearization is not only for determination of constants, but also for the detection of any deviation from linearity, which can be indicative of alternative mechanisms or artificial influences.

To return to the question about the appropriate substrate range it can now be understood that the K_m value as the substrate concentration at half saturation may be regarded as the medium amount of a range starting 1 order of magnitude below and ending 1 order of magnitude above this value.

2.2.5
Enzyme Inhibition

For characterization of a special enzyme, determination of the two kinetic constants is mostly sufficient, but sometimes further information is desirable. A more thorough analysis requires deeper knowledge of enzyme kinetics, which is outside the scope of this book; however, two important aspects, inhibition and multisubstrate reactions, are mentioned in the following text. The treatment of both mechanisms show various similarities. In comparison to the simple enzyme reaction discussed hitherto a second variable (inhibitor or cosubstrate) is introduced, so that for detailed studies all components involved in the reaction must be considered.

An **inhibitor** is defined as a substance, which specifically binds to the enzyme and reduces its activity. Actually, various influences can reduce enzyme activity, such as temperature, pH, ionic strength, or unspecific interactions with various substances, but inhibition in its strict sense requires binding of a substance to a specific site at the enzyme. Several modes of action of an inhibitor are known, and the main types are discussed.

To study the action of an inhibitor, its efficiency must be established at first. In an initial step the inhibitor binds to the enzyme, the intensity of the binding (**affinity**) is characterized by a binding or **dissociation constant** K_i[9] (dimension M^{-1}). Its value indicates the amount of inhibitor required to achieve half saturation, similar to the K_m value with respect to the substrate concentration. However, while K_m can be determined with the substrate alone, determination of K_i requires the presence of both the inhibitor and the substrate; the latter is needed to carry out the enzyme reaction, but it also influences the binding of the inhibitor. Therefore, it is easier to examine the degree of inhibition by practical testing. Because of the influence of substrate, its concentration should not be saturating for these experiments;

9) There exists some confusion about the respective terms. In fact dissociation expresses the opposite of binding, but the value of the dissociation constant is a measure of affinity, the lower the value, the higher the affinity. In physical chemistry literature the **association constant** (M), often designated as true binding constant, is preferred, but is only the reciprocal value of the dissociation constant, with higher values directly indicating higher affinity. The literature of enzyme kinetics prefers the dissociation constant; the Michaelis-Menten constant is derived from this too.

the inhibitory effect will become more clear at a half-saturating substrate concentration. At first the enzyme reaction should be tested in the absence of the inhibitor to yield the uninhibited reaction (100%) as a reference. In a second assay, a defined amount of the inhibiting compound is added and the velocity measured in comparison to the uninhibited reaction. If there is no change or only a small change, a higher amount of inhibitor must be added, while a lower amount is advisable if the reduction of velocity is too strong. Several concentrations should be tested to find out the amounts of inhibitor necessary to reduce the original activity to 25, 50, and 75%.

With this information, test series can be conducted, in which both the substrate and the inhibitor are varied. In the first series, the uninhibited reaction is tested with 10 different substrate concentrations around the K_m value substrate as already discussed. In the second series, the same substrate concentration series is tested again, but in the presence of a constant inhibitor concentration according to the 25% inhibition of the preliminary experiment. Thereafter, the same series is repeated with the inhibitor amount for 50% inhibition and finally for 75% inhibition. If these data are plotted in a linear diagram, such as the Lineweaver–Burk plot, all four series should yield straight lines, proving the universal validity of the Michaelis–Menten law, even in the presence of inhibitor. However, all four lines should take in different positions in the plot and if there is no unexpected deviation due to errors or special mechanisms (see below), a clear pattern must be revealed, displaying the respective inhibition mechanism. One pattern shows increasing steepness of the lines (in the Lineweaver–Burk plot) with increasing inhibitor concentration, with all lines intercepting at the same ordinate coordinate (Figure 2.16a, left plot). As already mentioned the ordinate intercept indicates the reciprocal maximum velocity $1/V$. Hence, all lines extrapolate to the same maximum velocity at infinite substrate concentration. This is just the feature of **competitive inhibition**. The inhibitor must bind to a specific site on the enzyme molecule and the simplest case is binding to the substrate site. This is the case especially for compounds, which are structurally similar to the substrate but cannot be converted to the product (*substrate analogs*). Upon binding they block the active site of the enzyme. However, as the substrate binds to the same site, very high (infinite) amounts displace the inhibitor completely and, therefore, the original maximum velocity is reached even in the presence of (limiting amounts of) the inhibitor. Conversely, high amounts of the inhibitor displace low amounts of the substrate.

Several modes for determination of the binding constant of the inhibitor, the **inhibition constant** $K_{ic} = [E][I]/[EI]$, exist. A favorable method is the **secondary diagram** (or replot), a graphical method, where the inhibitor concentrations (the uninhibited reaction being zero) are plotted against the slopes of the respective lines in the primary (Lineweaver–Burk) diagram. A straight line should result intersecting the abscissa exactly at the value of $-K_{ic}$ (Figure 2.17a). Deviation from linearity in this plot (if error can be excluded) is an indication of a more complicated inhibition type, while strict linearity is a further confirmation of the competitive mechanism.

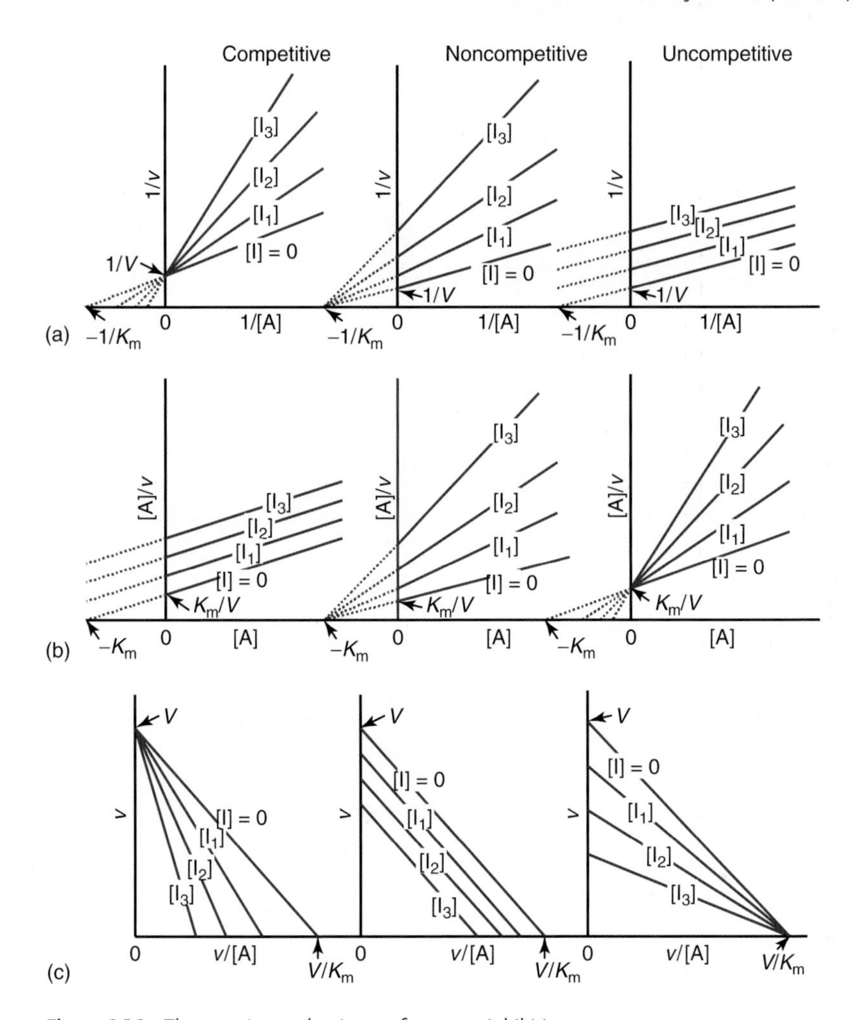

Figure 2.16 Three main mechanisms of enzyme inhibition in the linear presentations. (a) Lineweaver–Burk plot; (b) Hanes plot; and (c) Eadie–Hofstee plot; ——, lines in the experimental region, ……. lines in the extrapolated region.

If the lines in the Lineweaver–Burk diagram differ in their slope without common ordinate intersection, the inhibition is **noncompetitive**. But in this case too, all four lines *must* share a common intersection point, which now is left of the ordinate, either directly at the abscissa, or above or below it. The relative position of the intersecting point gives further information about the inhibition type. According to the meaning of the expression "noncompetitive" the inhibitor *does not* bind to the site of the substrate and cannot displace it. There must exist an additional site for the inhibitor and if both the substrate and the inhibitor bind to their respective sites completely independent of one another, the lines intersect directly on the abscissa (pure noncompetitive inhibition, Figure 2.16a,

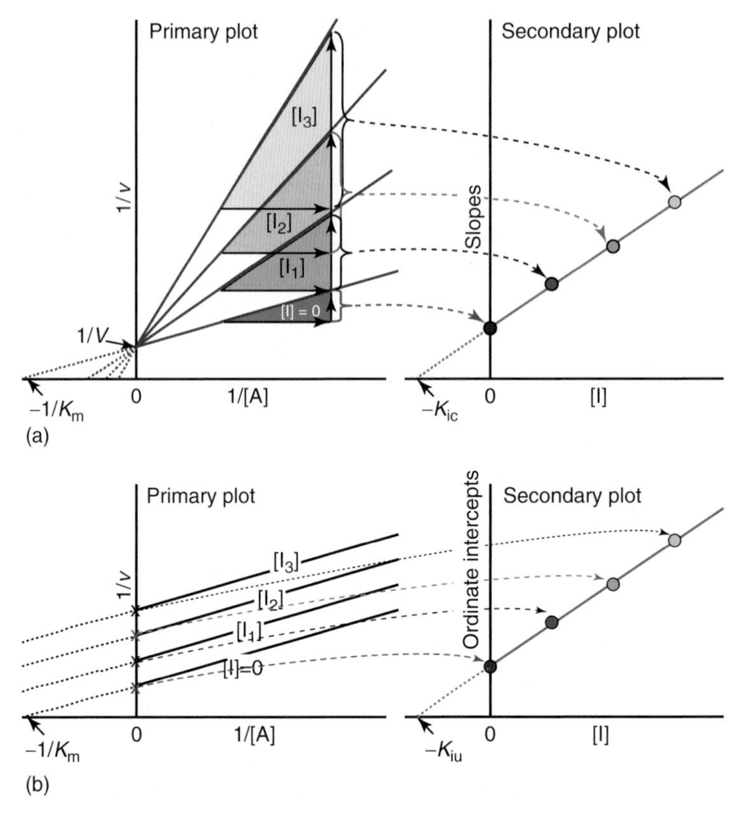

Figure 2.17 Secondary plot (right), derived from the slopes (a) and the ordinate intercepts (b) of the lines in the primary plot (Lineweaver–Burk plot, left). In the case of multisubstrate reactions on the abscissa 1/[B] (1/cosubstrate concentration) is plotted instead of [I], both in (a) and (b).

middle diagram). If, however, both compounds influence one another in their binding (in most cases this will be an impeding effect), they will bind better to the enzyme in the absence of the other component than in its presence. Consequently, two different inhibition constants must be considered, one for binding to the free enzyme (K_{ic}), the other for binding to the enzyme–substrate complex (K_{iu}), the former constant being smaller (higher affinity) than the latter: $K_{ic} < K_{iu}$. In this case, the lines intersect above the abscissa, while intersection is below, if binding to the free enzyme is weaker, that is, the already bound substrate helps the inhibitor in its binding (and *vice versa*): $K_{ic} > K_{iu}$, a more infrequent case. The inhibition types where both constants are unequal are sometimes called *mixed* inhibition.

If the inhibitor does not bind to the substrate site, where else does it bind? There may exist a particular, regulatory site for the inhibitor. This is the case with **allosteric enzymes**. Usually, these enzyme classes do not obey the Michaelis–Menten law, showing nonlinear curves in the linearized plots. Also in this case, experiments can be done as described above and the nonlinear curves should intersect also at a common point to the left of the ordinate. The typical noncompetitive inhibition, however, is observed when two or more substrates participate in the enzyme reaction. It can be imagined that an inhibitor competes with one substrate for its own binding site, but this inhibitor will not compete for the binding site of the second substrate. It will be competitive with respect to the first, but noncompetitive with respect to the second substrate. The second substrate cannot disclose the inhibitor from the binding site of the first substrate and, thus cannot restore the enzyme activity even at high concentrations.

If no common intersection point can be observed, that is, if the lines are strictly parallel (Figure 2.16a, right diagram), the inhibition is **uncompetitive** (consider: the inhibition is not "competitive," but it is also not "noncompetitive"!). For this infrequent inhibition type it can be imagined that the inhibitor cannot bind to the enzyme in the absence of substrate, but only to the enzyme–substrate complex: the substrate helps the inhibitor to bind, for example, by forming or completing a binding site for the inhibitor. Its affinity is quantified by the uncompetitive inhibition constant $K_{iu} = [EA][I]/[EAI]$, which in contrast to the competitive inhibition constant K_{ic} describes the binding exclusively to the enzyme–substrate complex. It can be determined from secondary plots similar to K_{ic}, but as the slope in the primary plot remains unchanged, the ordinate intercept is taken, with K_{iu} resulting from the abscissa intercept of the straight line (Figure 2.17b). Recalling the noncompetitive inhibition, both constants exist, since the inhibitor can bind both to the free enzyme and the enzyme–substrate complex. Consequently both secondary plots, one with the slope and the other with the ordinate intercept, can be drawn and the respective constants obtained. In the case of independent binding, resulting in a common abscissa intercept, both constants are equal.

Besides these three main inhibition mechanisms, two further types, namely, substrate and product inhibition should be mentioned. **Substrate inhibition** (or substrate surplus inhibition) is observed, when the reaction rate, after an initial increase, decreases with higher substrate concentrations, instead of approaching a saturation value (Figure 2.18a). It can be imagined that a first substrate molecule binds to the enzyme in quite a normal manner followed by binding of a second substrate molecule to a particular, inhibitory site. This can be the product binding site, so that the substrate can no longer be converted into the product. Formally, this is an uncompetitive inhibition type, since the inhibitory substrate molecule binds only to the EA complex. As, however, the same compound, the substrate, shares both functions, this one compound alone can be varied. In the low concentration range, the catalytic function prevails, while the inhibitory function predominates at higher concentrations. This causes characteristic deviations in linear plots. In the Lineweaver–Burk diagram, two regions can be differentiated (Figure 2.18b). The right part of the diagram (low substrate concentrations) shows linear behavior corresponding to the substrate function. From this part V and $-1/K_m$ can be obtained by extrapolation to the y and x axes, respectively. For higher concentrations, in the left part of the diagram, the curve deviates drastically upwards due to the inhibitory effect. A similar curve is obtained, when [A] is directly (instead of 1/[A])

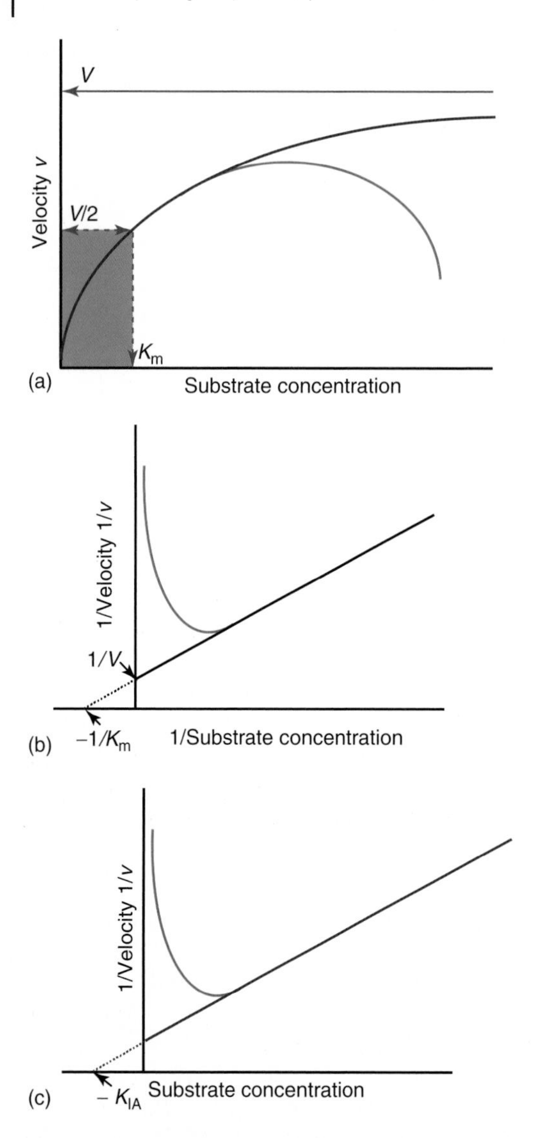

Figure 2.18 Substrate inhibition. (a) Michaelis–Menten plot; (b) Lineweaver–Burk plot; and (c) Dixon plot. Lines in the experimental region (blue, solid), deviation due to substrate inhibition (red), lines in the extrapolated region (blue, dotted). The determination of kinetic constants is shown.

plotted at the abscissa (Figure 2.18c). Here the linear range in the right part (high concentrations!) represents the inhibitory effect and extrapolation to the abscissa yields the dissociation constant K_{iA} for the inhibitory substrate.

Product inhibition is based on the general principle of reversibility of any chemical reaction and is, therefore, a feature of all enzyme reactions. When the product is formed during the catalytic process it remains bound to the enzyme for a short time before being released. When the product dissociates and its binding site becomes free, it can again be occupied by another product molecule from the surrounding solution. Because the substrate and the product bind principally to the same region of the catalytic center, binding of the product and the substrate is exclusive; both displace one another, and the inhibition is of the competitive type. If the product is added to the assay already at the start of the reaction, it behaves like a competitive inhibitor, and just the same pattern is observed, that is, straight lines with a common ordinate intercept at the position of V in the Lineweaver–Burk diagram (Figure 2.16).

Besides these chief inhibition types there exist even more, special mechanisms. If there is an indication that the observed inhibition pattern does not fit the rules described here, the special literature on enzyme kinetics should be consulted.

2.2.6
Multisubstrate Reactions

In the majority of all enzyme reactions, two, or sometimes even three different substrates are involved, and any general treatment must take note of this fact. In enzyme assays, all components, substrates, cofactors, and essential ions should be present in saturating amounts so that the maximum velocity will be obtained and no interfering influence from any component should occur. However, for a more detailed analysis, the dependence of the reaction rate on *all* substrates involved must be considered. Therefore, the most significant rules of such **multisubstrate reactions** are discussed in the subsequent text. The treatment resembles, in some respects, to that of inhibition mechanisms, with the main difference that inhibitors reduce the velocity, while the second substrate, the *cosubstrate*, accelerates it. Here too, three main mechanisms can be discerned. The description should begin with a mechanism similar to the noncompetitive inhibition, the *random mechanism*. We regard an enzyme reaction such as the alcohol dehydrogenase requiring two substrates, A (ethanol) and B (NAD^+). The reaction can only proceed in the presence of both substrates. If only one substrate (A) is added even in saturating amounts, nothing will happen and it is obvious now that the velocity will depend on the second substrate, if it is given in low concentrations in comparison to the first one. So a test series can be performed by varying A in the presence of saturating amounts of B, and, *vice versa*, that is, a series varying B at saturating concentrations of A. A hyperbolic curve obeying the Michaelis–Menten law should result and the kinetic constants can be determined as already described, for example, by linearization. Owing to the variation of both substrates separate Michaelis constants and maximum velocities are obtained for both, with K_m being characteristic for the respective substrate, whereby V as maximum velocity at saturating conditions with respect to all components must be the same for both substrates. For crude analysis this simple approach may be acceptable, while for an exact description it must be considered that the reaction

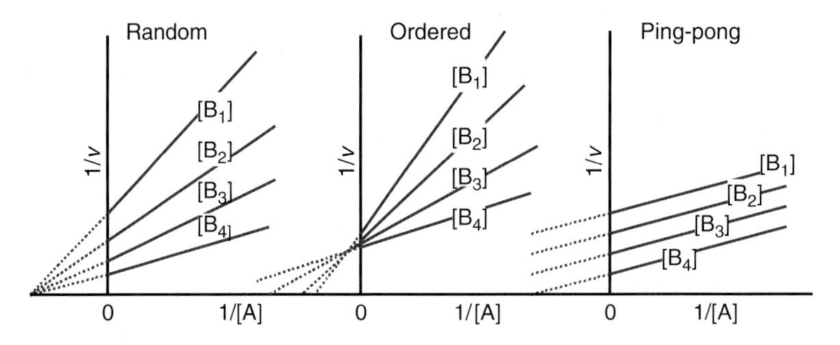

Figure 2.19 Multisubstrate mechanisms in the Lineweaver–Burk diagram. ——, lines in the experimental region; .... lines in the extrapolated region.

velocity depends now on two substrates and real saturating conditions cannot be attained. Therefore, a more thorough analysis should be undertaken, which will also provide information about the respective mechanism. The procedure is similar to the one described for inhibition, one substrate (e.g., A) will be varied in the presence of a constant (but not necessarily saturating) concentration of the other (B). Several (e.g., four) such series will be performed, modifying the constant amount of B between them. The outcomes are plotted in a linearized diagram, for example, the Lineweaver–Burk plot, which yields a pattern of straight lines with different slopes intercepting to the left of the ordinate, directly on the abscissa, or above, or below (Figure 2.19). This relative position depends on the case, whether both substrates bind independent of one another (intercept on the abscissa), or whether they impede or favor one another (intercept above or below the abscissa, respectively), as discussed for the noncompetitive inhibition. In the case of independent binding both substrates bind in random order (*random mechanism*); the one reaching the enzyme first will bind first. If both substrates interact with one another, an **ordered mechanism** results, the first substrate (A) binding to the free enzyme, the second one (B) to the EA complex. The common intercept of the straight lines will appear to the left of the ordinate above the x axis (Figure 2.19).

Also, similar to the noncompetitive mechanism, two binding constants for each substrate are defined, one (the **inhibition constant,** K_{iA} for substrate A, K_{iB} for substrate B) for binding to the free enzyme E, and the second one (the **Michaelis constant,** K_{mA} and K_{mB}) for binding to the enzyme substrate–complexes, EA and EB, respectively. These constants can also be determined from a secondary plot of the *reciprocal* concentrations (in contrast to direct plotting of the inhibitor concentration!) of the (constant) cosubstrate against the slopes and ordinate intercepts of the respective lines in the primary plot, as already described for the inhibition mechanisms (cf. Figure 2.17).

Parallel lines uncover a **ping-pong mechanism** (Figure 2.19) where the first substrate forms an intermediate with the enzyme, which reacts with the second substrate under restoration of the original enzyme form. An example is the transamination reaction, where an amino acid transfers its amino group to the active site of the enzyme, being released as an α-oxoacid, while another α-oxoacid removes this amino group becoming an amino acid, a mutual exchange as in the ball game mentioned above.

Multisubstrate reactions are more difficult to analyze, but they provide a lot of information about the respective enzyme. The coordinate intercepts of the linearized diagrams are complex expressions and the Michaelis constants and the maximum velocity cannot simply be taken from them. For more details, literature of enzyme kinetics should be referred.

2.2.7 Essential Conditions for Enzyme Assays

2.2.7.1 Dependence on Solvents and Ionic Strength

In the previous chapters, the dependence of enzyme reactions on the amounts of enzymes and substrates was described. Besides this, enzyme reactions are essentially dependent on environmental influences, such as solvent, pH, and temperature, as well as on unspecific influences of components that are not directly involved in the reaction. For the enzyme activity, the solvent plays a decisive role. Enzymes bound to or connected with the membrane prefer an apolar environment, as lipases, which are active in organic solvents. The great majority of enzymes, however, prefer the polar aqueous cell milieu. In organic solvents these enzymes are unstable and denature. Therefore, water is exclusively used as a solvent for enzyme assays. However, in some cases, the presence of organic solvents cannot be completely avoided. Several substrates and metabolites are water insoluble, especially in higher concentrations and must be dissolved as stock solutions in less polar solvents such as ethanol, acetone, tetrahydrofuran, or dimethylsulfoxide (DMSO). Aliquots from the organic stock solution are added to the aqueous assay solution, assuming that the diluted compound will remain resolved in water and the enzyme will tolerate the small amount of the apolar solvent added (only water mixable solvents should be used). Both preconditions must be established for each test, especially when the volume of the aliquot, and thus of the solvent, is changed to attain different final concentrations of the respective compound. Sometimes, organic solvents such as ethanol, are added to prevent microbial attack or to lower the freezing temperature.

Care must also be taken in enzyme assays containing different components, such that they are compatible with one another, and does not form precipitates or complexes (e.g., divalent cations in phosphate buffer) or influence the redox state (e.g., NAD, NADH, thiols).

Most enzymes are not only sensitive to high, but also very low ionic strength (the measure of the concentration of electrolytes). This must be considered with respect to all components, for example, high concentrations of substrates, or components not directly involved in the reaction, as protectives for thiol groups, or counterions for neutralization of acid or basic compounds.

Mono- and divalent ions are often necessary for the catalytic mechanism or for stabilization of the enzyme. On the other hand, they may cause unspecific effects, for example, promoting oxidative processes. To avoid such effects, frequently chelate forming substances, such as ethylenediaminetetraacetic acid (EDTA), are added providently, but they can also withdraw the stabilizing ions.

The ionic strength tolerated by the enzyme depends on its special requirements and must be individually be examined. Most enzymes prefer moderate ionic strength (~0.1 M) and will denature at considerably lower or higher values. Particular enzymes, especially from organisms growing in brine or at high temperatures prefer strong ionic strength (up to 1 M).

2.2.7.2 pH Dependency

Enzymes depend strictly on the pH in the medium. Responsible for this pH dependency are two different aspects: (i) charged groups are involved in the catalytic mechanisms and their state of protonation is decisive for the catalytic efficiency and (ii) enzymes are amphoteric substances with positive and negative charges on their surface and in the inner core, which stabilize the native structure. The catalytic effect can be investigated by an experiment, where the enzyme assay is carried out under otherwise identical conditions, but changing pH values. Usually a pH range between 3 and 12 should be covered with determinations in distances of at most 0.5 pH units. Usually an optimum curve results, increasing from the acid region, passing a maximum, and decreasing to the alkaline region (Figure 2.20). The maximum, the **pH optimum**, is a characteristic value for each

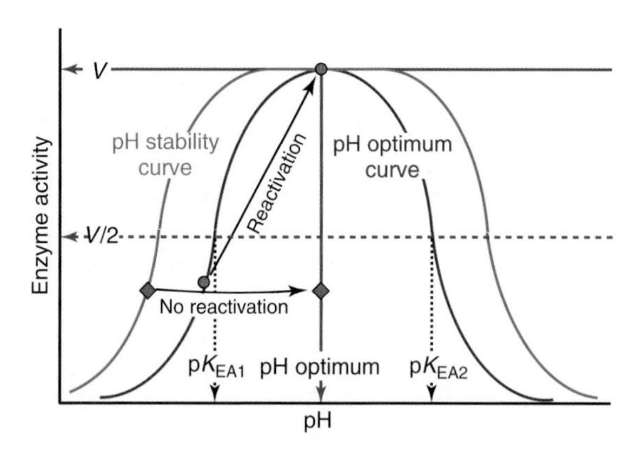

Figure 2.20 pH behavior of the enzyme activity. pH optimum curve: the enzyme is tested at the pH indicated. pH stability curve: the enzyme is preincubated for a particular time interval (e.g., 1 h) at the pH indicated and tested at the pH optimum. The p*K* values can be obtained from the inflexion points of the optimum curve. Shift of an enzyme sample from a marginal pH region of the optimum curve to the optimum pH restores the maximum activity (•), while no regain of activity occurs for a similar shift from the stability curve (♦).

Table 2.1 pK_a values of functional groups of various amino acids.

Amino acid	pK_a
Aspartic acid	3.86
Glutamic acid	4.32
Histidine	6.09
Cysteine	8.30
Serine	9.15
Tyrosine	10.10
Threonine	10.40
Lysine	10.53
Arginine	12.30

enzyme and is mostly in the neutral range between pH 6.5 and 8.5, but particular enzymes have extreme pH preferences, such as acid and alkaline phosphatases or pepsin, the latter with a pH optimum near 1. The whole optimum curve can be regarded as a combination of titration curves of charged groups, which are essential for catalysis, mostly amino acid residues, but cofactors can also be involved. One type of group will be active in the protonated state, others in the deprotonated state and their respective titration curves forming the sites of the pH optimum curve. For example, a carboxy group may be active in its charged state and the transition from the protonated ($-COOH$) to the deprotonated ($-COO^-$) state in the acid region will form the left side of the optimum curve, while deprotonation of an amino group ($-NH_3^+ \rightarrow -NH_2$) contributes to the right, declining site. If for each site only one single group is responsible, a pure titration curve results with an inflexion point corresponding to the pK_a value (the pH, where the respective group is just half protonated) of the titrated residue. Thus, the pK_a value can serve to identify the respective group involved in the catalytic mechanism (Table 2.1), but it must be considered, that pK_a values can be considerably changed (for ± 1–2 pH units) by the integration of the respective amino acid into the three-dimensional protein structure. Beyond that, often more than one residue is involved and the respective site of the optimum curve is an overlap of several titration curves.

The pH-dependent catalytic protonation processes are usually reversible. When the enzyme is incubated at a marginal pH with lower activity, it will regain its full activity when shifted to its pH optimum (Figure 2.20). In contrast to this, pH-dependent processes concerning the three-dimensional enzyme structure are mostly irreversible. A **pH stability curve** can discern between reversible and irreversible pH changes. Aliquots of the enzyme are preincubated for a particular time interval (e.g., 1 h) at various pH values, while the activity is tested thereafter at the optimum pH. Identical activities will be obtained as long as the pH-dependent changes are reversible, but after an irreversible change the enzyme cannot return to its optimum activity (Figure 2.20). In comparison to the pH optimum curve, the pH stability curve expands more toward extreme pH values and has a broad plateau of similar activities around the pH optimum. As the enzyme has its highest activity at its pH optimum this is taken as the actual pH for enzyme assays. Sometimes, however, special aspects require a deviation from the optimum pH, as in the case of the alcohol dehydrogenase, where an alkaline pH is applied to push the reaction in the direction of the product against the reaction equilibrium, which favors the substrate.

2.2.7.3 Isoelectric Point

Besides the pH optimum, the **isoelectric point** (IP) is a characteristic value for each enzyme. It is the pH value where the positive and the negative charges of the enzyme or protein are counterbalanced, that is, the protein is neutral without positive or negative surplus charges. At this pH the enzyme does not move in the electric field (this feature is used to determine the IP) and it has its lowest solubility in water.

2.2.7.4 Buffers: What Must Be Regarded?

According to the importance of the pH for enzyme activity and stability it is necessary to make sure that the enzyme exists in its optimum pH always. Only when water is the solvent, this cannot be guaranteed. Additions, such as substrate or cofactors, if they are not strictly neutral, can change the pH dramatically. To stabilize the pH, **buffers** are used. They consist of a weak acid and a strong basic component. The relationship between the concentration of the buffer components and the pH is described by the **Henderson–Hasselbalch equation** (Box 2.7). The efficiency to stabilize the pH, the **buffer capacity**, depends on the concentration and on the interval between the actual pH and the pK_a value of the buffer. As a rule, it can be assumed that the buffer capacity is within one pH unit below and above the pK_a value, that is, two pH units together. This is not a very broad range, but there exist various buffer systems, such that for each pH an appropriate buffer can be found (Table 2.2). As long as the same pH is required always, such as for enzyme assays, this is not a problem; but if the pH needs to be changed, for example, to prepare a pH optimum or stability curve, a broader range is needed. This can be achieved by combination of different buffer systems. It must, however, be considered, that due to ionic strength and the nature of ions, enzyme activity can vary considerably in different buffer systems even at identical pH. Such deviations may be approximated by determinations at overlapping pH values, but this is not really satisfying. Instead it is recommended to use universal buffers consisting of more components, such as the Teorell–Stenhagen buffer, covering broader pH ranges (Box 2.9).

Box 2.7: Henderson–Hasselbalch Equation

Mass action law for the acid (Ac)

$$K_a = \frac{[H^+][Ac^-]}{[HAc]}$$

K_a: ionization constant
~ dissociation constant for the acid

Rearranging

$$[H^+] = K_a \frac{[HAc]}{[Ac^-]}$$

Definitions
$-\log [H^+] = pH$ = negative logarithm of hydrogen concentration
$-\log K_a = pK_a$

Transforming into the logarithm

$$-\log[H^+] = -\log K_a - \log \frac{[HAc]}{[Ac^-]}$$

Final form of the Henderson–Hasselbalch equation

$$pH = pK_a - \log \frac{[HAc]}{[Ac^-]}$$

Even in the presence of buffers addition of acid or basic compounds can considerably modify the pH. Some compounds, for example, NAD^+, are available in an acid form and must be neutralized before usage. Also the enzyme reaction itself can change the pH, like lipases by releasing fatty acids.

To choose a particular buffer system besides the desired pH other criteria must also be considered. Some buffers, such as phosphate buffers, and especially diphosphate buffers possess complex-forming capacities, which may withdraw essential metal ions and can form precipitates with them. Nevertheless, phosphate buffers are not only inexpensive but also compatible with most enzymes. Particular buffer ions can exert activating or inhibiting effects on enzymes. There are "biological buffers" available, such as 3-(*N*-morpholino)propanesulfonic acid (MOPS), *N*-(2-hydroxyethyl)piperazine-*N'*-(ethanesulfonic acid (HEPES), *N*-tris(hydroxymethyl)methyl-2-aminoethane-sulfonic acid (TES) ("Good buffers," Good *et al.*, 1966; Good and Izawa, 1972; Stoll and Blanchard, 1990, cf. Table 2.2), which may be advantageous for special enzymes. Frequently Tris/HCl buffers are used, but in some cases detrimental effects, even covalent reactions with proteins have been reported. From the chemical viewpoint most buffers are stable for months or years, but microbial attack, perceptible by a progressing turbidity, limits their stability. Therefore, for longer storage freezing is recommended.

Attention must be paid to the temperature dependency of the pH. Instructions for buffer preparation refer usually to room temperature (25 °C), but enzyme studies are often carried out in the cold (~4°) and the pH must be adjusted at this temperature, according to the appropriate correction function of the pH meter.

Table 2.2 Biological buffers (pK_a values and pH range refer to 25°C).

Short name	Full name	pK_a	pH range
MES	2-(*N*-Morpholino)ethanesulfonic acid	6.15	5.5–6.7
BIS-TRIS	Bis(2-hydroxyethyl)iminotris (hydroxymethyl)-methane	6.5	5.8–7.2
ADA	*N*-(2-Acetamido)-2-imididiacetic acid	6.6	6.0–7.2
PIPES	Piperazine-*N*,*N*′-bis (2-ethanesulfonic acid)	6.88	6.1–7.5
ACES	*N*-(Carbamoylmethyl)-2-aminoethanesulfonic acid	6.9	6.2–7.6
MOPSO	3-(*N*-Morpholino)-2-hydroxypropanesulfonic acid	–	–
BIS-TRIS PROPANE	1,3-Bis(tris[hydroxymethyl] methylamino)propane	6.8 (pK_{a1}) 9.0 (pK_{a2})	6.3–9.5
BES	*N*,*N*′-Bis(2-hydroxyethyl) 2-aminoethanesulfonic acid	7.15	6.4–7.8
MOPS	3-(*N*-Morpholino)propanesulfonic acid	7.20	6.5–7.9
P_i	Phosphate	7.21 (pK_{a2})	6.5–8.0
TES	*N*-Tris(hydroxymethyl)methyl-2-aminoethane-sulfonic acid	7.50	6.8–8.2
HEPES	*N*-(2-Hydroxyethyl)piperazine-*N*′-(ethanesulfonic acid)	7.55	6.8–8.2
DIPSO	3-(*N*,*N*′-Bis[2-hydroxyethyl]amino)-2-hydroxy-propanesulfonic acid	7.6	7.0–8.2
MOBS	4-(*N*-Morpholino)butanesulfonic acid	7.6	6.9–8.3
TAPSO	3-(*N*-Tris[hydroxymethyl] methylamino)-2-hydroxy-propanesulfonic acid	7.7	7.0–8.3
TEA	Triethanolamine	7.8	7.3–8.3
POPSO	Piperazine-*N*,*N*′-bis (2-hydroxypropanesulfonic acid)	7.85	7.2–85
HEPPSO	*N*-(2-Hydroxyethyl)piperazine-*N*′-(2-hydroxy-propanesulfonic acid)	7.9	7.2–8.6
EPPS (HEPPS)	*N*-2-Hydroxyethylpiperazine-*N*′-3-propanesulfonic acid	8.0	7.3–8.7
TRIS	Tris(hydroxymethyl)aminomethane	8.1	7.0–9.0
TRICINE	*N*-Tris(hydroxymethyl) methylglycine	8.15	7.4–8.8
Gly-amide	Glycinamide, hydrochloride	8.20	7.4–8.8
BICINE	*N*,*N*-Bis(2-hydroxyethylglycine)	8.35	7.6–9.0
Gly-Gly	Glycylglycine	8.40	7.7–9.1
HEPBS	*N*-(2-Hydroxyethyl)piperazine-*N*′-(4-butanesulfonic acid)	8.3	7.6–9.0
TAPS	*N*-Tris(hydroxymethyl)methyl-3-aminopropane-sulfonic acid	8.4	7.7–9.1

Table 2.2 (*continued*)

Short name	Full name	pK_a	pH range
AMPD	2-Amino-2-methyl-1,3-propanediol	8.8	7.8–9.7
TABS	*N*-Tris(hydroxymethyl)methyl-4-aminobutane-sulfonic acid	8.9	8.2–9.6
AMPSO	3-[(1,1-Dimethy-2-hydroxyethyl) amino]-2-hydroxy-propanesulfonic acid	9.0	8.3–9.7
CHES	2-(*N*-Cyclohexylamino) ethanesulfonic acid)	9.3	8.6–10.0
EA	Ethanolamine, hydrochloride	9.5	8.9–10.2
CAPSO	3-(Cyclohexylamino)-2-hydroxy-1-propanesulfonic acid	9.6	8.9–10.3
AMP	2-Amino-2-methyl-1-propanol	9.7	9.0–10.5
CAPS	2-(Cyclohexylamino)-1-propanesulfonic acid	10.4	9.7–11.1
CABS	4-(Cyclohexylamino)-1-butanesulfonic acid	10.7	10.0–11.4

Box 2.8: How to Prepare Buffers?

1) Common procedure
   Example: 1 l, 0.1 M Tris/HCl pH 8.1
   - Step one: Dissolve the basic component in part of the final volume: 12.11 g Tris base (tris(hydroxymethyl)aminomethane, $M_r = 121.14$) in 600 ml H_2O
   - Step two: Add the acid component (1 N HCl) dropwise under pH control until the desired pH (8.1) is reached
   - Step three: Fill up to 1 l with H_2O
2) Phosphate buffer
   Example: 0.1 M potassium phosphate pH 7.5
   - Step one: Prepare 1 M stock solutions (concentrated solutions are more stable)
     - 136.1 g KH_2PO_4 fill up to 1 l with H_2O
     - 228.2 g $K_2HPO_4 \cdot 3H_2O$, fill up to 1 l with H_2O
     - Store both solutions in the refrigerator
   - Step two: Dilute aliquots of both solutions 10-fold (e.g., fill up 100 ml to 1 l)
   - Step three: Adjust at the pH meter, the acid solution (KH_2PO_4) to pH 7.5 by pouring the basic solution (K_2HPO_4) under permanent stirring

2.2.7.5 How to Prepare Buffers?

Box 2.8 describes the general procedure for preparation of buffers. The weak component (e.g., Tris) is dissolved in lesser (60–80%) than the final volume, and a (e.g., 10 times) concentrated solution of the strong component (e.g., HCl) is added slowly under permanent stirring until the desired pH is reached. Now the solution is adjusted to the final volume with water. If large quantities of the same buffer are needed, for example, for routine tests, it may be favorable to prepare a concentrated stock solution and to dilute it just before usage. This also has the advantage that concentrated solutions are less susceptible to microbial growth. It must, however, be considered that the pH of the buffer depends on its concentration, and after dilution, the actual pH must be controlled. It is therefore advisable, instead of storing the concentrated buffer to prepare separate concentrated stock solutions for each component and to dilute and adjust the pH just before usage. Special buffer systems and their preparations are described in Box 2.9.

Box 2.9: Special Buffers

Phosphate buffered saline (PBS, physiological salt solution) 0.02 M sodium/potassium phosphate pH 7.2, 0.9% NaCl

Procedure
Dissolve 1.09 g KH_2PO_4, 2.14 g $Na_2PO_4 \cdot H_2O$, and 9.0 g NaCl in 1 l H_2O

Universal buffer according to Teorell and Stenhagen, buffer range pH 2.0–12.0.

Solutions

- 100 ml 0.33 M citric acid monohydrate, $M_r = 210.1$; 7 g in 100 ml H_2O
- 100 ml 0.33 M phosphoric acid, 85%, $M_r = 98.0$; 2.2 ml in 100 ml H_2O
- 100 ml boric acid, $M_r = 61.8$; 3.54 g in 343 ml 1 M NaOH

Procedure

- Mix the solutions and fill up to 900 ml with H_2O
- Adjust to the desired pH (between 3 and 12) with 1 N HCl
- Fill up to 1 l with H_2O

Universal buffer according to Britton and Robinson, buffer range pH 2.6–11.8
Dissolve in 800 ml

- 6.004 g citric acid, monohydrate (28.6 mM)
- 3.89 g KH_2PO_4 (28.6 mM)
- 5.263 g barbital (28.6 mM)
- 1.77 g boric acid (28.6 mM)

Adjust to the respective pH with 1 N NaOH and bring to 1 l.

2.2.7.6 Temperature Dependency

According to the Van't Hoff rule the reaction velocity accelerates with increasing temperature by a factor of 2–3 per 10 °C. This is also valid for enzyme reactions. Thus, the activity of enzymes depends essentially on the assay temperature and it is an important question which temperature should be used. As enzymes are valuable compounds, assay conditions yielding highest activities are preferred. This holds for the already discussed pH optimum and in comparison to this, a temperature optimum is often mentioned, although no such optimum exists. According to the Van't Hoff rule, the enzyme activity should increase with the temperature without passing any optimum. However, practical reasons, such as boiling of water, limit the assay temperature. A more severe limitation is the thermosensitivity of the protein structure. At elevated temperature, enzymes become irreversibly denatured and precipitate, losing their catalytic capacity. Denaturation does not occur at the same temperature for all enzymes, rather it depends on various factors determining the enzyme stability. Some enzymes are very thermosensitive, while others, especially from thermophilic organisms, maintain their stability even at the temperature of boiling water. It is difficult to predict the relative temperature stability of a particular enzyme from its structure and the same type of enzymes can be thermosensitive in one organism and thermostable in another, with only minor structural modifications. It is argued that thermostability is not really a problem for proteins. They have been temperature resistant in earlier times when the temperature on the earth was considerably higher. Concomitant with the general lowering of temperature, thermostability was no longer required and became lost. Because of the diverse temperature preferences of enzymes a common assay temperature for all enzymes cannot be defined.

When considering the temperature dependency of the enzyme activity, both effects, the increase of the reaction velocity due to the Van't Hoff rule, and the decrease because of thermal denaturation overlap so that an apparent optimum curve results (Figure 2.21). However, even this apparent optimum is not constant for a particular enzyme, but depends on the incubation conditions. Incubated at a higher temperature, the enzyme denatures in a time-dependent process, which usually follows exponential first-order kinetics, becoming linear in a half logarithmic representation (Figure 2.22). Only at low and moderate temperatures will the enzyme be stable. A diagram based on the Arrhenius equation (Box 2.10) makes this behavior more obvious. Plotting $\ln v$ against $1/T$ (absolute temperature, Kelvin, $0\,^{\circ}\mathrm{C} = \sim 273.15$ K) yields a straight line indicating the range of stability of the enzyme, that is, the validity of the Van't Hoff rule (Figure 2.21). At higher temperatures, when the enzyme is thermal destabilized, the curve deviates from linearity, the reaction rate slows down, and finally reaches zero. This denaturation process is time dependent and more pronounced, the longer the enzyme remains exposed to the high temperature. Therefore, it makes a severe difference whether the assay is carried out immediately after addition of the enzyme or whether the enzyme is preincubated for a particular time interval at this temperature. This is already the case if, after addition of the enzyme to the assay mixture, some

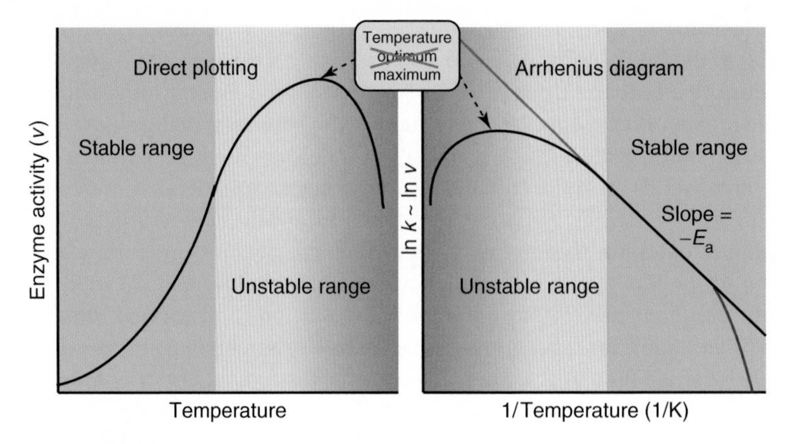

Figure 2.21 Temperature behavior of the enzyme activity in direct plotting and the Arrhenius diagram. Experimental region (black); hypothetical progression in the absence of enzyme denaturation (red); example for a physiological, temperature-dependent transition of the enzyme (blue).

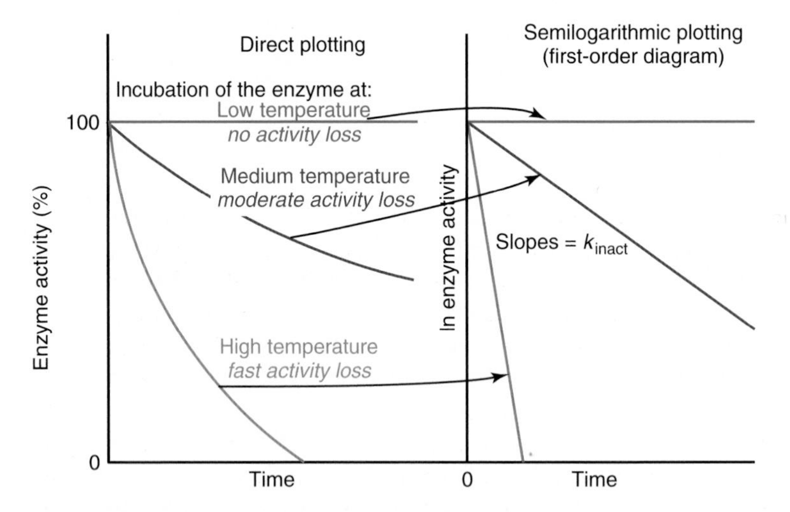

Figure 2.22 Temperature stability of the enzyme activity in direct and semilogarithmic diagrams. The enzyme is preincubated in different temperatures and tested after particular time intervals at the normal assay temperature.

time is needed to start the test. The actual velocity will decrease and the activity maximum in Figure 2.21 shifts to lower temperatures. This occurs only in the higher temperature range of denaturation; in the lower range, preincubation should not reduce the enzyme activity (however, other reasons for destabilization of the enzyme can exist, cf. Section 2.2.7.7).

Box 2.10: Arrhenius Equation

$$k = Ae^{\frac{-E_a}{RT}}$$

Transforming into the logarithm

$$\ln k = \ln A - \frac{E_a}{RT}$$

$$\log k = \log A - \frac{E_a}{2.3RT}$$

$$\log \frac{k_2}{k_1} = \frac{E_a}{2.3R}\left(\frac{T_2 - T_1}{T_1 T_2}\right)$$

Definitions

E_a: activation energy
A: collision factor
T: absolute temperature
R: gas constant
k: catalytic rate constant (can be replaced by the velocity v)

Sometimes deviations from linearity in the Arrhenius plot are observed even in the lower (stable) range, caused by transitions between different active states of the enzyme. Also, the presence of isoenzymes with the same activity, but differing in their structure can be responsible for such inhomogeneities (Figure 2.21).

Owing to these reasons it is not recommended to take the thermal activity maximum as the assay temperature. Rather a temperature in the stable, linear range should be chosen; evidently, the highest temperature just before onset of denaturation. Because this temperature depends on the respective enzyme, a universal standard temperature cannot be defined, but at least a preferential temperature suitable for the majority of the enzymes may be taken. Dependent on the importance of the respective criteria, one of three potential standard temperatures can be chosen, 25, 30, or 37 °C (Box 2.11). Tempering of the assay is inevitably required to maintain constant temperature. Room temperature is not reliable, as it depends not only on the actual room climate, but also on the influences from the outside, such as solar radiation and open windows or doors. Particular enzymes, such as thermophilic or thermosensitive enzymes have special temperature requirements. Some enzymes are even cold sensitive and become destabilized at very low temperatures. The most significant consequences of the temperature dependency of enzymes are summarized in Box 2.12.

Box 2.11: Standard Temperatures for Enzyme Assays

Temperature (°C)	Reasons in favor	Reasons against
25	Slightly above room temperature Easy to maintain No need for preincubation of the assay mixture Suited especially for thermosensitive enzymes	Low enzyme activities Relatively high enzyme amounts required per assay
30	Only modest tempering necessary Closer to physiological conditions Medium enzyme activity Reasonable enzyme requirement	Preincubation of assay mixture required
37	Physiological temperature High enzyme activities Low enzyme amounts required	Preincubation of assay inevitably Long warming up intervals Risk of incipient inactivation

Box 2.12: Consequences of Temperature Dependency of Enzymes

- The Arrhenius diagram should show a linear region, within which the enzyme is stable
- Activation energy calculated from the slope is a measure of the catalytic efficiency of the enzyme ranging mostly between 40 and 60 kJ mol^{-1}
- Inhomogeneities in the linear region can be indications for conformational changes or isoenzymes
- At high temperatures enzymes become instable
- This works against the van't Hoff rule (steady increase of reaction rate with temperature)
- Consequently, the temperature curve declines, passing a maximum temperature (falsely designated as optimum)
- The maximum temperature is not constant; it depends on time the enzyme is exposed to this temperature (the longer the exposure the lower the temperature maximum)
- Hence the temperature maximum is not suitable as assay temperature, although the enzyme develops its highest activity there
- Assay temperature should be within the linear range

2.2.7.7 Why Are Enzymes Unstable?

It is generally observed, that the stability of enzymes is limited. In aqueous solution, they lose their activity within weeks or even days, and some become inactivated even within hours. The chemical nature of enzymes and their protein structure, gives no direct indication for such instability. Actually, proteins can be considerably stable. Therefore, knowledge of the reasons for the instability can help to prolong the lifetime of a particular enzyme. In the previous chapters instabilities due to extreme temperatures, pH values, or ionic strength have been mentioned, and such conditions must be avoided. Keeping the enzyme at low temperature (+4 °C) in a buffered solution at its pH optimum is generally recommended. Besides influences due to inappropriate conditions, proteolytic attacks and chemical modifications caused by oxidative processes or reactive components in the solution are frequent reasons for inactivation (Box 2.13).

Box 2.13: Reasons for Inactivation of Enzymes: General Rules; Special Enzymes Can Have Different Requirements

Effect	Conditions of inactivation	Protection
Temperature	High temperature Very low temperature (seldom)	Low temperature (~4 °C)
pH	Extreme (low and high) pH values	pH optimum
Ionic strength	Very high (>1 M) and very low (<10 mM) electrolyte concentrations[a]	Medium electrolyte concentration (~0.05–0.2 M)
Proteolysis	Contaminations with proteases, not completely removed by purification procedure	Protease inhibitors (cf. Table 2.3) Use of recombinant enzymes
Chemical modification	Contamination with reactive reagents (SH-active compounds)	High purity of all components Avoidance of oxidative conditions Chelating reagents
SH poisoning	Oxidative conditions, promoted by heavy metal ions	SH reagents, mercaptoethanol, dithiothreitol (DTT), dithioerythritol (DTE)

[a] Enzymes from halophilic and thermophilic organisms prefer concentrated milieu.

Proteolytic attack is mostly due to contaminations with proteases from the same source from which the enzyme was isolated. Even very low amounts, not detectable in electrophoresis, can have detrimental effects, proteases are also enzymes acting in catalytic amounts. The action of a protease becomes evident with the disappearance of the enzyme band during electrophoresis, concomitant with the appearance of new, smaller bands. Protease inhibitors can prevent proteolysis. Some special inhibitors are efficient against most proteases, while others react with only one protease type (Table 2.3). This unintended copurification of proteases

Table 2.3 Frequently applied additives for enzyme assays.

Additive	Mostly applied substance	Concentration range	Application and remarks
Monovalent cations	K^+, Na^+	Dependent on special enzyme requirements, ~0.1 M	Essential for several enzyme reactions and for protein structure (counterions of surplus charges) Components of several buffers
Bivalent cations	Mg^{2+}, Ca^{2+}	Dependent on special enzyme requirements, ~1 mM	Cofactors of several enzyme reactions (metalloproteases, kinases, neutralization of di- and triphosphates (e.g., ATP, GTP, ThDP) Stabilization of protein structure
Chelate former	EDTA, EGTA	1–2 mM	Capture divalent cations to protect from oxidative processes, however also removal of cations essential for catalysis and structure
Thiol reagents	DTT, DTE mercaptoethanol	0.1–0.2 mM	Protects SH group from oxidation and formation of –S–S-bridges
Protease inhibitors	α_2-macroglobulin, leupeptin	~0.5 μg ml^{-1}	General protease inhibitors
	TCLK	~0.05 μg ml^{-1}	Inhibitors for serine and cysteine proteases
	TPCK	0.1 μg ml^{-1}	
	PMSF	0.1 mg ml^{-1}	Inhibits serine proteases, instable in aqueous solution (0.5 h half-life!)
	E-64, calpain inhibitor I	~0.01 mg ml^{-1}	Inhibits cysteine proteases
	EDTA	1 mM	Inhibit metalloproteases by complexing of divalent cations
Proteins	Serum albumin	1 mg ml^{-1}	Protects against denaturation especially in diluted solutions
Oxygen traps	GOD/catalase	cf. assay in Section 3.3.1.14	Prevent oxidative modifications
	sodium dithionite	~0.1 mM	Strong reducing reagent

Calpain inhibitor I, *N*-acetyl-leu-leu-norleucinal; DTE, dithioerythritol; DTT, dithiothreitol; E-64, *N*-[*N*-(L-3-trans-carboxirane-2-carbonyl)L-leucyl]-agmatine; EDTA, ethylenediaminetetraacetic acid; EGTA, ethylene glycol-bis(β-aminoethyl ether)-*N,N,N′,N′*-tetraacetic acid; GOD, glucose oxidase; PMSF, phenylmethylsulfonyl fluoride; TCLK, L-1-chloro-3-(4-tosylamido)-7-amino-2-heptanone; ThDP, thiamine diphosphate; TPCK, L-1-chloro-3-(4-tosylamido)-4-phenyl-2-butanone.

with the desired enzyme is a further reason to gain recombinant enzymes instead of preparing them from natural sources.

2.2.7.8 How Can Enzymes Be Stabilized?

Enzymes feel best in their natural environment in the cell. The cell medium is highly concentrated especially with proteins and, accordingly, enzymes prefer high protein concentrations, and dislike strong dilution. This can be achieved by a high concentration of the enzyme itself (10–20 mg ml^{-1}). If high amounts of the enzyme are not available, addition of other proteins has a similar effect. To add foreign proteins to highly purified enzyme appears strange; therefore only inert proteins are applied, mostly serum albumin. This protein also tempers harmful proteolytic or oxidative effects.

When treating enzymes, for example, during purification procedures, unsuitable conditions cannot always be completely avoided and temporarily the structure of the enzyme may became partially deformed. During such occurences the enzyme can lose some of its activity. Often structural areas from the inside, especially hydrophobic regions, become exposed and can either be a target for proteases (while native proteins are mostly protected, at least against proteases of the own cell), or cause aggregation with other proteins, producing insoluble, irreversible precipitates.

To avoid inactivation of the enzyme various additives can be introduced into the assay solution, depending on the special features of the respective enzyme. If not disturbing for the assay it is generally advantageous to add substrates and/or cofactors. Their binding stabilizes the three-dimensional enzyme structure. Mono- and bivalent metal ions are frequently added. They act as counterions of surplus charges and, bivalent cations, especially, have structure stabilizing effects with particular enzymes. On the other hand, heavy metal ions support oxidative modification, especially, of thiol groups and are trapped with chelate forming substances such as EDTA, but care must be taken in the removal of functional cations. To avoid poisoning of SH groups, thiol active reagents are added. Oxygen can be removed from the assay solution by applying a vacuum or degassing with a nitrogen stream. Sodium dithionite is very efficient in capturing traces of oxygen, but it is a strong reducing reagent, which can damage the enzyme. A powerful oxygen trap is the coupled enzyme system, glucose oxidase (GOD) and peroxidase (POD) or catalase (*cf.* assay in Section 3.3.1.14). Additives for stabilizing enzymes are summarized in Table 2.3.

2.2.7.9 How to Store Enzymes?

A further problem of limited enzyme stability is the storage of enzymes, especially for longer periods or for shipping. Several methods, summarized in Box 2.14, are suggested. Their application depends on the compatibility with the respective enzyme. As already mentioned enzymes do not like to exist in diluted solutions, but even in concentrated solutions at low temperature their stability is limited, for example, because of proteolytic attack. Such solutions are also susceptible to the growth of microorganisms, which use the enzyme as nutriment. A simple

Box 2.14: How to Store Enzymes?

Storage method	Conditions[a]	Advantages	Disadvantages
Aqueous solution	Buffered solution (pH optimum, 4 °C), protectives (protease inhibitors, chelating, and SH reagents)	Gentle method No special treatment For short-term storage only (few days)	Danger of proteolytic attack, oxidative processes, microbial growth
Freezing	−20 to −80 °C, buffered solution, with 20–30% glycerol or sucrose	Fast and simple method	Shearing forces (water crystals) are harmful for protein structure Permanent freezing required (risk of freezer defect, difficult to transport), avoid repetitive freezing
Lyophilization	Solutions with volatile buffers to avoid high ionic strength, addition of stabilizers (e.g., glycerol) recommended	Stable preparation, easy to handle (e.g., mailing) relatively resistant to temperature fluctuations	Lyophilization procedure often harmful for protein structure Risk of denaturation Contaminant substances (buffer ions) accumulate
Crystallization	Special crystallization conditions (ammonium sulfate, polyethylene glycol)	Stable preparation for long-term storage (years)	Laborious method, applicable only for few enzymes, which are easy to crystallize
Precipitation	Ammonium sulfate or polyethylene glycol precipitation	Simple method for long-term storage	Rough method, tolerated only by stable enzymes

[a]High protein concentration (10–20 mg ml^{-1}) recommended for all methods.

preservation method is freezing, and in the frozen state the enzyme may be stable for long time, but the main problem is the conversion into this state (and back into solution), which is stressful for the enzyme, especially due to the shearing forces of the ice structure. Additives such as glycerol or sucrose moderate this effect, but repetitive freezing and thawing of the same enzyme sample should be avoided. It is advisable to divide the enzyme preparation into small samples before freezing. Similar problems of conversion into another state exist with the other methods, such as precipitation and lyophilization. If the enzyme tolerates lyophilization it is a practical method. Storage in the refrigerator (4 °C) is sufficient and enzyme samples can easily be shipped in this state. For any special enzyme the most tolerated method must be tried out.

References

Good, N.E. and Izawa, S. (1972) *Meth. Enzymol.*, **24B**, 53–68.

Good, N.E., Winget, G.D., Winter, W., Connolly, T.N., Izawa, S., and Singh, R.M.M. (1966) *Biochemistry*, **5**, 467–477.

Stoll, V.S. and Blanchard, J.S. (1990) *Meth. Enzymol.*, **182**, 24–38.

Teorell, T. and Stenhagen, E. (1939) *Biochem. Z.*, **299**, 416–419.

2.3 Instrumental Aspects

2.3.1 Spectroscopic Methods

2.3.1.1 Absorption (UV/Vis) Photometry

The different spectroscopic techniques are valuable tools for enzyme analysis and enzyme assays. Unquestionably, the most applied method is absorption photometry in the UV/Vis range. Advantages as well as disadvantages are summarized in Box 2.15. Besides the relatively low costs and easy handling, its broad applicability making the absorption photometer indispensable for any biochemical laboratory, is emphasized. For enzyme assays the possibility of continuous measurements is of decisive importance. On the other hand, its sensitivity and selectivity is limited and, therefore, this method is not very suitable to identify special substances. Since the chemical differences between enzyme substrates and their products is often minor, they mostly cannot be discerned by absorption spectroscopy and due to this fact many reactions are difficult to observe by this technique. But because of its convenience, assays are often modified to render them accessible for photometric studies.

A broad variety of instruments is offered and it is not always easy to select the appropriate apparatus. If finance is not a constraint, a universal spectrophotometer equipped with all extras may be purchased, but this makes operation often more complicated. An instrument possessing just the features required serves in many cases the same purpose, especially for routine tests.

For the selection of a suitable instrument and for appropriate operation it is essential to understand the features and the mode of action of absorption photometers. The fundamental principle of absorption photometers is simple: light penetrates a solution, where part of the light is absorbed by a dissolved substance. The photometer quantifies the light intensity lost as measure of the amount of the respective substance.

Box 2.15: Absorption Spectroscopy (UV/Vis)

Criteria for application

- Generally applicable (all substances absorb within the accessible range)
- Easy handling
- Moderate price
- Variety of applications
  - Protein determination
  - Nucleic acid determination
  - Metabolite concentration
  - Enzyme assays
  - Binding measurements
  - Conformational analysis

Disadvantages

- Moderate sensitivity
- Poor selectivity

The first essential part of an absorption photometer is the **light source**. Principally, any light source can be taken; the German inventors of the photometer, Elster and Geitel from Göttingen (1891) used sunlight. Since the sun is not always shining, and even if it shines, its intensity is not constant, a more stable light source is preferred. A further requirement is the selection of particular wavelengths and the ability to scan spectra. For this a light source covering the whole spectral range from 200 to 800 nm with the same intensity for each wavelength is required. Unfortunately, such a light source does not exist. Commercially available lamps cover only part of the desired spectral range, and even within this limited spectral range the light intensity is not constant. One type, deuterium or hydrogen lamps, emits UV light, while a second type, tungsten and halogen lamps, covers the visible range. So for the whole spectral range both lamp types must be combined (Figure 2.23). A movable mirror directs light from the respective lamp onto the sample. Newer instruments perform this automatically, while for older ones this must be done manually. Wrong mirror positions cause operating errors (cf. Box 2.20). A third type of lamps (Hg and Cd lamps) emits lines of particular wavelengths, instead of continuous spectral light and accordingly measurements are only possible at these fixed wavelengths. The desired wavelength is selected by an interference filter, which absorbs all other wavelengths. This principle of a **filter photometer** (Figure 2.23) is suitable for routine tests, when only a fixed wavelength is needed, provided that the lamp emits a line within the interesting range. Since absorption bands are usually broad and it is not always necessary to measure exactly at the maximum, a suitable wavelength can be found most of the time. Box 2.17 and Figure 2.27 demonstrate this for NAD(P)H, where two lines of a Hg lamp can be chosen, if it is not possible to measure directly at the maximum at 340 nm.

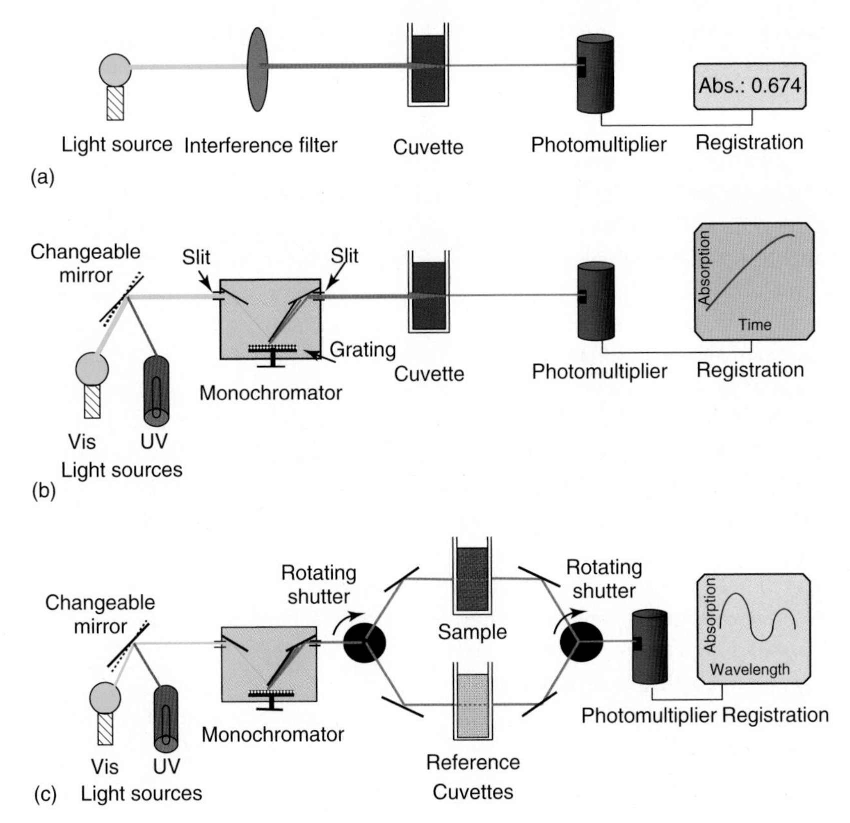

Figure 2.23 Construction schemes of absorption photometers. (a) Filter photometer. (b) UV/Vis spectrophotometer. (c) Double beam spectrophotometer.

Although the whole spectral range can be covered with the UV and Vis lamps, the fact remains, that the emitted light intensity is not the same for all wavelengths, rather each lamp possesses a special characteristic comparable to an optimum curve. When measuring at a single wavelength, the respective intensity can be adjusted to 100%, but through scanning spectra the variable intensities will superimpose and deform the spectra. To avoid this, spectrophotometers possess correction functions so that the operator need not take care of this; but it would be helpful to be aware of this fact.

The device for selection of a particular wavelength is the second essential part of a photometer. Unlike the filter photometer, **spectrophotometers** use a **monochromator** for wavelength selection. Older instruments dispersed the light through prisms, while newer ones prefer gratings. They show linear progression of wavelength when the grating is rotated with constant velocity by a motor to scan spectra. Prisms, in contrast, show no linear relationship, but they produce purer monochromatic light, while gratings show a considerably high stray light portion. Special coatings reduce this effect, and the groove density is also decisive for the quality of the grating. In fact, the quality of the monochromator determines, essentially, the value of the photometer. It must be kept in mind that the monochromator does not provide a single wavelength, but cuts out a part of the whole spectrum, which is broader for the basic apparatus and sharper for high-quality instruments. Constriction of the entrance and the exit slits of the monochromator sharpens the cut-out window, but this only makes sense with high-resolution instruments. The sharper the light beam, the lower the light intensity and the stability of the instrument. With normal enzyme assays usually, the higher intensity of a broader spectral range is more favorable than purer wavelengths at the expense of lower intensity.

The monochromatic light passes the **cuvette** containing the sample solution, which is placed in a cuvette holder, a light dense closable compartment. According to the mode of application various types of cuvettes are available, which will be described together with instructions for handling them, in the following section. Special instruments possess cuvette holders with several (e.g., six) cuvette positions. They can, one after the other, be automatically exposed to the light beam allowing simultaneous determinations of several samples. This is advantageous especially for routine assays, while for accurate studies recording of the complete progress curve of one sample is preferable. Enzyme assays must be carried out at a constant temperature, and tempering of the cuvette holder by a special heating device or an external water bath is indispensable.

The **double-beam photometer** (Figure 2.23) possesses two separate cuvette holders, one for the sample, and the second for a reference. After leaving the monochromator the light is split into two identical beams, one penetrating the sample, and the other penetrating the reference cuvette on exactly parallel paths. The difference between both cuvettes is displayed as photometric signal; the value of the sample is automatically reduced by the blank value, and drifts in sample and blank (e.g., due to instability of a component) are compensated. Such instruments allow the detection of small differences within strongly absorbing systems, if the absorbing component is present both in the sample and the reference cuvette in the same concentration. This feature serves to observe the formation of a product even if its spectrum differs only slightly from those of the substrate.

A **photomultiplier** detects the intensity of the light beam leaving the cuvette and transfers this signal to a display or a **recording device**, a recorder with pen and paper in older instruments and a monitor in newer instruments. Modern computer-controlled instruments carry out the adjustments of lamps and baseline automatically after switching on, while the operator must do this for older instruments. The absorption in the cuvette is displayed either as a number or as a time-dependent progression. Computer-controlled instruments calculate the velocity directly, but care must be taken, so that for enzyme assays only the initial linear range is taken. Usually the instrument calculates a regression analysis over the whole time range irrespective of whether the curve is linear or shows systematic or artificial deviations. This is demonstrated in Figure 2.24 for a progress curve, which is linear only in its initial part, while the computer calculates the velocity for

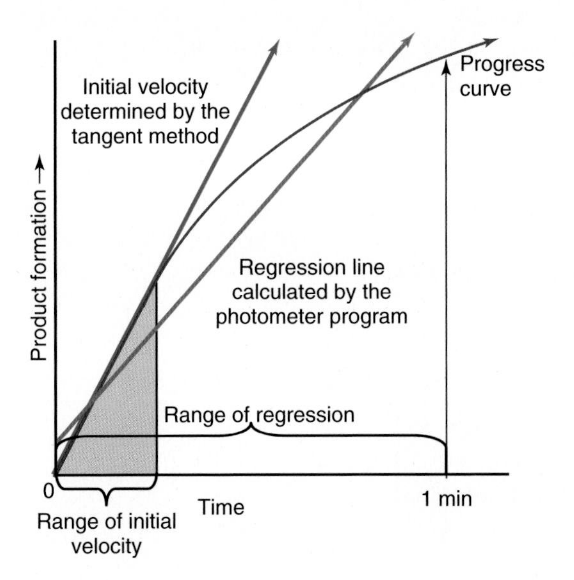

Figure 2.24 Determination of the velocity from a progress curve by regression analysis of the photometric computer program compared with the tangent method.

the whole range. Usually the program allows concentrating to the linear range, but this must be commanded by the operator.

Attention must also be paid to the absorption range, which is likewise adjusted by the computer program, usually for a wide range, for example, between 0 and 2, to cover all measurements. However, for common enzyme assays such a range is far too broad and will give poor sensitivity for the results. Sensitive measurements, such as initial rates, occur in a narrow range within an absorption of 0.1. The consequences of such an adjustment are demonstrated in Figure 2.25, which is an example of a typical progress curve for a dehydrogenase reaction. It attains an absorption of about 0.3 within 1 min. The whole progress curve is viewed on the screen, but it fills only the lower part (Figure 2.25, left diagram). If the display is adjusted to 0.1 as maximum absorption the initial part of the progress curve expands over the whole screen, the initial reaction becomes clearly visible, and now it is obvious that the reaction proceeds too fast and the linear range runs off within few seconds, that is, too much enzyme has been applied. Tenfold dilution yields just the initial part of the progress curve within a convenient time (Figure 2.25, right diagram).

Another point of attention concerns the baseline. It may be assumed, that a reaction starts always at zero absorption and, accordingly, programs adjust the baseline to zero. In reality this must not be the case. The absorption is usually adjusted to zero *before* addition of the component, which initiates the reaction, the enzyme, or the substrate. The added substance can itself contribute to the absorption and the reaction start deviates from zero. Within a broad absorption

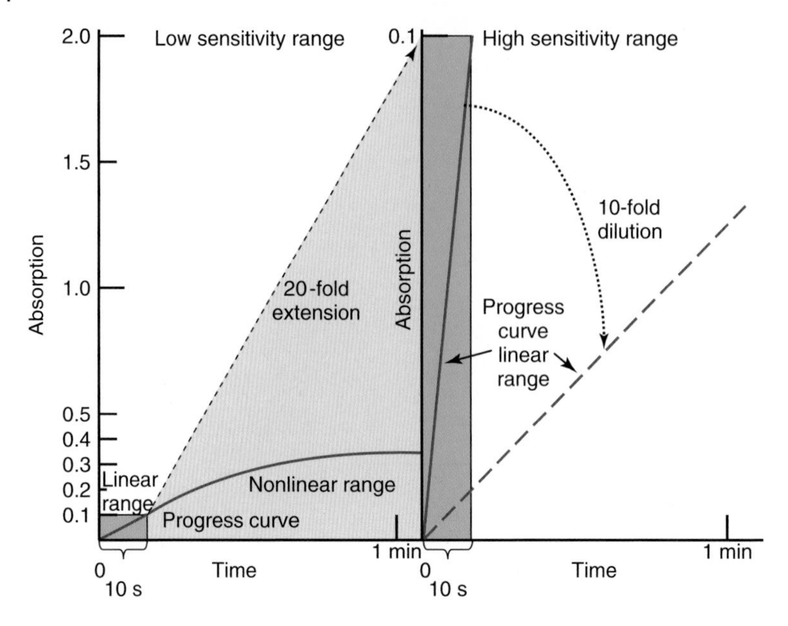

Figure 2.25 Progress curves of an enzyme-catalyzed reaction. The left diagram shows the adjustment of the photometric computer program to a low sensitive range; the right diagram with the sensitivity increased by a factor of 20, shows the same progress curve and a progress curve of a 10-fold diluted enzyme solution.

range this may be tolerated, but it can easily exceed a more sensitive range, so that the reaction proceeds outside the screen (Figure 2.26). However, to return to the less sensitive range is not the appropriate consequence. Rather, by control experiments the extent of the absorption jump caused by the added component should be estimated and considered for baseline adjustment. This is also the case if the initial absorption falls below zero, for example, due to dilution of the assay mixture by the added sample. Finally, it must be considered, that several reactions proceed with an absorption decrease (if substrate consumption is measured) and, if started at zero absorption, the reaction will immediately run out of the screen. In this case the baseline should be adjusted to the upper part of the screen. Fortunately, the instruments record reactions even outside the screen (if not too far apart) and the result can be visualized after the end of the measurement, but it is inconvenient if the actual reaction cannot directly be observed.

A newer development is the **diode array photometer**, where the total light beam passes the cuvette directly, with the monochromator being mounted behind the cuvette. After leaving the monochromator, the spectral dispersed light impinges on an array of diodes, so that the instrument measures the whole spectrum instantly and time-dependent spectral changes can be observed.

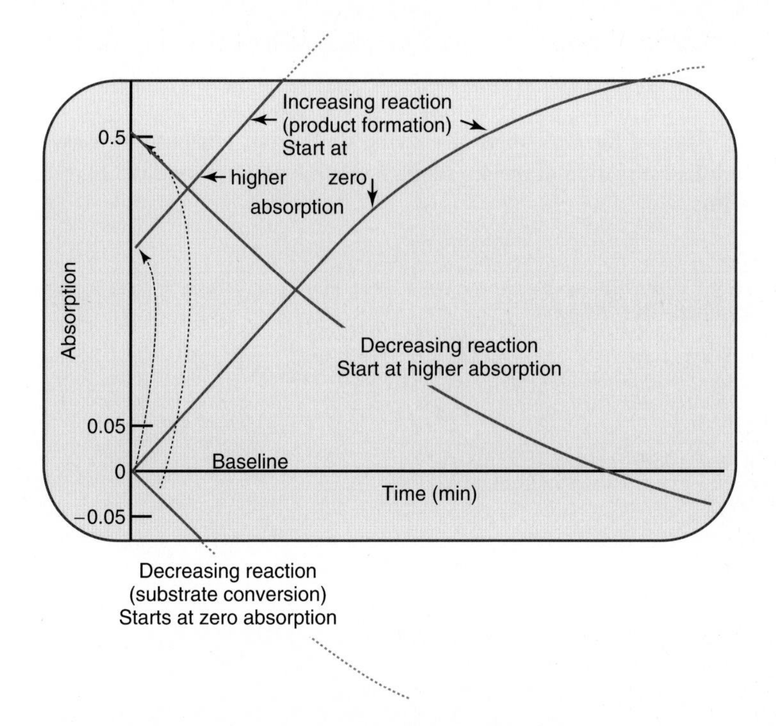

Figure 2.26 Progress curves of enzyme reactions displayed on a photometric screen. The dotted lines show the progression of the reaction traces outside the screen.

Another special device is the **ELISA or microplate reader**, which is mostly based on the filter photometer technique, so that only particular wavelengths can be chosen. It measures consecutively, the absorption in the (96) wells of a microtiter plate. This instrument is preferentially applied for the ELISA method, but other tests, such as enzyme assays or protein determinations can also be adapted to microtiter plates, which allow fast determinations of serial tests with small volumes, such as variation of substrate or inhibitor concentrations. ELISA readers are available also for fluorescence measurements. In Chapter 3 some examples for enzyme assays with the microplate reader are presented, but most other assays can also be adapted to this technique.

As already mentioned the (dimensionless) photometric units are called **absorption** (or **absorbance**, the previous expression "extinction" should not be used), although the direct photometric signal is the **transmittance** T, the ratio of the light intensity I reduced by the absorbing substance to the intensity of the blank I_0 (Box 2.16). The transmittance is exponentially related to the concentration c of the absorbing substance. A linear relationship between the negative logarithm of the transmittance, the absorption A, and the concentration, is expressed by the

Lambert–Beer's law (Box 2.16). The factor ε, the **absorption coefficient** (previously extinction coefficient) is a material constant of the absorbing substance and allows the calculation of its molar concentration from the absorption (Box 2.17). Its knowledge is essential for evaluation of enzyme units from photometric assays (Box 2.18) and the estimation of the appropriate enzyme amount required for the assay (Box 2.19).

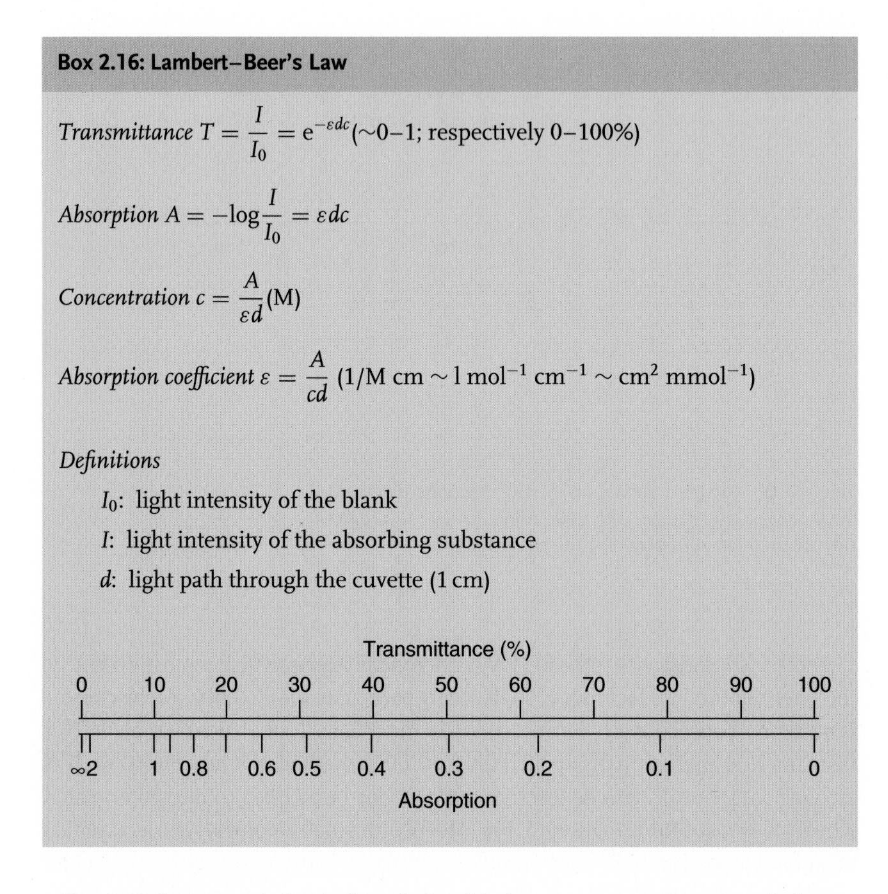

Box 2.16: Lambert–Beer's Law

Transmittance $T = \frac{I}{I_0} = e^{-\varepsilon dc}$ (~0–1; respectively 0–100%)

Absorption $A = -\log\frac{I}{I_0} = \varepsilon dc$

Concentration $c = \frac{A}{\varepsilon d}$ (M)

Absorption coefficient $\varepsilon = \frac{A}{cd}$ ($1/\mathrm{M\,cm} \sim 1\,\mathrm{mol^{-1}\,cm^{-1}} \sim \mathrm{cm^2\,mmol^{-1}}$)

Definitions

I_0: light intensity of the blank

I: light intensity of the absorbing substance

d: light path through the cuvette (1 cm)

Box 2.16 (lower part) shows the relationship between transmittance and absorption. At maximum transmittance (100%) absorption is zero. Absorption increases with the concentration of the absorbing substance to infinity, finally, while transmittance declines to zero. The scale demonstrates that the main absorption region ranges between 0 and 1, while the region between 1 and ∞ is strongly compressed and gives no reliable values. Actually, the validity of the Lambert–Beer's law is restricted to the lower region; absorption values above 1 are no more strictly proportional to the concentration of the absorbing substance. Quantitative measurements are only reliable within the lower region; if the absorption exceeds 1 the solution must be diluted accordingly. This must also be considered when using a double-beam photometer, which displays absorption differences between samples and blanks. If both possess a high basic absorption (e.g., 1.8), but only a small difference (e.g., 0.2), the measurement is outside the validity of the Lambert–Beer's law, although the instrument displays a low value.

Box 2.17: Absorption Coefficient

Dimensions of the molar absorption coefficient ε

$$1\,\text{mol}^{-1}\,\text{cm}^{-1} = \text{M}^{-1}\,\text{cm}^{-1} \sim \text{cm}^2\,\text{mmol}^{-1}$$

$1\,\text{mol}^{-1}\,\text{mm}^{-1}$ according to the SI system (the value of ε must be divided by 10)

Molar absorption coefficients of substances used for enzyme assays:

Compound	nm	ε ($M^{-1}cm^{-1}$)	Remarks
NAD(P)H	340	6300	Maximum in the visible range
	334	6180	Wavelength of filter photometer
	365	3400	Wavelength of filter photometer
	260	14 300	Maximum in the UV range
NAD(P)	260	17 400	
ABTS	414	24 600	
BAPNA	405	9620	
Cytochrome *c*	550	21 000	Difference oxidized – reduced
Dianisidine	436	8300	
Ellman's reagent	412	14 150	
H_2O_2	240	40 000	
$K_3[Fe(CN)_6]$	436	755	
4-Nitroanilide	405	10 200	
4-Nitrophenol	405	18 500	
	348	5500	pH independent

Absorption photometry is principally an easy technique and not very susceptible to disturbances. Nevertheless, even with this method various mistakes can occur (Box 2.20) and careful performance is required. The cuvettes, especially the windows exposed to the light path, must be completely clean (not touched by hand!), and solutions must be perfectly clear, as air bubbles or dust disturbs the measurement. Accordingly, absorption cannot be determined in turbid solutions or suspensions. Special devices enable measurements in homogenates with debris of cell walls or membranes.

Box 2.18: Calculation of Enzyme Activity for a Photometric Assay

- Volume activity:

$$\text{units/ml} = \frac{\text{measured value} \cdot \text{assay volume} \cdot \text{dilution factor}}{\text{time} \cdot \text{concentration constant} \cdot \text{enzyme volume}}$$

Example: Absorption

$$\text{units/ml} = \frac{\text{absorption/ min} \cdot \text{assay volume} \cdot \text{dilution factor}}{\text{absorption coefficient} \cdot \text{path length} \cdot \text{enzyme volume}}$$

Expressed in:
International units

$$\text{IU/ml} \sim \mu\text{mol/ min /ml}$$

$$= \frac{\text{absorption/ min} \cdot \text{assay volume (ml)} \cdot \text{dilution factor} \cdot 10\,000}{\varepsilon_{\text{nm}}(\text{l} \cdot \text{mol}^{-1}\text{cm}^{-1}) \cdot 1\ \text{cm} \cdot \text{enzyme volume (ml)}}$$

Katal (SI system)

$$\text{kat/ml} \sim \text{mol/s/ml}$$

$$= \frac{\text{absorption/s} \cdot \text{assay volume (ml)} \cdot \text{dilution factor}}{\varepsilon_{\text{nm}}(\text{l} \cdot \text{mol}^{-1}\text{cm}^{-1}) \cdot 1\ \text{cm} \cdot \text{enzyme volume (ml)} \cdot 1000}$$

- Total activity: volume activity × total volume (ml)
- Specific activity: volume activity/mg protein × ml^{-1}
  ~ total activity/mg total protein

Box 2.19: Calculation of the Enzyme Amount for a Photometric Assay: Lactate Dehydrogenase (LDH) as an Example

Recommended absorption difference: ~0.1/min

- 1 IU = 1 μmol conversion/ml/min
- Absorption of 1 M NADH in 1 ml cuvette (1 mmol NADH/ml) ~ 6300
- 1 μmol NADH/ml/min (1 IU) = 6.3 absorption/min
- Absorption of 0.1 ~ 0.016 IU ~ 0.016 μmol NADH/ml/min
- LDH (commercial preparation) 5500 IU ml^{-1} (10 mg ml^{-1})
- 5000-fold dilution (2 μg ml^{-1}) ~ 1.1 IU ml^{-1} = 0.0165 IU in 15 μl

Conclusion:

15 μl LDH of a 5000-fold dilution (0.03 μg ~ 0.2×10^{-12} mol) in 1 ml assay solution produces an absorption change of *0.103/min.*

Box 2.20: Frequent Pitfalls of Photometric Measurements

Error	Detection and elimination
Unexpected high absorption value in the display (even of blank)	Remove cuvette and adjust photometer to zero. If no improvement check the photometer: incorrect wavelength wrong lamp on wrong mirror position slit width too narrow false cuvette orientation (frosted sites in the light beam) If photometer ok, insert a cuvette with water and observe the display: if still unexpectedly high: wrong cuvette (glass or plastic in the UV range?) If ok insert the sample cuvette and read the displayed value. If still too high, check solution, one component may be too concentrated
Unexpectedly low display	Any additive forgotten? Cuvette not in the place of the light beam (if cuvette holder with multiple positions)
Turbid solutions	Centrifuge or filter (sterile filter) the solution

A prerequisite for absorption measurements is the presence of an absorbing substance. Practically, each substance shows absorption at least in the UV range, but, as already discussed, for enzyme assays the absolute absorption of a substrate or a product is not decisive; rather a detectable difference must exist. There are only few examples of really strong differences. The most prominent example is NAD (and NADP), the substrate of many dehydrogenase reactions, revealing clear differences between its oxidized and reduced form (Figure 2.27, Box 2.17). In the UV region at 260 nm there is a strong absorption peak, which is more intense for the oxidized than for the reduced form, the difference signal between both absorptions is $\Delta\varepsilon_{260} = 3.1 \times 10^3\ \mathrm{l\,mol^{-1}\,cm^{-1}}$. A second peak at 340 nm is less intense ($\varepsilon_{340} = 6.3 \times 10^3\ \mathrm{l\,mol^{-1}\,cm^{-1}}$), but as it appears only with the reduced form, the difference signal at this wavelength is twice compared with the more intense UV peak. Another advantage is that this wavelength can be measured with both the visible and the UV device of the photometer. Since most of the other enzyme reactions show no such striking differences, they are, as far as possible, coupled with a dehydrogenase reaction (*coupled assay*, see Section 2.4). Another way to render reactions accessible for photometric measurements is to convert either the product formed or the remaining substrate into a strongly absorbing form by a subsequent chemical reaction. However, this procedure allows no continuous observation of the enzyme reaction, rather it must be stopped after a particular time

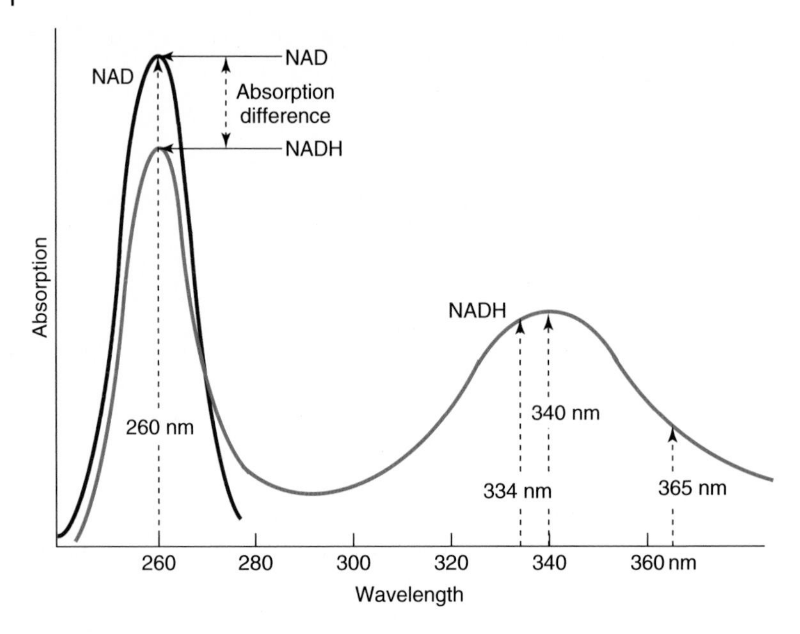

Figure 2.27 Spectra of oxidized and reduced NAD. The wavelengths of the absorption maxima in the UV and visible range and of convenient interference filter are indicated.

to execute the chemical conversion. On the other hand this procedure requires only a simple photometric device, a fixed wavelength, mostly in the visible range, without the need for time-dependent measurements. This technique is called **colorimetry**, and the appropriate instruments are colorimeters.

Computer-controlled instruments possess programs for various applications, which are helpful especially for routine tests, when the same procedure is always performed. But for research studies, when there exist no schematic procedures, the program complicates manipulation and rapid adaptation to varying assay conditions, because for each change the program must be supplied with the actual data.

References

Cantor, C.R. and Schimmel, P.R. (1980) *Biophysical Chemistry*, W.H. Freeman, New York.
Harris, D.H. and Bashford, D.I. (1987) *Spectrophotometry*, IRL Press, Oxford.
Chance, B. (1991) *Annu. Rev. Biophys. Biophys. Chem.*, **20**, 1–28.
Haid, E., Lehmann, P., and Ziegenhorn, J. (1975) *Clin. Chem.*, **21**, 884–887.

2.3.1.2 Cuvettes

Spectral methods require cuvettes for the sample solution and according to the particular technique and application various types of cuvettes are commercially available. The main aspects are discussed taking cuvettes for absorption photometry as example.

Shape For rough determinations, normal circular glass tubes can be used, but usually standard cuvettes are used. They are 3 cm in height and have a quadratic basis, 10 × 10 mm, according to 1 cm light path (Figure 2.23). Front and back sides are plain polished (and should not be touched), and right and left sides are frosted for handling. The light beam penetrates the front and back windows in a rectangular direction to minimize reflections (which occur actually with circular tubes). Up to 3 ml of solution can be filled. For smaller volumes it must established, that the light beam passes completely through the solution. If the volume is too small the beam will touch the meniscus yielding deficient values. The shape and position of the light beam depends on the respective instrument; some instruments tolerate smaller volumes, while others do not. If the minimal volume is not known it must be determined.[10)]

Since 3 ml is a rather large quantity for many enzyme assays, reduced cuvettes for smaller volumes exist. Principally, the path length of 1 cm should be maintained, smaller light paths reduce the sensitivity of the method. Frequently applied are reduced cuvettes with thickened left and right sides, and front and back sides corresponding still to the standard cuvette. The total volume reduces to 1 ml (half-microcuvettes), but also cuvettes for smaller volumes are available (microcuvettes). For these the above discussion about the position of the light path must be referred to, which is usually adjusted to standard and not to reduced cuvettes. Deviations will result if the light path touches the thickened sides, especially if the cuvette is repeatedly removed and replaced into the cuvette holder. To avoid such errors reduced cuvettes with black sides are recommended. Even cuvettes for volumes down to 0.1 ml are available, but their sites must obligatorily be blackened. For special applications such as diluted samples, cuvettes with longer path lengths (2–10 cm) are used, but they require a suitable cuvette holder.

Cuvettes for other spectroscopic techniques are adapted to the respective method. Fluorescence cuvettes are similar to absorption cuvettes, but all sides must be plain polished, as fluorescence is emitted in all directions. Thus, care must be taken in handling these cuvettes. Gloves should be used and only the upper part above the light path should be touched. For CD and ORD cylindrical cuvettes are usually used, which are in different shapes and path lengths.

Material Glass is the suitable material for cuvettes, but it has two essential disadvantages: it breaks easily and it is impermeable for UV light. So its application is limited to the visible region. Unbreakable plastic cuvettes are available. They possess the additional advantage that they can be discarded after usage, rendering cleaning unnecessary, but producing wastes. Also, cuvettes made from UV

10) The position of the light beam can be visualized, selecting a wavelength in the visible range and inserting a white paper in the place of the cuvette.

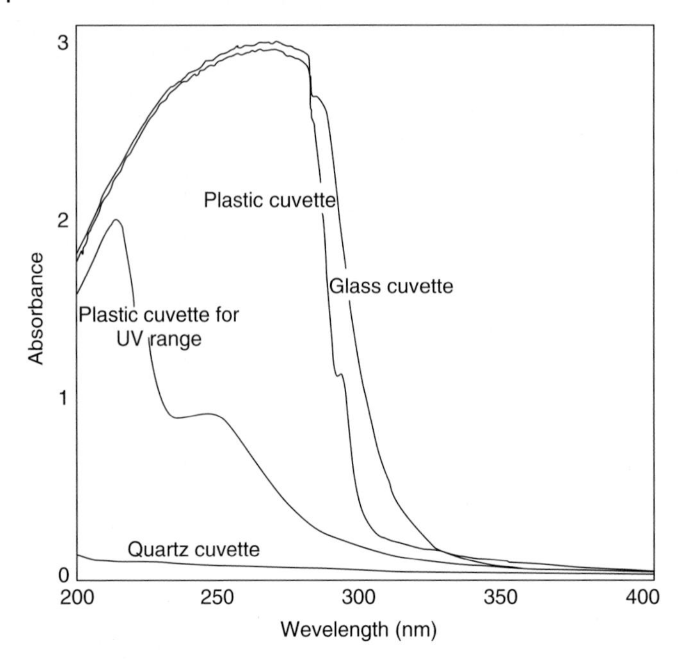

Figure 2.28 Spectral features of cuvettes made from different materials.

permeable plastic material are offered (Figure 2.28). Otherwise expensive quartz cuvettes must be used for the UV range.

Cleaning Cleaning of cuvettes is a special technique. Accurate spectroscopic experiments certainly require the highest claims for cleanliness. Disposable plastic cuvettes serve this criterion, although they do not satisfy higher optical demands. Glass and quartz cuvettes can be cleaned with distilled water, but especially after repeated usage obstinate contaminations may remain, which can be removed by dishwashing detergents. Suppliers of cuvettes offer special cleaning agents. Sometimes concentrated hydrochloric acid is used, but this is harmful and can damage the glued edges of the cuvette. Protein contaminations can be removed with alkaline solutions. If the same cuvette is repeatedly used, it should be dried after each use. This is often achieved by rinsing with methanol or acetone, but such a procedure is not recommended, since contaminations are even more adsorbed to the glass walls. By drying with filter paper fibers may remain in the cuvette. Sometimes it is sufficient to beat out the remaining liquid drops carefully before refilling; the adherent liquid traces still present are not really disturbing, if the same solution is used repeatedly. Special cuvette centrifuges, which instantly remove residual liquids together with contaminations, are very advantageous.

2.3.1.3 Turbidity Measurements

Incident light is scattered in suspensions, emulsions, or colloids. This *Tyndall effect* can be used for assays, where larger particles such as lysozyme hydrolyzing bacterial cell walls, cellulases digesting cellulose particles, or lipase acting on emulsified lipids are enzymatically degraded. Principally, two different parameters can be measured, the reduction of the incident light (**turbidimetry**), or the intensity of the scattered light (**nephelometry**).

Determination of turbidity is comparable to absorption photometry, with the main difference, that the light attenuation is not dependent on the wavelengths. Therefore, these measurements can be carried out in common photometers, but special instruments are also offered. The general photometer structure, light path, slit width, and cuvette volume is well suited for turbidity measurements also. The cuvette position should be distant from the photomultiplier to exclude scattering light as far as possible from the incident beam. A monochromator is not really necessary and any wavelength (e.g., 450 nm) can be chosen.

For nephelometric measurements just the scattering light is determined and the incident beam must be excluded. This is achieved, as with fluorescence photometry (see following section), by measuring the light intensity in the right angle from the incident beam. Hence fluorimeter and fluorescence cuvettes, polished on all sides, can be used. Unlike real fluorescence measurements, the wavelengths for excitation and emission must be identical. As with turbidity measurements, scatter is independent of the wavelength, which may be selected within the medium visible range. For quantification, a calibration curve with a standard suspension, for example, with formazin (formaldehyde azine) should be measured.

Reference

Bock, R. (1980) *Methoden der Analytischen Chemie*, vol. 2, part 1, Verlag Chemie, Weinheim, pp. 323–328.

2.3.1.4 Fluorescence Photometry

Compared with UV/Vis photometry, fluorimetric measurements are significantly more sensitive (by about a factor of 100), an important feature for enzyme assays. Nevertheless, this method is seldom applied, due to various disturbances and the expensive equipment, which needs thorough experience (Box 2.21). The most serious limitation, however, is the fact that only few compounds demonstrate the phenomenon of light emission (**fluorophores**), although just this feature makes fluorescence more selective than absorption spectroscopy. A distinct spectrum can more easily be associated with a special compound. According to theory, an electron of the absorbing substance occupies an excited state upon absorbing a photon, and fluorescence light is emitted when the electron returns to its ground state. Since energy is lost during this process, the emitted light is always of longer wavelength than the absorbed light. However, emission can only occur if the excited state persists for a period of at least some nanoseconds, but during this time other processes compete for its energy. If these processes are faster than the lifetime of the excited state, it becomes deactivated and no fluorescence (only absorption) will be observed. Only few compounds, preferentially those with conjugated, aromatic, or heterocyclic structures, maintain the exited state for a longer time and show fluorescence.

Box 2.21: Fluorescence Measurements

Advantages

- Linear dependency on concentration
- High sensitivity
- High specifity
- Sensitive against environmental influences (e.g., polarity)

Disadvantages

- Elaborate instruments
- Quenching
  - Impurities
  - Concentration quenching
  - Inner filter effect
  - Temperature
  - Scattering
  - Raman effect
- No absolute signal (relative fluorescence intensity depending on the instrument)
- Accumulation of different effects

As fluorescence light is emitted in all directions, it is measured at right angles to avoid overlap with the strong incident, exciting beam. The construction scheme of a **spectrofluorimeter** (Figure 2.29) resembles principally that of an absorption photometer (cf. Figure 2.23). One essential difference concerns the lamp, which must emit intense light for excitation. Mostly a xenon arc lamp is used, where a luminous arc is generated between two electrodes applying high voltage. Because this arc can move during the measurement, the signal is not completely stable and high-quality instruments have special devices to compensate for such fluctuations. Since the emission wavelengths are in the lower spectral range, only one lamp is used, covering both the UV and shorter visible range. As already discussed for the lamps used in absorption photometry, the intensity of the xenon lamp also has a wavelength-dependent characteristic; it is weak in the far UV range below 250 nm as well as in the visible range above 500 nm with a maximum intensity between 300 and 400 nm. Because this characteristic deforms the fluorescence spectra, correction functions are provided with high-quality instruments.

Before entering the cuvette, the light beam is dispersed in an excitation monochromator. Because the light emitted by the fluorophore in the right angle and not the

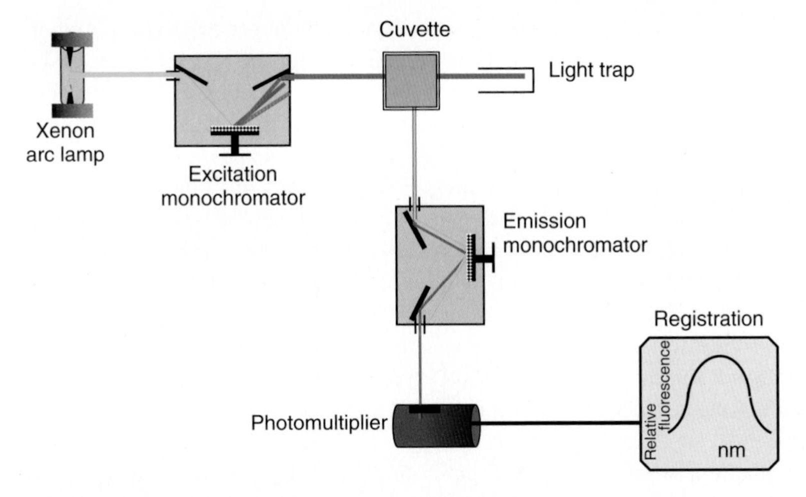

Figure 2.29 Construction scheme of a spectrofluorimeter.

intensity of the incident beam is measured, fluorescence cuvettes are plain polished on all sides. For the study of the emitted light, a second monochromator, the emission monochromator, is mounted at right angles to the incident beam. Accordingly, two types of spectra are obtained. The **excitation spectrum**, obtained by scanning the excitation monochromator, corresponds principally to the absorption spectrum of the emitting substance. The **emission spectrum** is always at higher wavelengths compared to the excitation. Thus, for assays both the excitation and the emission wavelength must be known and usually the maximum values of both spectra are taken.[11] If experiments are carried out only at fixed wavelengths, monochromators are not necessary and can be replaced by interference filters. Simple fluorimeters spare even the selection for the emission wavelength, detecting the total emission, which is often sufficient and has the advantage of higher intensity. However, care must be taken to exclude the scattering peak. Part of the incident beam is scattered, with various effects contributing to this scattering: reflections at the cuvette windows or at particles (dust, air bubbles, precipitates, macromolecules) in the solution, or Rayleigh scattering of molecules. Scattering light must be excluded by stray filters or the emission monochromator.

An important difference between fluorescence and absorption photometry is the fact, that in the absence of a fluorescent substance no light reaches the photomultiplier, while in the absence of an absorbing substance the total light intensity arrives at the absorption photomultiplier. Therefore, highly sensitive photomultipliers detecting even traces of light can be used only for fluorescence spectroscopy, rendering this a very sensitive technique. Unlike the exponential dependency of

11) Sometimes disturbances, like Raman or scattering bands force deviation from the maximum.

the absorption, emission is directly proportional to the fluorophore concentration and the fluorescence intensity should rise linearly with the concentration of the emitting substance. This is true, indeed, for very low amounts, while at higher concentrations a deviation of the fluorescence intensity from linearity and finally a decrease, generally called **quenching**, is observed. This happens in the presence of any high concentration, also of contaminating compounds. Therefore, fluorescence should be measured only in diluted solutions and only substances of highest purity should be taken. Quenching effects also have oxygen dissolved in aqueous solution and increase of temperature (Box 2.21). The fluorescence intensity is also reduced by the inner filter effect, occurring when a compound in the cuvette reabsorbs the emitted light. When measuring at high sensitivity, sharp symmetric **Raman bands** appear at constant distance from the excitation wavelength, already in water in the absence of a fluorophore, and can easily be misinterpreted as real fluorescence.

The **quantum yield**, the ratio of photons emitted to photons absorbed, is a characteristic material constant, comparable to the absorption coefficient in UV/Vis spectroscopy. It is a measure of the emission intensity of the compound and ranges from 0 to 1 (respectively, 0–100%). It is, however, not frequently used, because it is more difficult to determine the absolute concentration of a compound from its fluorescence. A difficulty of fluorescence measurements is that there is apparently no upper limit for emission, unlike UV/Vis photometry, which ranges between 0 and 1 transmission. Therefore, fluorescence cannot be calibrated by the instrument, and the yield of emitted light depends essentially on the construction pattern of the fluorimeter, the lamp intensity, the light path, and the photomultiplier sensitivity. Owing to the instability of the arc lamp, even fluorescence values measured on the same instrument under identical conditions can suffer time-dependent variations. Therefore, the fluorescence signal is displayed only in relative values (*relative fluorescence*). For absolute determinations, calibration with inner standards, such as quinine sulfate as a stable fluorophore, is required.

Since only few substances emit fluorescence light, fluorescence photometry may be regarded as a method suitable only for special applications. However, looking more closely, a variety of applications exist. Natural fluorescent compounds and metabolites are summarized in Box 2.22. Since the three aromatic amino acids, especially tryptophan, show fluorescence emission, practically all proteins are susceptible to this effect, but this signal serves less for enzyme assays. More important is the fact that NAD(P) emits fluorescence, but only in its reduced form. Owing to this feature, all NAD(P)-dependent dehydrogenases can be tested with this technique with higher sensitivity compared to the absorption method. However, besides NAD(P)H, flavins and a few other metabolites, such as anthranilic acid and bile acids, the majority of enzyme substrates and products are nonfluorescent. In such cases fluorescent chromophores can be introduced. Prominent examples are ethenoadenosine nucleotides, which possess an additional five-membered ring from the nitrogen at position 1 to the 6-NH_2 group of adenine. This modification converts the nonfluorescent compounds AMP, ADP, ATP, coenzyme A (CoA), and also the oxidized NAD into fluorophores. It causes only a small structural change, so that the modified compounds are mostly recognized like the natural ones. Frequently, bioconjugates are used, where a fluorescent group is conjugated to a biomolecule, for example, umbelliferone (7-hydroxycoumarin), bound to lipid or carbohydrate substrates.

Box 2.22: Fluorescent Substances for Enzyme Assays

Compound	Maxima (nm) Excitation	Emission
NADH, NADPH	340	470
FMN, FAD	445	536
Tryptophan	280	350
Tyrosin	273	303
Pyridoxal phosphate	295	392
Fluorescein	495	518
Rhodamine B	575	630
Anilinonaphthalene sulfonate (ANS)	350	515
Dimethylaminonaphthalene sulfonylchloride (DANSYL-Cl)	320	580
4-Methylumbelliferone	358	448
Etheno-ATP	265 300	410

Fluorescence is sensitive to polarity and many fluorophores show a characteristic increase in emission intensity, often connected with a blue shift upon transfer from a polar solution, such as water to an apolar milieu. Release of fluorophoric groups bound to a less polar macromolecular structure, such as membrane lipids or even proteins, into aqueous solution is accompanied by loss of fluorescence intensity, so that this process can directly be followed fluorimetrically. Various enzyme assays are based on this principle. A great variety of fluorescent substances is available. Fluorescein isothiocyanate (FITC), which conjugates with amino groups, for example, from proteins, is broadly used. Alexa Fluor dyes are very efficient fluorescent bioconjugates, available for various wavelengths.[12)]

Fluorescence resonance energy transfer (*FRET*) is a frequently applied method, where a pair of fluorophores is fixed to a biological system, for example, an enzyme. The emission wavelength of one fluorophore (*donor*) must overlap with the absorption band of the other (*acceptor*). If the donor is excited, the (not excited) acceptor emits light. This happens by radiationless energy transfer and is strictly dependent on the distance between both fluorophores, which must be shorter than 8 nm. With this method neighboring positions (such as catalytic and regulatory

12) Molecular Probes Inc., R.G. Haugland, Eugene, Oregon, USA.

centers of enzymes) can be detected and the distance between both fluorophores can be estimated from the intensity of the acceptor fluorescence.

References

Cantor, C.R. and Schimmel, P.R. (1980) *Biophysical Chemistry*, W.H. Freeman, New York.
Guibault, G.G. (1990) *Practical Fluorescence*, Marcel Dekker, New York.
Dewey, T.G. (1991) *Biophysical and Biochemical Aspects of Fluorescence Spectroscopy*, Plenum, New York.

2.3.1.5 Luminometry

One of the most sensitive enzymatic test methods is the luminescence reaction. It is based on the luciferase reaction (EC 1.13.12.7)

$$\text{luciferin} + \text{ATP} + \text{O}_2 \xrightarrow{\text{luciferase}} \text{oxyluciferin} + \text{PPi} + \text{H}_2\text{O} + \text{light}$$

a reaction proceeding by emission of light. It is responsible for the glow of the firefly. ATP is determined using luciferin as substrate. The emission maximum is at 562 nm. ATP concentrations down to 1×10^{-15} mol can be detected (cf. Section 3.3.1.33). This high sensitivity is utilized to couple the luciferase reaction with ATP generating reactions, for example, the creatine kinase (CK, EC 2.7.3.2)

$$\text{creatine phosphate} + \text{ADP} \xrightarrow{\text{CK}} \text{creatine} + \text{ATP}$$

or the pyruvate kinase (PK, EC 2.7.1.40)

$$\text{phosphoenolpyruvate} + \text{ADP} \overset{\text{PK}}{\rightleftharpoons} \text{pyruvate} + \text{ATP}$$

With the bacterial luciferase from *Vibrio (Photobacterium) fischeri*, NAD(P)H concentrations as low as 10^{-10} M can be measured. The first step in the reaction sequence is catalyzed by a flavine mononucleotide (FMN) reductase (NAD(P)H dehydrogenase, EC 1.6.8.1)

$$\text{NADPH} + \text{H}^+ + \text{FMN} \longrightarrow \text{NADP}^+ + \text{FMNH}_2$$

followed by alkanal monooxygenase (EC 1.14.14.3)

$$\text{FMNH}_2 + \text{RCHO} + \text{O}_2 \longrightarrow \text{FMN} + \text{RCOOH} + \text{H}_2\text{O} + \text{light}$$

RCHO is a long-chain aliphatic aldehyde (8–14 C atoms), and light is emitted at a maximum wavelength of 493 nm.

Bioluminescence can be measured in a liquid scintillation counter or with a fluorimeter, but special instruments (*luminometer*), are also available. They are about 1×10^5 times more sensitive than absorption photometers. Because light is emitted by the enzyme reaction itself, no light source is needed; rather it is important to shield the instrument from any infiltrating light. Also a monochromator is not required; the total light emitted from the sample is measured with a photomultiplier. The sample should be close to the detector, a feature, which is not sufficiently realized in other instruments. The signal is given as *relative light unit* (RLU). Some instruments count *photons per second*.

The light output is directly proportional to the concentration-limiting component, for example, ATP in the luciferase assay. After starting the reaction the light impulses are measured for a particular time interval, for example, 15 s. The light output depends on various factors, especially on the reaction volume and must be carefully standardized. A stable light source, a defined radioactive sample in a scintillation cocktail or a chemiluminescent reaction, placed instead of the cuvette into the sample chamber, is used for standardization of the system.

References

Campbell, A.K. (1989) in *Essays in Biochemistry*, vol. 24 (eds R.D. Marshall and K.F. Tipton), Academic Press, London, pp. 41–81.

DeLuca, M. and McElroy, W.D. (1978) *Meth. Enzymol.*, **57**, 3–15.

Seitz, W.R. and Neary, M.P. (1976) *Methods Biochem. Anal.*, **23**, 161–188.

Stehler, B.L. (1968) *Methods Biochem. Anal.*, **16**, 99–179.

2.3.1.6 Polarimetry

This method makes use of the fact that most biological substances possess at least one asymmetric center, a carbon atom carrying four different ligands or an asymmetric three-dimensional structure, which causes a deflexion of the plane of polarized light. Especially for carbohydrates, polarimetry often yields the clearest optical signal. Actually, the first thoroughly investigated enzymatic reaction, that of invertase cleaving saccharose into glucose and fructose, was investigated in this manner.

The light in the polarimeter is linearly polarized by a polarizing filter (*Nicol's prism*) before entering the cuvette. A solution containing an asymmetric compound causes a deviation of the plane of polarization from the original orientation. A second polarization filter, the *analyzer*, is mounted behind the cuvette. In the absence of an optical active substance, passage of polarized light is only possible if the filter is oriented parallel to the polarizing filter (baseline), but in the presence of an asymmetric compound it must be rotated in the direction of the deviation until the photodetector detects the maximum light intensity. The angle of rotation α depends on the nature of the compound, its concentration (c), the path length (l), and the wavelength. Rotation is stronger at shorter than at longer wavelengths. In contrast to the rectangular cuvettes in absorption photometry, cylindrical cuvettes, mostly with path length of 10 cm for dilute solutions, are used. The *specific rotation* $[\alpha]$ is an intrinsic constant of the respective compound and is defined as the rotation α in degrees (°) of a solution of 1 g/ml in a cuvette of 10 cm path length: $\alpha = [\alpha]lc$.

With a simple polarimeter, the rotation is directly observed by eye through a polarimeter tube. The analyzing filter is manually rotated to approach the maximum of the incident light, when the position of the filter matches just the deviation of the polarization plane. The degree of rotation can be read from the analyzer position. With modern ORD spectrometer the rotation of the plane of linear polarized light can be automatically followed with dependence on both time and wavelength.

2.3.2 Electrochemical Methods

2.3.2.1 pH Meter and Glass Electrodes

Since the activity of enzymes depends strictly on the pH, in any study with enzymes the pH must carefully be controlled. For this purpose as well as for preparation of buffers and for special enzyme assays, pH meters are indispensable tools in biochemical laboratories. Various types of pH meters and electrodes are available. The most essential criterion is the accuracy; further advantageous features are a plastic jacket to protect against breakage, especially caused by magnetic stirrers, and small diameters for measuring directly in cuvettes. The fact that glass electrodes as single-rod measuring cells are meanwhile standard, allows one to easily forget that the principle of pH measuring is the determination of the potential difference, the **electromotive force** E, between an indicator E_i and a reference electrode E_r, which is directly proportional to the pH value

$$E = \{E_i + (RT \times F^{-1}) \ln a_{H+}\} - E_r = (-59.15 \text{ mV}) \text{ pH} - E_r$$

R is the gas constant, F the Faraday constant ($2.3026 \times RT \times F^{-1} = 59.15$ (at 25 °C) is the steepness, i.e., the voltage change per pH unit), a_{H+} is the activity of hydrogen ions. In water, the glass electrode generates a thin gel layer at its surface allowing only hydrogen ions to penetrate (Figure 2.30). Commercial devices are

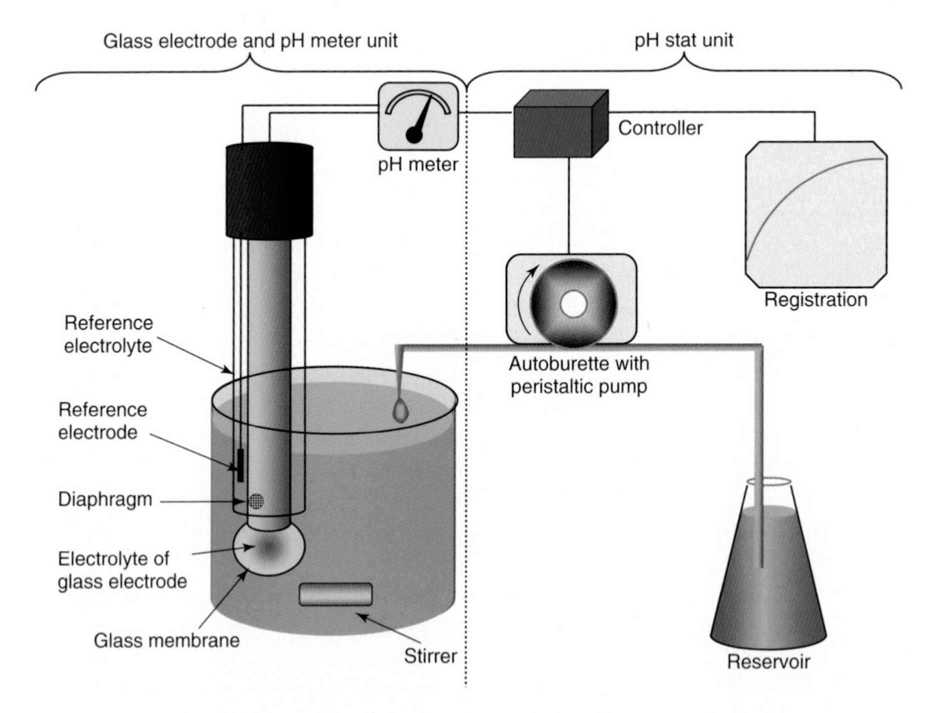

Figure 2.30 Glass electrode with single-rod measuring cell, connected with a pH meter and a pH stat.

manufactured as single-rod measuring cells, where the reference electrode encloses concentrically the glass electrode. The pH meter acts like a voltmeter; it determines the potential between glass and reference electrode. Measurements between pH 2 and 12 can be performed with the glass electrode. The reference electrode (except from electrodes with solid or gel electrolytes) must be refilled with 3 M KCl (23 g KCl in 100 ml H_2O) from time to time and the glass electrode should be stored in this solution. Special attention must be drawn to the diaphragm of the reference electrolyte, which mediates the contact with the solution. Protein solutions often block the diaphragm. Treatment in an ultrasonic bath, short immersing in alkaline and acid solutions, or digestion with pepsin (0.5 g in 100 ml 0.4% HCl) for some hours is recommended for cleaning.

Before usage the pH meter must be calibrated. The zero point is controlled with a pH 7.0 buffer solution (at this pH glass electrodes have usually a potential of zero), and, thereafter, the steepness is adjusted with an acid or alkaline buffer (pH 4.0 or 9.0). Calibration must be done at a defined temperature, for example, 25 °C.

The precision of pH measurements depends essentially on the quality of the glass electrode and is normally in the range of 0.03–0.05 pH units, but depends also on external factors such as temperature, accuracy of calibration buffer, state of the diaphragm, and the reference electrolyte.

Under normal conditions glass electrodes can be used for one to three years. Aging is perceptible by a delay in the response time and a decrease of the steepness in the alkaline range. The calibration becomes increasingly difficult. Treatment with 4% NaOH and 4% HCl for 5 min, respectively, can have a regenerating effect.

2.3.2.2 pH Stat

The strong preference for photometric and colorimetric measurements largely displaced the pH stat, although it is a valuable method for continuously pursuing enzyme reactions. However, this relatively complicated device is useful only for special applications, where the pH stat reactions producing a pH change are observed. Such changes can principally be measured with a pH meter equipped with a glass electrode, but the activity of enzymes depends strongly on the pH and becomes affected by the progressing pH change during the reaction, and constant assay conditions cannot be maintained. The pH stat allows continuous observing of pH changes at constant pH. A pH meter with a glass electrode is connected to a controller, which regulates an autotitrator system (Figure 2.30). Change of the pH in the solution triggers an impulse and the titrator supplies acid or alkaline solution until the preadjusted pH is restored. The amount of reagent added by the titrator is registered in dependence of the time. So the progression of the reaction can be followed at constant assay conditions. Examples for enzyme tests with the pH stat are lipase and choline esterase.

2.3.2.3 Potentiometry

This method serves to study redox reactions involving redox pairs, such as $NAD(P)^+/NAD(P)H$, $FAD/FADH_2$, or cytochrome Fe^{2+}/Fe^{3+}. The potential difference between an indicator electrode (platinum electrode) and a reference electrode (calomel electrode) is determined. The potential of the reference electrode is constant and independent of the components, while the indicator electrode detects changes of the active substances. The potential refers to a standard potential of zero of a hydrogen streamed platinum electrode plunging in a 1.228 M HCl solution. Reducing or oxidizing reagents are used to modify the redox state of the redox pair.

2.3.2.4 Oxygen and Carbon Dioxide Electrodes

Many enzyme-catalyzed reactions proceed by consumption or release of gases, mostly oxygen and carbon dioxide. Various methods have been developed to study these reactions by determining the amount of gas converted, such as the manometric technique of Otto Warburg or radioactive labeling. Most convenient is the use of gas-specific electrodes, which are available both for oxygen and carbon dioxide. The **oxygen electrode** developed in 1953 by Clark consists of a platinum wire inside a glass tube as cathode, and a silver/silver chloride anode. Both plunge into a saturated potassium chloride solution, maintaining a constant voltage of 0.5–0.8 V between them. The sample solution must be airtightly sealed within the electrodes compartment, a Teflon membrane separating both the sample and the electrode compartment. The sample is filled and removed with a syringe through a small airtight canal. The oxygen from the sample penetrates the membrane and becomes reduced at the cathode

$$O_2 + 4H^+ + 4e^- \longrightarrow 2H_2O$$

while at the anode silver is oxidized

$$4Ag + 4Cl^- \longrightarrow 4AgCl + 4e^-$$

The current generated is proportional to the oxygen concentration. Owing to the diffusion through the membrane, the electrode shows a rather long delay time of some minutes. For quantification the method is calibrated with a solution completely liberated from oxygen (by a nitrogen stream) and a solution saturated with oxygen.

The **carbon dioxide electrode** resembles a glass electrode for pH measurements, wrapped by a rubber or silicon layer. The electrode plunges into the sample solution and dissolved carbon dioxide penetrates through the layer and becomes hydrated to carbonic acid. The resulting pH change is related to the CO_2 concentration in the solution

$$\Delta pH = S\,\Delta \log pCO_2$$

S is a constant. CO_2 measurements need strict pH control. Quantification can be done with standard hydrogen carbonate solutions or defined partial CO_2 pressures.

References

Clark, L.C., Wolf, R., Granger, D., and Taylor, Z. (1953) *J. Appl. Physiol.*, **6**, 189–193.
Lessler, M.A. and Bierley, G.P. (1969) *Methods Biochem. Anal.*, **17**, 1–29.

Beechey, R.B. and Ribbons, D.W. (1972) *Meth. Microbiol.*, **6B**, 25–53.
Nicolls, D.G. and Garland, P.B. (1972) *Meth. Microbiol.*, **6B**, 55–63.

2.3.3 Radioactive Labeling

Radioactive labeling has a broad field of applications in biochemistry, cell and molecular biology, and in medical diagnosis. However, it is hazardous because of direct radiation and of contaminating materials and equipment, producing wastes, which are difficult to depollute. Therefore, radioactive labeling is replaced more increasingly by nonradioactive methods; a prominent example is the introduction of fluorescence marker for DNA sequencing.

For enzyme assays the enzymatic cleavage of nucleoside triphosphates, such as ATP and GTP, labeled with ^{32}P, can be studied, whereby the released inorganic phosphate $^{32}P_i$ must be separated from the reaction solution. A relative advantage is the short half-life of ^{32}P. Another application is the release of carbon dioxide, where the respective carbon atom of the substrate is labeled with ^{14}C (e.g., C1 of pyruvate for the pyruvate decarboxylase reaction). The reaction, proceeding in airproof vessels, is stopped after a particular time by pouring an acid from a separate compartment inside the vessel into the reaction compartment. A filter disc mounted above the reaction solution and soaked with an alkaline medium, such as hyamin, adsorbs the released $^{14}CO_2$. The filter disc is transferred into a scintillation mixture and the radioactivity is measured in a scintillation counter. Escape of any gas from the vessel must be carefully avoided, since $^{14}CO_2$ as a long-lived isotope can be inhaled and incorporated into the organism.

2.3.4 Diverse Methods

In the previous sections, the most convenient techniques for enzyme assays have been discussed. If, for a special enzyme reaction, none of these methods is applicable, other techniques must be taken into consideration, but they are generally more laborious and less suited for routine tests, although under special circumstances installation of a more complex technique even for routine methods can be necessary. Since such methods usually serve other applications, such as bioanalytics or protein purification and are treated in the pertinent literature, they are only briefly described here.

Referring to the previous consideration, any enzyme assay must identify either the product formed or the disappearance of the substrate. If optical or electrochemical methods do not work, there remain principally, two possibilities. The substrate and the product differ either in their chemical reactivity or in their structure (otherwise both compounds should be identical).

To make use of a difference in the reactivity, the enzyme reaction may be stopped after a particular time, joining an **indicator reaction**. Three aspects must be considered:

1) For termination of the enzyme reaction, inactivating conditions for the enzyme may be chosen, such as high temperature by transferring the assay solution into a boiling temperature bath, or extreme pH change by addition of a strong acid. Frequently trichloroacetic acid or perchloric acid are used, both of which also act by precipitating proteins. Then the concentrations of substrate or product in the assay mixture can be analyzed. Alternatively, a chemical agent trapping the product or the remaining substrate and initiating the chemical indicator reaction is directly added to the assay mixture.
2) It must be established that the respective substance takes part exclusively in the indicator reaction, for example, the product, but not the substrate and also no other compound of the assay solution. Otherwise, the respective substance must be isolated, for example, by extracting with organic solvents or chromatographic methods, before starting the indicator reaction.
3) The resulting product of the indicator reaction should easily be determined, for example, by the colorimetric method (cf. Section 2.3.1.1).

Substrates and products differing in structural features (seize, configuration, charges) can be separated and identified by chromatographic techniques, such as ion exchange, gel filtration, reversed phase or enantiomer selective materials. Conventional chromatographic equipment can be used, but **HPLC** (high performance liquid chromatography) and **FPLC** (fast protein liquid chromatography) devices are strongly recommended. They combine high resolution and precise reproducibility with the advantage of high sensitivity and, thus, requirement for very small samples and are, therefore, suitable also for routine assays. Other separation methods, such as ultracentrifugation, ultrafiltration, precipitation, extraction with organic solvents, can also be used, but resolution and sensitivity are limited and larger amounts of the sample are required.

Although gas-specific electrodes have essentially replaced the **manometric method** of Otto Warburg for the study of gas exchange reactions, it is described in the following text. The reaction vessel manufactured from glass contains a central cylindrical compartment melted into its bottom and a laterally attached pear-shaped compartment, so that the enzyme, the substrate, and the stop solution can be applied separately. The system is connected to a manometer in an airtight manner. By appropriately inclining the device, enzyme and substrate solutions are mixed to start the reaction. During the reaction the vessel is permanently shaken and tempered. After a particular reaction time the stop solution is added, which also serves to release any dissolved gas from the aqueous solution, such as acid for CO_2. The volume change due to gas exchange is read from the manometer scale.

Finally, the course of an enzyme reaction can be followed by measuring the reaction heat with a **calorimeter**. Reactions proceed with either heat release (*exothermic*, negative reaction enthalpy) or heat consumption (*endothermic*, positive reaction enthalpy), and the change in heat can be observed in a calorimeter. Microcalorimeters with sample volumes down to 0.2 ml are available.

2.4
Theory of Coupled Enzyme Reactions

2.4.1
Two Coupled Reactions

If an enzyme reaction is not accessible to a convenient detection method it can be coupled to an easily measurable reaction. A prerequisite is that the coupled reactions must either accept the product of the test reaction as its own substrate, or the product of the coupled reaction must be accepted as substrate by the test reaction. If there is no such reaction directly available, even three reactions can be coupled. Hexokinase (HK) catalyzing the phosphorylation of D-glucose is an example for coupling

$$\text{D-glucose} + \text{ATP} \xrightarrow{\text{HK}} \text{glucose-6-P} + \text{ADP}$$

Substrates and products exhibit no significant absorption difference. If the reaction is coupled with the enzyme glucose-6-phosphate dehydrogenase (G6PDH), it takes glucose-6-P as substrate

$$\text{glucose-6-P} + \text{NADP}^+ \xrightarrow{\text{G6PDH}} \text{gluconate-6-P} + \text{NADPH} + \text{H}^+$$

and reduces $NADP^+$, so that the progress of the reaction can be followed through the absorption increase at 340 nm. The coupled reaction is called **indicator reaction** (G6PDH in our example), and the reaction under study (HK), the **test** (or helper) **reaction.**

Some important aspects must be considered for coupled assays. Optimum conditions should be maintained for both enzymes, but this is problematic if they differ in essential features, such as pH optimum or temperature behavior. For instance, if one enzyme is thermophilic it will show its highest activity at a temperature where the other one denatures. Also of importance is the state of the equilibrium of the reactions involved. At least the equilibrium of the final reaction must favor the product site to guarantee quantitative conversion. If this is not the case, the final reaction cannot be a reliable measure for the test reaction, and such a combination is inappropriate. On the other hand, the equilibrium of the primary reaction favoring its substrate is not a problem, as long as the subsequent reaction proceeds quantitatively to the product site, because this reaction eliminates all the products of the preceding reaction and forces it also to proceed quantitatively – against its own equilibrium.

Thus, for the development of a coupled enzyme assay the first requirement is to determine test conditions compatible for all involved enzymes. If such conditions cannot be found, the assay can only be conducted stepwise, starting with the first reaction until it comes to the end, joining the second reaction thereafter.

The indicator reaction should never become limiting; its enzymes, cofactors, and cosubstrates, such as NADP in the above example, must be present in a large surplus. One should also be aware of the fact that any change of the test conditions during the experiment (e.g., substrate or cofactor concentration, addition of inhibitors) may alter the conditions, making this prerequisite no longer valid so that the indicator reaction can become rate limiting, yielding incorrect results.

Two distinct arrangements, requiring different treatments, are possible with coupled enzyme assays. In most cases the indicator reaction follows the reaction to be analyzed, as in the upper example, and the detectable component develops at the end of the reaction sequence. The reverse case, where the indicator reaction precedes the test reaction is seldom.

After the start of a coupled assay with the indicator reaction at the end of the reaction sequence, it needs some time to reach a stationary phase, where formation and conversion of the intermediate substrate (P_{int}) is constant

$$S \xrightarrow{E_{test}} P_{int} \xrightarrow{E_{in}} P_{end}$$

This stationary phase is the optimum state for the coupled assay and should be reached as fast as possible. To realize this, the test enzyme E_{test} must be limiting and must work under conditions of substrate saturation $[S] \gg K_m$, so that $v_{test} \cong V_{test}$. The indicator reaction works with a surplus of the indicator enzyme E_{in}. When the reaction starts a lag phase[13] is observed, during which the steady-state concentration of P_{int} increases (Figure 2.31). As P_{int} is supplemented by the primary reaction at constant velocity, the indicator enzyme also reacts with constant velocity v_{in}, which obeys the Michaelis–Menten equation with $[P_{int}]$ as substrate concentration. Because of $v_{test} \cong V_{test}$, the effective velocity of the indicator reaction must be equal to the maximum velocity of the test reaction V_{test}

$$v_{in} = \frac{V_{in}[P_{int}]}{K_{m/in} + [P_{int}]} = V_{test} \tag{2.20}$$

The concentration of the intermediate P_{int} under these conditions is

$$[P_{int}] = \frac{K_{m/in} V_{test}}{V_{in} - V_{test}} \tag{2.21}$$

Obviously, the intermediate concentration $[P_{int}]$ behaves in a reverse manner to the maximum velocity of the indicator reaction and thus to the amount of the indicator enzyme $V_{in} = k_{in}[E_{in}]$. Large amounts of the indicator enzyme thus reduce $[P_{int}]$ and improve the conditions for the coupled assay for linearity. Because of the versatile connections between the two reactions under various relationships of components involved, there are different possibilities of order of the resulting reaction. Zero order, that is, linearity, is achieved if the test reaction proceeds with its maximum velocity V_{test}. In the stationary state, if $[P_{int}]$ = constant

$$v_{in} = \frac{V_{in}[P_{int}]}{K_{m/in} + [P_{int}]} = \frac{V_{in}}{\frac{K_{m/in}}{[P_{int}]} + 1} \tag{2.22}$$

the denominator $K_{m/in}[P_{int}]^{-1} + 1$ will also be constant and the reaction proceeds in a linear manner with constant velocity. It is important, that the stationary (zero-order)

13) Lag phases can sometimes be observed with normal reactions, when the enzyme is subject to a slow activation process, but can have also artificial reasons, such as warming up of the assay solution.

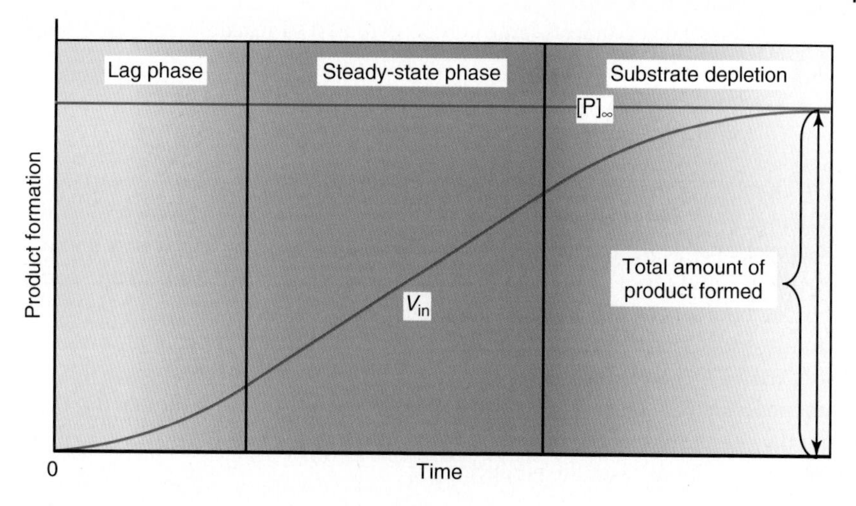

Figure 2.31 Progress curve of a coupled enzyme assay. For substrate determination the concentration of the product at the end of the reaction is measured, as indicated.

state is reached quickly, since the initial lag phase approaches asymptotically and the substrate concentration decreases constantly so that the time period, during which substrate saturation prevails, is limited.

Coupled reactions with two substrates are treated in a similar manner. As long as both substrates can be regarded as saturating, the condition $v_{test} \cong V_{test}$ should hold. But it is often difficult to fulfill conditions of saturation for both substrates simultaneously, for example, with dehydrogenases, NADH cannot be added in too high concentration due to its high absorbance.

The indicator reaction follows usually the reaction of the enzyme to be tested. In special cases, for example, when the substrate of the test enzyme is unstable and must be provided by the preceding reaction, the situation can be reversed. If the preceding reaction serves only as provider for the substrate and the test reaction can directly be measured, there is no principal problem, as long as the first enzyme is present in a surplus, so that the test reaction becomes rate limiting. This is the case for the malate dehydrogenase (MDH) reaction

$$\text{oxaloacetate} + \text{NADH} + \text{H}^+ \xrightarrow{\text{MDH}} \text{malate} + \text{NAD}^+$$

when the unstable oxaloacetate is supplemented by the aspartate aminotransferase (AAT) reaction

$$\text{aspartate} + \text{2-oxoglutarate} \xrightarrow{\text{AAT}} \text{oxaloacetate} + \text{glutamate}$$

More problematic is the situation, where the provider reaction is at the same time the indicator reaction to be determined, as in the case of citrate synthase (CS), where oxaloacetate becomes supplemented by the MDH reaction

$$\text{malate} + \text{NAD}^+ \xrightarrow{\text{MDH}} \text{oxaloacetate} + \text{NADH} + \text{H}^+$$
$$\text{acetylCoA} + \text{oxaloacetate} \xrightarrow{\text{CS}} \text{citrate} + \text{CoA}$$

This reaction sequence is used for the determination of acetate (respectively, acetyl-CoA). However, the amount of NADH formed in the first reaction must not be proportional to the acetate turnover in the second reaction. This is only the case when the equilibrium of the first (indicator) reaction favors the substrate. Under this condition only small amounts of NADH will be formed, until the CS reaction captures oxaloacetate and thus, forces its formation, and consequently also that of NADH. To enable this, the equilibrium of the CS reaction must favor the end product. This is a general rule for coupled assays with initial indicator reactions that the equilibrium of the first reaction should favor the substrate, and that of the test enzyme should favor the product.

2.4.2 Three Coupled Reactions

The situation with three coupled enzyme reactions is principally similar. The first reaction carried out by the test enzyme is followed by a second enzyme reaction (E_{con}), connecting the test reaction with the indicator reaction

$$\text{S} \xrightarrow{\text{E}_{\text{test}}} \text{P}_{\text{int1}} \xrightarrow{\text{E}_{\text{con}}} \text{P}_{\text{int2}} \xrightarrow{\text{E}_{\text{in}}} \text{P}_{\text{end}}$$

Here both intermediate products P_{int1} and P_{int2} must be kept very small to reduce the initial lag phase. The effective velocities of both the connecting and the indicator enzyme should be similar and equal to the maximum velocity of the test reaction

$$V_{\text{test}} = v_{\text{con}} = v_{\text{in}}$$

2.5 Substrate Determination

The high specificity of enzymes can be used to determine precisely concentrations of compounds such as metabolites even in crude cellular extracts. The respective compound must be accepted by an enzyme as its substrate and converted to a product. The relative change due to the enzyme reaction indicates the presence and the amount of the respective compound. The fact that the respective compound is not identified by its absolute absorption, but by its relative change enables the detection out of a variety of other metabolites. Care must be taken especially in crude extracts so that the determination will not be disturbed by side reactions. The reaction will either be directly followed if substrate or product can be detected by an appropriate measuring method, or coupled to a further enzymatic reaction with a detectable product, such as a dehydrogenase. Two principally different methods for enzymatic substrate determination are applied, namely, the end point and the kinetic methods.

2.5.1 End Point Method

The end point method is the best and simplest procedure, provided that the substrate or the product can directly be determined and the reaction proceeds irreversibly. The reaction is started with the addition of the enzyme, employing the respective test conditions. The course of the reaction must be followed, either continuously (e.g., photometrically), or by removing and analyzing samples after particular time intervals. When reaching its end and no more turnover can be detected, the reaction will attain a positive plateau value if product formation (Figure 2.31), a negative value if substrate consumption is measured. The plateau value is directly related to the amount of product formed and can be calculated from its absorption coefficient or a calibration curve of a known standard solution. As long as the reaction proceeds irreversibly, substrate and product concentrations should be identical. It must, however, be considered that the real plateau value will be reached only *ad infinitum* and that the value will be underestimated if the experiment is terminated too early. Since slow enzyme reactions require a long time to reach the plateau, high enzyme amounts should be taken, which can also serve to avoid disturbing influences, such as inactivation or side reactions.

With two-substrate reactions also, the concentration of the cosubstrate can be determined instead of that of the substrate (if it is easier to detect, e.g., NADH), which reacts with the same stoichiometry. For this, the cosubstrate (as well as all other components) must be present in a higher concentration than the substrate; otherwise, the amount of the cosubstrate will be measured unintentionally.

For substrate determination linear initial velocities are not essential as for the determination of the enzyme activity and thus steady-state conditions must not be regarded. Of greater interest is the final phase of the reaction, which, in its simplest case obeys first-order (or pseudo-first-order with two-substrate reactions, the cosubstrate present in a surplus)

$$\ln[\mathrm{A}] = \ln[\mathrm{A}]_0 - k_1 t \tag{2.23}$$

Table 2.4 shows that the time required for a nearly complete conversion of substrate is 10-fold compared to that needed to reach 50% turnover. Real conditions can severely deviate, since the actual reaction must not obey first-order and other influences may also be considered, especially product inhibition, which can essentially elongate the time required to reach equilibrium.

Many enzyme reactions are not irreversible; they rather approach a defined equilibrium state and in such cases, substrates will not quantitatively be converted to products. Under these conditions two different approaches for substrate determination can be applied. Even if the substrate is not converted quantitatively, the relative amount of product formed at the end of the reaction is determined by the state of the equilibrium and can be calculated from the equilibrium constant of

Table 2.4 Time required for the conversion of substrate to product assuming an irreversible first-order reaction and a half-life time of 10 min.

Product formed (% of $[A]_0$)	Reaction time (min)
50	10.0
80	23.3
90	33.3
95	43.4
98	56.5
99	66.7
99.9	100.0

the reaction. Likewise, the percentage of product formed from a particular initial substrate amount can be determined in a reference experiment, and from the data obtained the amount of an unknown sample can be analyzed.

Alternatively, quantitative conversion of the substrate can be enforced even in the case of an unfavorable state of equilibrium by trapping the product with the aid of chemical or enzymatic reactions. This is applied, for instance, in the alcohol dehydrogenase reaction where aldehydes or ketones formed are trapped with semicarbazide or hydrazine. Protons can be withdrawn from the equilibrium in the presence of alkaline pH. When inorganic phosphate in the glyceraldehyde-3-phosphate dehydrogenase reaction is substituted by arsenate, the unstable 3-phosphoglycerate-1-arsenate is formed in place of the 1,3-bisphosphoglycerate. Its rapid decay drives the reaction quantitatively to the product site.

2.5.2
Substrate Determination by Coupled Enzyme Reactions

Coupled enzyme reactions can also be applied for substrate determination. Here the conditions are not so stringent as with determination of enzyme activity, since only the value at the end of the reaction must be obtained. The essential prerequisite is that the final indicator reaction must react quantitatively with the product. One or more reactions can precede the indicator reaction. They need not necessarily be irreversible, since the indicator reaction removes the intermediate from the equilibrium to form the final product. If two or more enzymes are coupled in a reaction sequence, only one must be completely specific to its substrate. An example is the relatively unspecific conversion of glucose by HK

$$\text{D-glucose} + \text{ATP} \xrightarrow{\text{HK}} \text{ADP} + \text{D-glucose-6-phosphate}$$

Besides D-glucose, the enzyme accepts also other hexoses, such as fructose and mannose. For selective determination of glucose, instead of HK, the more specific, but also more (about 250-fold) expensive glucokinase may be taken. But since the following indicator reaction of the D-glucose-6-phosphate dehydrogenase

$$\text{D-glucose-6-P} + \text{NADP}^+ \xrightarrow{\text{G6PDH}} \text{gluconate-6-P} + \text{NADPH} + \text{H}^+$$

is highly specific for D-glucose-6-phosphate, by-products of HK originating from other substrates do not disturb the quantitative determination of glucose in the coupled assay.

2.5.3 Kinetic Method for Substrate Determination

The initial substrate concentration of an enzyme-catalyzed reaction is related to the reaction velocity according to the (rearranged) Michaelis–Menten equation

$$[\text{A}] = \frac{K_\text{m}v}{V - v} \tag{2.24}$$

Thus knowing the kinetic constants K_m and V the actual substrate concentration can be derived from the initial velocity. Likewise, a standard curve, following the hyperbolic Michaelis–Menten curve can be prepared by determining the reaction velocity at different substrate concentrations. With this method, substrates can be determined only in the lower concentration range ($[\text{A}] \leq K_\text{m}$), while in the higher range (nearer to saturation) even strong variations in the substrate concentration cause only slight the changes in the velocity.

A more stringent relationship holds for first-order reactions. Here the substrate conversion within a defined time period $\Delta t = t_2 - t_1$ is directly proportional to the initial substrate concentration $[\text{A}]_0$

$$-\frac{\text{d}[\text{A}]}{\text{d}t} = k_1[\text{A}] \tag{2.25}$$

$$[\text{A}] = [\text{A}]_0\text{e}^{-k_1t} \tag{2.26}$$

$$-\frac{\text{d}[\text{A}]}{\text{d}t} = k_1[\text{A}]_0\text{e}^{-k_1t} \tag{2.27}$$

$$[\text{A}]_0 = -\frac{\Delta[\text{A}]}{\text{e}^{-k_1t_1} - \text{e}^{-k_1t_2}} \tag{2.28}$$

If k_1 is known a concentration change between a defined time interval $t_2 - t_1$ is measured from which [A] can be determined. For this fixed-time procedure the time interval $t_2 - t_1$ must be the same for all measurements. With the Michaelis–Menten equation, first-order conditions can only be achieved at low substrate concentrations ($[\text{A}] \ll K_\text{m}$). Therefore, the amounts of substrate to be determined by this procedure must be rather low. Alternatively, the Michaelis constant should be high. This condition can be achieved by the addition of a competitive inhibitor, which increases the apparent Michaelis constant. An example for the substrate determination with the kinetic method is the glucose determination with the coupled assay of glucose oxidase and peroxidase (cf. Section 3.3.1.14).

2.5.4 Enzymatic Cycling

Very low amounts of metabolites can be detected by enzymatic cycling. The metabolite to be determined functions as an intermediate within a reaction sequence and remains constant in its concentration. Coenzyme A (CoA) is an example of a metabolite occurring in the cell at very low concentration. In a coupled reaction it can be formed from acetyl phosphate by phosphotransferase (PTA). Malate dehydrogenase (MDH) serves as an indicator reaction and provides oxaloacetate for the citrate formation by the citrate synthase (CS):

$$\text{acetylphosphate} + \text{CoA} \xrightarrow{\text{PTA}} \text{acetyl-CoA} + \text{P}_i$$

$$\text{malate} + \text{NAD}^+ \xrightarrow{\text{MDH}} \text{oxaloacetate} + \text{NADH} + \text{H}^+$$

$$\text{acetyl-CoA} + \text{oxaloacetate} + \text{H}_2\text{O} \xrightarrow{\text{CS}} \text{citrate} + \text{CoA}$$

Low amounts of NAD or NADP can be determined by enzymatic cycling, for example, the regeneration of NADPH by coupling of the glutamate dehydrogenase (GluDH) and glucose-6-phosphate dehydrogenase (G6PDH).

$$\text{2-ketoglutarate} + \text{NH}_3 \xrightarrow{\text{GluDH}} \text{glutamate} + \text{H}_2\text{O}$$

$$\text{NADPH} + \text{H}^+ \qquad \text{NADP}^+$$

$$\text{6-phosphogluconate} \xleftarrow[\text{G6PDH}]{} \text{glucose-6-phosphate}$$

In this case the constant remaining NADPH intermediate cannot be used as an indicator reaction. Instead the reaction is stopped after a defined time, for example, 30 min. The total amount of 6-phosphogluconate formed is analyzed separately with the 6-phosphogluconate dehydrogenase reaction applying the end point method. With a high number of passages through the cycle a more than million-fold increase in sensitivity can be achieved. Since the amount of the intermediate, which should be determined, remains constant during the reaction, it can be regarded like a catalyst.

Generally the reaction sequence for enzymatic cycling can be formulated as

$$\text{A} + \text{B} \xrightarrow{\text{E1}} \text{P} + \text{Q}$$

$$\text{P} + \text{C} \xrightarrow{\text{E2}} \text{A} + \text{R}$$

The sum of the reaction sequence is then

$$B + C \longrightarrow Q + R$$

A is the substance to be determined, either a cosubstrate or a coenzyme. The concentration of A must be limiting and smaller than its own K_m value, and the concentrations of the other two substrates B and C must be large. Also, the activities of both enzymes E1 and E2 should be high. When the cyclic system reaches the steady state, the rate of formation of P must be equal to the back reaction for reformation of A

$$k_1[A] = k_2[P]$$

The first-order rate constant for the overall reaction k is

$$k = \frac{V}{K_m} = \frac{k_1 k_2}{k_1 + k_2} \tag{2.29}$$

V is the velocity with saturating levels of A, which is the substrate or cofactor to be determined.

References

Ånggård, E. and Samuelsson, B. (1974) in *Methods of Enzymatic Analysis*, 3rd edn (ed. H.U. Bergmeyer), Verlag Chemie, Weinheim, pp. 1925–1933.

Bergmeyer, H.U. (1977) *Grundlagen der Enzymatischen Analyse*, Verlag Chemie, Weinheim.

Bergmeyer, H.U. (1983) *Methods of Enzymatic Analysis*, 3rd edn, vol. 1, Verlag Chemie, Weinheim.

Copeland, R.A. (2000) *Enzymes, A Practical Approach to Structure, Mechanism and Data Analysis*, 2nd edn, John Wiley & Sons, New York.

Lowry, O.H., Passonneau, J.V., Schultz, D.W., and Rock, M.K. (1961) *J. Biol. Chem.*, **236**, 2746–2753.

Passonneau, J.V. and Lowry, O.H. (1978) in *Principles of Enzymatic Analysis* (ed. H.U. Bergmeyer), Verlag Chemie, Weinheim, pp. 86–87.

3
Enzyme Assays

In chapter 2 the fundamental requirements for enzyme assays have been described while in the current chapter protocols for various enzyme assays and general methods for dealing with enzymes – such as protein assays and concentration methods – are presented. The protocols can be treated like cookery recipes and reproduced exactly step by step, but it is understood that the rules discussed earlier should be kept in mind, because any description can mediate only a general pattern, and according to the particular system, modifications are often required. The discussed rules should enable the reader to independently develop an assay; for example, for a newly isolated enzyme. Conditions essential for any enzyme assay are summarized in Box 3.1 and in Box 3.2 general hints for concentration ranges of the assay components are given. Table 3.1 lists stock solutions frequently needed for enzyme assays, while buffers are listed in Table 2.2 and their preparation is described in Boxes 2.8 and 2.9.

3.1
Enzyme Nomenclature

The following enzyme assays are described in the order of the respective EC numbers. The significance of these numbers will be explained in this section. When first investigating enzymes, the discoverers named the new enzymes and many of these historical names are still in usage, such as *diaphorase, Zwischenferment,* and *old yellow enzyme;* however, in the course of time, many have lost their meaning. With an increasing number of enzymes being discovered, the need for a clear system of nomenclature became urgent. In 1956 the *International Union of Biochemistry* (IUB) established an *International Commission of Enzymes* under the leadership of Prof. M. Florkin; the commission prepared a classification and nomenclature system for all known enzymes. In 1961 the commission was substituted by a *Standing Committee on Enzymes* and from 1969, by an *Expert Committee on Enzymes,* which consistently updates the nomenclature and publishes supplements (see literature below). The present enzyme list contains over 3000 entries. Each enzyme is designated by a four-figure code and a systematic enzyme name.

Practical Enzymology, Second Edition. Hans Bisswanger.

Box 3.1: Essential Conditions for Enzyme Assays

Condition	Recommended range[a]
Solvent	Polar (aqueous)
Temperature	25–37 °C
pH	6.5–8.5
Ionic strength	0.05–0.2 M
Enzyme amount	As low as possible
Substrate amount	Saturating (~50-fold K_m)
Cofactors	Saturating (~50-fold K_d)

[a] For most assays, distinct enzymes require deviating conditions.

Box 3.2: Dependence of Enzyme Reaction on Concentrations of Assay Components

Component	Mode of dependency
Enzyme	Strictly linear, no saturation
Specific binding components, directly involved in the reaction (substrates, products, cofactors, essential metal ions)	Hyperbolic saturation behavior according to Michaelis–Menten equation (Eq. (2.13))
Nonspecific components, not directly involved in the reaction (ions, buffer, stabilizing and protecting substances) (cf. Table 2.3)	Weak influences, no saturating effects

The **systematic name** (SN) should characterize the reaction catalyzed. At first, the direction of the reaction must be defined. For a distinct class of enzymes the same direction is taken for all reactions, even for cases where this is not the physiological direction. The systematic name consists of two parts, the first denotes the substrate(s), separated by a colon (:), while the second describes the nature of the reaction, ending with *–ase;* for example, alcohol:NAD^+ oxidoreductase. The systematic names of oxidoreductases follow the general pattern donor:acceptor oxidoreductase.

It turned out that the systematic names are too circumstantial for practical usage and therefore, have been simplified into **recommended names** (RNs), which are also specified in the enzyme list. For instance, the recommended name for alcohol:NAD^+ oxidoreductase is alcohol dehydrogenase. It implies an oxidation of the hydroxyl group of the alcohol, but unlike the systematic name, gives no information about the acceptor. For the recommended names

Table 3.1 Frequently used solutions for enzyme assays. For the given substances, the most common forms are indicated but other forms are also available, for example with (or without) crystal water, as free acid or with different counterions.

Solution	Molecular mass (M_r)	Concentration (M)	Preparation	Remarks
NaOH, sodium hydroxide, caustic soda	40.0	1	4.0 g, resolve and adjust to 100 ml with H_2O	Etching! Vigorous warming up upon resolution Store in PE flasks
KOH, potassium hydroxide, potash lye	56.1	1	5.61 g, resolve and adjust to 100 ml with H_2O	Etching! Store in PE flasks
HCl, hydrochloric acid	34.5	1	82.8 ml concentrated HCl (37%, density 1.18) fill up with H_2O to 1 l	Etching liquid and vapors
$MgCl_2 \cdot 6H_2O$	203.3	0.1	203 mg in 10 ml	–
TCA, trichloroacetic acid, $C_2HCl_3O_2$	163.4	3	49 g, adjust to 100 ml with H_2O	Protein precipitating reagent for protein determination (biuret) and stopped enzyme tests
ADP, disodium salt	471.2	0.1	471 mg, adjust to 10 ml with H_2O	Keep frozen
ATP, disodium salt, trihydrate	605.2	0.1	605 mg, adjust to 10 ml with H_2O	Keep frozen
NAD^+, free acid	663.4	0.1	663 mg, adjust to 10 ml with H_2O	Keep frozen
NADH, disodium salt	709.4	0.01	71 mg, adjust to 10 ml with H_2O	Unstable, even in frozen state
$NADP^+$, sodium salt	765.4	0.1	765 mg, adjust to 10 ml with H_2O	Keep frozen
NADPH, tetrasodium salt	833.4	0.01	83 mg, adjust to 10 ml with H_2O	Unstable, even in frozen state
DTE, dithioerythritol, Cleland's reagent	154.2	0.1	154 mg, adjust to 10 ml with H_2O	Similar to DTT, dithiothreitol, both can be used alternatively

the term oxygenase is used if molecular oxygen is incorporated: monooxygenases incorporating one atom and dioxygenases both atoms of O_2. D and L, applied to define asymmetric C-atoms, e.g. in D-sugars and L-amino acids, are ignored as long as no ambiguity exists, otherwise they must be mentioned as in L-lactate dehydrogenase and D-lactate dehydrogenase. Anionic substrates should end with *–ate* (mal*ate* dehydrogenase) instead of *-ic* (malic enzyme). Directly annexing *–ase* to the substrate name indicates its hydrolysis (lactase). Recommended names should generally end in *–ase*, with the exception of some proteolytic enzymes where the ending *-in* (trypsin) is retained. In contrast to systematic names recommended names usually refer to the direction of the reaction that has been demonstrated. Fantasy names (reparase, caspase) should not be used.

If an enzyme catalyzes more than one reaction, for example if it is composed of various subunits performing successive partial reactions, it should be denoted as a *system*, but often a *complex* is preferred for such functional units (e.g., pyruvate dehydrogenase system (complex), fatty acid synthase system (complex)). Since enzymes are classified on the basis of the chemical reaction they catalyze, enzymes catalyzing the same reaction, for example those from different organisms or isoenzymes, are classified by one systematic name and one code number; but some mostly historically founded exceptions exist, such as acid and alkaline phosphatase or cholinesterase and acetylcholinesterase. Enzymes not performing a chemical reaction are not classified in the enzyme list.

The **code numbers** for enzyme classification (**EC numbers** from **E**nzyme **C**ommission) contain four figures, separated by points. The first figure indicates the number of one of the six main enzyme classes to which the particular enzyme belongs:

1) Oxidoreductases
2) Transferases
3) Hydrolases
4) Lyases
5) Isomerases
6) Ligases (synthetases)

The following two figures indicate subclasses and sub-subclasses, respectively. The fourth figure is the serial number of the individual enzyme within its sub-subclass. An excerpt of the enzyme list with the main, the sub, and the sub-subclasses including some characteristic examples of individual enzymes is presented at the end of the book.

References

Enzyme Nomenclature (1976) *Biochim. Biophys. Acta*, **429**, 1–45.
Webb, E.C. (ed.) (1992) *Enzyme Nomenclature*, Academic Press Inc., San Diego.
Schomburg, D. and Salzmann, M. (1990) *Enzyme Handbook*, Springer-Verlag, Berlin.

3.2 Practical Considerations for Enzyme Assays

The assay protocols are ordered according to the EC numbers and start with the recommended name and the EC number of the respective enzyme as headings, followed by the systematic name (SN), abbreviation, and common (trivial) names (if in use), and the reaction formula. A short description of enzyme features follows, especially of those that are of some significance for the assay, such as cofactors, molecular mass, Michaelis constants, and pH optimum. This information is given only for orientation. Enzymes of the same type but from different organisms often differ remarkably in their features and it is not the intention of this chapter to collect all the known data. For instance, Michaelis constants reported in separate publications, even for enzymes from the same organism, can differ by more than a factor of 10. Actually, it is not possible to determine such constants with too high an accuracy, so very precise values (e.g., 2.163 mM) must be regarded with caution. From the variety of organisms, mammalian (human) enzymes are preferentially considered; in some cases thoroughly investigated examples, for example from bacteria, are also considered.

It was not possible to consider all enzyme assays described so far. But from comparing distinct enzyme assays, general principles can be observed. For instance, various hydrolyzing enzymes are tested with substrates connected with a chromophoric residue, like the nitrophenyl or umbelliferyl groups. Upon cleavage a strongly absorbing or fluorescent compound is released, which can be detected by the appropriate method. Various assays of such types are described here, but combined substrates of related enzymes are also available and can be tested by adapting the assay conditions.

Originally, enzymes were isolated directly from the respective organisms by various purification procedures, but this approach became increasingly displaced by the advent of **recombinant enzymes**. With the methods of gene technology, the respective DNA region from any organism can be cloned and expressed, for example in a bacterial host with an *N*-terminal polyhistidine tag, consisting of six successive histidine residues (*6xHis-tag*). This tail enables easy purification using a nickel affinity column (Ni-Separose®, Ni-NTA-Agarose®) and specific elution with a 20 mM imidazole solution. The bound polyhistidine tag can change the features of the enzyme, but if this is not too extensive, the enzyme may be used in this form. Otherwise the polyhistidine tag can be removed with exoproteases. A further advantage of recombinant enzymes is the application of side-directed mutagenesis and the expression of mutant enzymes with various modifications.

A list of all required **assay solutions** with instructions for their preparation is given (for the preparation of buffers see Boxes 2.8 and 2.9). The respective substances are often available in various forms, as free acid or as salt with differing counterions, anhydrous or hydrated. Usually the most common form is indicated together with its molecular mass (M_r) and the amount needed (in grams) for preparation of the solution. For routine assays larger quantities may be prepared and stored, either in the cold, or for sensitive substances such as substrates and cofactors, in the frozen state (∼ −20 °C or less). Usually solid substances are dissolved in water (H_2O, principally understood as *distilled* or *deionized*) or buffer (especially for acid or alkaline substances). Enzymes must always be dissolved or diluted in buffer, the pH set according to their pH optimum. Water-insoluble substances must be dissolved in an appropriate solvent, for example dimethylsulfoxide (DMSO), acetone, ethanol or methanol. A small volume of such a solution is diluted with the aqueous assay mixture assuming that the substance may remain dissolved, but this condition must be controlled.

Each assay can be prepared separately by mixing the respective aliquots of the assay solutions directly in a test tube or a cuvette, but such a procedure is prone to errors and hence, more asssays may need to be performed. It is therefore recommended that one **assay mixture** be prepared to be used for all assays. This saves time, reduces errors, and establishes equal composition for all assays. In the following descriptions, the quantity is calculated for 10 assays, more or less can be prepared accordingly. The assay solutions should contain all required components, with the exception of one essential component with which the assay will be started. This is usually an enzyme; therefore, it is omitted from the assay mixture. However, if there is reason, one may start with another component, the substrate or a cofactor. In this case, the enzyme may be added directly to the assay mixture instead of the particular substance. It must be remembered that components of the assay mixture should not influence one another (e.g., by forming precipitates or inducing oxidative processes). This holds true especially for enzymes. As already discussed, they are not very stable in diluted solution, but on the other hand, cofactors and other additions of the assay mixture can act as stabilizing influences. Substrates or cofactors can induce slow activation processes with special enzymes (*hysteretic enzymes*) and preincubation with the respective components is required in such cases. However, conversion of the substrate by the enzyme must strictly be avoided (e.g., by omitting another essential component). The assay mixture is prepared only for the actual test series and should not be stored for a long time. It must be kept on ice and be warmed up before starting the assay.

For the assay **procedure**, the respective quantity of the assay mixture for a single assay is filled in a test tube or cuvette and the reaction is started by addition of the missing component from the assay mixture, usually the enzyme. The final assay volume depends on the respective detection method; here it is generally estimated for 1 ml. Upon addition, the progression of the reaction must be observed immediately. It is essential that it occurs within the observable range, that is not too slow and not too fast. This can be regulated by the amount of enzyme as discussed in the previous sections (2.2.3ff, Box 2.3). Because of the varying activities of enzyme preparations, suggestions for the quantity of enzyme are given only in exceptional cases. Usually 20 μl of diluted enzyme solution is considered and the appropriate activity should be tested out in preliminary assays. The suitable amount can be obtained by changing the dilution factor and/or the added volume. In the latter case, the buffer quantity must be modified to achieve the final assay volume (1 ml). An absorption change of 0.1 in 1 min is favorable for photometric assays (cf. Boxes 2.18 and 2.19).

Any enzyme assay requires a **blank** or reference, the type of blank depends on the type of assay. Stopped assays refer to the blank as zero value and it is obvious that any deviation of the blank will distort all data. Therefore, the blank must be measured with high accuracy and during longer series, repeated determinations of the blank should be carried out. Usually the assay mixture can be taken as blank, the starting component (enzyme sample) replaced by the buffer. It must, however, be made sure, that the starting component does not change the features of the assay mixture; for example, by contributing a distinct absorption in photometric assays. Otherwise, this contribution must be considered or a more indifferent component (e.g., substrate) is used as the starting component. It is also obvious that the blank must be stable, neither a spontaneous reaction nor an enzyme reaction dare occur. Usually, the photometer is adjusted to zero with the blank. This is convenient, because the sample values will be obtained directly, already reduced by the blank value. However, this is not really necessary and it is often advantageous to adjust the photometer to zero without blank and to afterwards measure the absorption, both of the blank and the samples. This appears circumstantial, because the blank must be subtracted from the sample values. However, examination of the blank absorption is a valuable control. Unusual deviations can be detected and if necessary, a new blank can be prepared for correction of the samples, while all sample values are of no use if the instrument was adjusted to zero with a wrong blank value. A further control for the blank is the fact that limiting values of a measured series must tend to zero (respectively the blank value); that is, a calibration curve must extrapolate to zero absorption for zero concentration of the component determined.

For continuous assays such as the progress curve of an enzyme reaction, a blank is not really necessary because the velocity is taken from the steepness of the curve, an absolute value does not have to be determined. Nevertheless, absorption of the assay mixture before starting the reaction is a control for correct composition. But for time-dependent measurements, it must be made sure that the assay solution shows no time-dependent change in the absence of the enzyme reaction. Sometimes spontaneous drifts are observed; for example, due to instability of a component or oxidative processes. If such effects cannot be suppressed, the drift must be subtracted from the velocity of the sample.

The **calculation** of enzyme activity is generally described in Section 2.2.3 and in Boxes 2.18 and 2.19: in special cases it is indicated at the end of the assay procedure.

The knowledge of hazardousness of chemicals is presupposed. Hints are given in special cases, but the reader is generally referred to the security rules indicated in the relevant literature; instructions provided together with the chemicals by the respective companies are recommended. All data and instructions presented in the following protocols have been presented carefully, but due to the large amount of data complete guarantee cannot be given.

Reference

Hengen, P. (1995) *Trends Biochem. Sci.*, **20**, 285–286.

3.3
Special Enzyme Assays

3.3.1
Oxidoreductases, EC 1

3.3.1.1 Optical Assay
NAD(P)-dependent dehydrogenases (oxidoreductases) belong to one of the most frequently tested enzyme classes, therefore some general aspects are mentioned beforehand. These enzymes differ in their substrate specificity; for example, for lactate or malate, but all share a similar cosubstrate, only differing in a phosphate group (NAD^+ or $NADP^+$). Because both forms possess the same spectroscopic feature of an additional absorption band at 340 nm appearing only in the reduced state, a general assay – the *optical assay* based on this feature – can be performed for all NAD(P)-dependent dehydrogenases. Not only the substrate must be adapted to the respective enzyme specificity, but also other features such as affinity (K_m), pH optimum, temperature behavior, and in particular, the state of equilibrium must be considered. The latter aspect determines whether the reaction proceeds in the reductive (absorption increase) or the oxidative (absorption decrease) direction. The assay can be carried out in any conventional UV or visible photometer and in a filter photometer; the respective wavelengths are indicated in Figure 2.27. As has been discussed earlier, it is not really necessary to measure just at the maximum wavelength; the absorption at 339 nm is nearly the same as that at 340 nm, but is more stable against temperature fluctuations.

The state of the equilibrium is not the only decisive factor for the direction of the assay reaction. A further aspect is the availability of the respective substrate and the fact that the absorption is considerably high if started from the reductive site; for example, an absorption of 0.63 for 0.1 mM NADH. Usually, enzyme assays should be performed at saturation concentrations of all substrates, practically at least 10-fold K_m. This cannot always be realized under such conditions. For the alcohol dehydrogenase reaction, the unfavored reductive reaction is enforced by trapping the products (acetaldehyde and H^+) with a chemical reagent and an alkaline pH (cf. Section 3.3.1.3).

Reference

Ziegenhorn, J., Senn, M., and Bücher, T. (1976) *Clin. Chem.*, **22**, 151–160.

3.3.1.2 Fluorimetric Assay
Fluorimetric assays have the advantage of high sensitivity and are recommended for detection of low activity levels. However, the method requires special equipment (fluorimeter), high accuracy, and purity of the solutions. Only the reduced form of NAD(P), when excited at 260 nm (or 340 nm), emits light at 470 nm with about 2% quantum yield and the fluorescence increase at this wavelength is a signal for the enzyme reaction. The fluorimetric assay is principally applicable to all

dehydrogenase reactions forming or oxidizing NAD(P)H. The same solutions and concentrations described for the respective photometric test can be used, but very high concentrations of NAD(P)H cause quenching effects, so that the saturation range is not completely accessible. Therefore, the direction of reduction is preferred for this assay.

3.3.1.3 Alcohol Dehydrogenase, EC 1.1.1.1

SN: alcohol:NAD oxidoreductase. aldehyde reductase, ADH

$$\text{aldehyde} + \text{NADH} + \text{H}^+ \rightleftharpoons \text{alcohol} + \text{NAD}^+$$

The reaction is reversible and can be tested from either side. The equilibrium constant for the reaction with acetaldehyde is 8×10^{-12} M and favors the formation of ethanol. Alcohol dehydrogenase, a Zn enzyme, exists in different forms depending on the source. The commonly used ADHs are from yeast or horse liver, they differ in structure and specificity. ADH from yeast ($M_r = 148\,000$), homotetramer, is most active with ethanol. K_m (mM): 16.7 (ethanol), 2.8 (acetaldehyde), 0.12 (NAD), 0.06 (NADH), pH optimum: 8.3. Mammalian ADH ($M_r = 79\,000$), a homodimer, has a preference for higher alcohols. K_m (mM): 7.0 (ethanol), 3.4 (acetaldehyde), 0.25 (NAD), 0.01 (NADH), pH optimum: 7.5. Inhibitors: 1,10-phenanthroline (competitive with NAD, Zn-complexing agent); fluoroethanol ($K_i = 2.5$ mM), inhibitor of mammalian ADH, competitive with ethanol. ADH is not stable in dilute solution, addition of BSA and a thiol reagent (DTE, DTT) is recommended for stabilization.

A. Reduction Assay The reaction is tested in the physiological direction:

$$\text{aldehyde} + \text{NADH} + \text{H}^+ \rightleftharpoons \text{ethanol} + \text{NAD}^+$$

A disadvantage is the volatility and toxicity of acetaldehyde.

Assay solutions

0.1 M potassium phosphate pH 7.5

BSA solution: 0.1 g BSA in 100 ml 0.1 M potassium phosphate pH 7.5

0.01 M NADH (disodium salt, $M_r = 709.4$; 71 mg 10 ml^{-1} H_2O)

0.5 M acetaldehyde ($M_r = 44.0$, $d = 0.78$ g ml^{-1}; 2.2 g (2.82 ml) in 100 ml H_2O)

ADH diluted in BSA solution: commercially available preparations must be diluted according to their activity: 0.02 IU ADH convert 0.02 μmol NADH/min, corresponding to an absorption difference of 0.126; from a commercial ADH with 400 IU mg^{-1} prepare a stock solution of 10 mg ml^{-1} (~4000 IU ml^{-1}) in 0.1 M potassium phosphate pH 7.5. A 4000-fold dilution (2.5 μl in 10 ml BSA solution) corresponds to 1 IU ml^{-1}, 20 μl for the enzyme assay ~0.02 IU

Assay mixture

Components	*Concentration (mM)*
9.5 ml 0.1 M potassium phosphate pH 7.5	95
0.1 ml 10 mM NADH	0.1
0.2 ml 0.1 M acetaldehyde	2.0

Procedure

0.98 ml assay mixture

0.02 ml ADH (0.02 IU)

The absorption decrease at 340 nm is measured at 25 °C, absorption coefficient for NADH: $\varepsilon_{340} = 6.3 \times 10^3 \, \text{l}\,\text{mol}^{-1}\,\text{cm}^{-1}$.

B. Oxidation Assay To test the unfavored back reaction

$$\text{ethanol} + NAD^+ \rightleftharpoons \text{acetaldehyde} + NADH + H^+$$

the products must be removed from equilibrium: acetaldehyde by semicarbazide and the protons by alkaline pH.

Assay solutions

75 mM glycine/sodium diphosphate pH 9.0 (10 g $Na_4P_2O_7 \cdot 10H_2O$ + 0.5 g glycine in 300 ml H_2O, adjust to pH 9.0 with 1 N HCl)

0.1 M potassium phosphate, pH 7.5

Ethanol p.A. ($M_r = 46.1, d = 0.81\,\text{g}\,\text{ml}^{-1}$)

0.1 M NAD ($M_r = 663.4$; 0.663 g in 10 ml 0.1 M potassium phosphate pH 7.5)

0.1 M 1,4-dithioerythritol (DTE, $M_r = 154.3$; 154 mg in 10 ml 0.1 M potassium phosphate pH 7.5)

BSA solution: 0.1 g BSA in 100 ml 0.1 M potassium phosphate, pH 7.5

2.2 M semicarbazide (semicarbazide·HCl, $M_r = 111.5$; 2.5 g in 10 ml 2 N NAOH, adjust to pH 6.3–6.5 with 5 N NaOH)

ADH, for dilution see above

Assay mixture

Components	*Concentration (mM)*
8.8 ml 75 mM glycine/sodium diphosphate pH 9.0	66 mM
0.3 ml ethanol	0.53 M
0.2 ml 0.1 M NAD	2.0 mM
0.2 ml 0.1 M DTE	2.0 mM
0.3 ml 2.2 M semicarbazide	66 mM

Procedure

0.98 ml assay mixture

0.02 ml diluted ADH

The increase in the absorption at 340 nm is measured at 25 °C. The absorption coefficient for NADH is $\varepsilon_{340} = 6.3 \times 10^3\ \mathrm{l\,mol^{-1}\,cm^{-1}}$.

References

Andersson, L. and Mosbach, K. (1982) *Meth. Enzymol.*, **89**, 435–445.
Bergmeyer, H.U. (1983) *Methods of Enzymatic Analysis*, 3rd edn, vol. **2**, Verlag Chemie, Weinheim, pp. 139–141.
Dafeldecker, W.P., Meadow, P.E., Pares, X., and Vallee, B.L. (1981) *Biochemistry*, **20**, 6729–6734.
Pietruszko, R. (1982) *Meth. Enzymol.*, **89**, 429–435.

3.3.1.4 Alcohol Dehydrogenase ($NADP^+$), EC 1.1.1.2

SN: alcohol:$NADP^+$ oxidoreductase, aldehyde reductase (NADPH)

$$\text{aldehyde} + \text{NADPH} + H^+ \rightleftharpoons \text{alcohol} + \text{NADP}^+$$

From different sources, mammalian enzyme $M_r = 74\,000$, dimeric, K_m (acetaldehyde) 0.25 mM; identical with glucuronate dehydrogenase, EC 1.1.1.19, mevaldate reductase EC 1.1.1.33, lactaldehyde reductase, EC 1.1.1.55; can be tested with alcohol dehydrogenase assays, EC 1.1.1.1 (Section 3.3.1.3), replacing NAD (respectively NADH) by NADP (or NADPH).

References

Das, B. and Srivastava, S.K. (1985) *Biochim. Biophys. Acta*, **840**, 324–333.
Turner, A.J. and Hryszko, J. (1980) *Biochim. Biophys. Acta*, **613**, 256–265.
Wartburg, V.J.-P. and Wermuth, B. (1982) *Meth. Enzymol.*, **89**, 506–513.

3.3.1.5 Homoserine Dehydrogenase, EC 1.1.1.3

SN: homoserine:NAD(P)$^+$ oxidoreductase, aspartate kinase–homoserine dehydrogenase I, AK–HDH I

$$\text{L-homoserine} + \text{NAD(P)}^+ \rightleftharpoons \text{L-aspartate-4-semialdehyde} + \text{NAD(P)H} + \text{H}^+$$

Multifunctional allosteric enzyme (*Escherichia coli*), fused with aspartate kinase (cf. assay in Section 3.3.2.11), M_r: 360 000, homotetramer. (4 × 84 000), K_m (mM): 0.013 (homoserine), 0.17 (L-aspartate-4-semialdehyde), 0.073 (NADP), 0.09 (NADPH).

Assay solutions

0.1 M CHES/HCl, 0.4 M KCl pH 9.0 (2-(*N*-cyclohexylamino)ethanesulfonic acid, $M_r = 207.3$, 20.7 g, KCl, $M_r = 74.6$, 29.8 g, dissolve in 600 ml, adjust to pH 9.0 with 1 M HCl, fill up to 1 l)

0.25 M L-homoserine ($M_r = 119.1$, 298 mg in 10 ml H_2O)

0.1 M NADP (disodium salt, $M_r = 787.4$; 787 mg in 10 ml)

0.1 M DTT (dithiotreitol, $M_r = 154.2$, 154 mg in 10 ml H_2O)

Assay mixture

Components	*Concentration (mM)*
9.26 ml 0.1 M CHES/HCl, 0.4 M KCl pH 9.0	93/370
0.5 ml 0.25 M L-homoserine	12.5
0.03 ml 0.1 M NADP	0.3
0.01 ml 0.1 M DTT	0.1

Procedure

0.98 ml assay mixture

0.02 ml enzyme solution

The absorption increase at 340 nm is measured at 30 °C, absorption coefficient for NADPH: $\varepsilon_{340} = 6.3 \times 10^3\ \text{l mol}^{-1}\ \text{cm}^{-1}$.

References

Angeles, T.S. *et al.* (1989) *Biochemistry*, **28**, 8771–8777.
James, C.L. and Viola, R.E. (2002) *Biochemistry*, **41**, 3720–3725.

Truffa-Bachi, P. (1973) in *The Enzymes* (ed. P.D. Boyer), 3rd edn, vol. 8, Academic Press, New York, pp. 509–553.
Wedler, F.C. and Ley, B.W. (1993) *J. Biol. Chem.*, **268**, 4880–4888.

3.3.1.6 Shikimate Dehydrogenase, EC 1.1.1.25

SN: shikimate:NADP$^+$ 3-oxidoreductase

$$\text{3-dehydroshikimate} + \text{NADPH} + \text{H}^+ \rightleftharpoons \text{shikimate} + \text{NADP}^+$$

This enzyme catalyzes a step of the biosynthetic pathway of aromatic amino acids. For the assay, the reverse reaction is observed. In *Escherichia coli* the enzyme exists as a monomer ($M_r = 29\,400$), in other organisms as a dimer. K_m : 1.25×10^{-5} M (NAD), pH optimum 8.5.

Assay solutions

0.1 M sodium carbonate pH 10.6 (Na_2CO_3, $M_r = 106$; 10.6 g in 1 l, adjust to pH 10.6 with a 0.1 M (8.4 g l^{-1}) $NaHCO_3$ solution)

0.1 M shikimic acid ($M_r = 174.2$; 174 mg in 10 ml)

0.1 M NADP (disodium salt, $M_r = 787.4$; 787 mg in 10 ml)

Assay mixture

Components	*Concentration (mM)*
9.2 ml 0.1 M sodium carbonate pH 10.6	92
0.4 ml 0.1 M shikimic acid	4
0.2 ml 0.1 mM NADP	2

Procedure

0.98 ml assay mixture

0.02 ml diluted enzyme

The increase in the absorption at 340 nm is measured at 25 °C. The absorption coefficient for NADPH is $\varepsilon_{340} = 6.3 \times 10^3$ l mol^{-1} cm^{-1}.

References

Chaudhuri, S. and Coggins, J.R. (1985) *Biochem. J.*, **226**, 217–223.
Coggins, J.R., Bookock, M.R., Chaudhuri, S., Lambert, J.M., Lumsden, J., Nimmo, G.A., and Smith, D.D.S. (1987) *Meth. Enzymol.*, **143**, 325–341.
Lumsden, J. and Coggins, J.R. (1977) *Biochem. J.*, **161**, 599–607.

3.3.1.7 L-Lactate Dehydrogenase, EC 1.1.1.27

SN: *S*-lactate:NAD^+ oxidoreductate, L-lactic dehydrogenase, LDH

$$\text{L-lactate} + NAD^+ \rightleftharpoons \text{pyruvate} + NADH + H^+$$

The reaction can be tested in both directions, but the equilibrium favors the reduction of pyruvate, although high absorption of NADH must be taken into account. For routine assays, LDH from pig heart is suitable. In mammals (bovine, pig, rabbit) a homotetramer, $M_r = 140\,000$, two isoenzymes, M_4 type prevailing in anaerobic tissue (skeleton muscle), H_4 type predominant in aerobic tissues (heart muscle, liver, kidney); hybrids composed of subunits from both types (e.g., H_3M) exist depending upon the oxygen supply. LDH from bacteria is specific for NAD, but oxidizes besides *S*-lactate other *S*-2-hydroxymonocarboxylic acids; mammalian LDH accepts NADP also, but with reduced efficiency. K_m (mM, pig, mammalian): 6.7 (lactate), 0.16 (pyruvate, strong substrate inhibition at pyruvate concentrations > 0.2 mM), 0.25 (NAD), 0.011 (NADH); inhibitors: oxalate ($K_i = 0.2$ mM), competitive with lactate, noncompetitive with pyruvate, malonate, tartrate; pH optimum 7.0.

A. Spectrophotometric Reduction Assay

$$\text{pyruvate} + NADH + H^+ \rightleftharpoons \text{L-lactate} + NAD^+$$

Assay solutions

0.1 M potassium phosphate pH 7.0

0.01 M NADH (disodium salt, $M_r = 709.4$; 71 mg 10 ml^{-1} H_2O)

0.1 M pyruvate (sodium salt, $M_r = 110$; 110 mg in 10 ml H_2O)

LDH solution (e.g., 10 mg ml^{-1}, 550 U ml^{-1}; dilute to 1 IU ml^{-1} with 0.1 M potassium phosphate pH 7.0)

Assay mixture

Components	*Concentration (mM)*
9.4 ml 0.1 M potassium phosphate pH 7.0	94
0.2 ml 0.01 M NADH	0.2
0.2 ml 0.1 M pyruvate	2.0

Procedure

0.98 ml assay mixture

0.02 ml diluted LDH

The absorption decrease at 340 nm is measured at 25 °C; absorption coefficient for NADH: $\varepsilon_{340} = 6.3 \times 10^3\ \mathrm{l\,mol^{-1}\,cm^{-1}}$.

B. Fluorimetric Reduction Assay The fluorimetric assay for LDH is described representatively for all reactions where NAD(P)H is formed from or converted to NAD(P). The fluorimetric assay is about 100-fold more sensitive than the spectrophotometric assay, depending on the sensitivity of the used fluorimeter.

Assay solutions

0.1 M potassium phosphate pH 7.0

0.01 M NADH (disodium salt, $M_r = 709.4$; 71 mg 10 ml^{-1} H_2O)

0.1 M pyruvate (sodium salt, $M_r = 110$; 110 mg in 10 ml H_2O)

LDH solution (e.g., 10 mg ml^{-1}, 550 U ml^{-1}; dilute to 1 IU ml^{-1} with 0.1 M potassium phosphate pH 7.0)

Assay mixture

Components	*Concentration (mM)*
19.35 ml 0.1 M potassium phosphate pH 7.0	97
0.05 ml 0.01 M NADH	0.025
0.4 ml 0.1 M pyruvate	2.0

Procedure

1.98 ml assay mixture

0.02 ml LDH dilution

The relative fluorescence intensity per minute, excited at 260 nm and emitted at 470 nm (ΔFl · min^{-1}) and 25 °C is determined. A fluorescence standard curve with NADH in the respective concentration range is performed for quantification.

Calculation

Volume activity

$$\mathrm{IU\ ml^{-1}} = \frac{\Delta \mathrm{Fl\ min^{-1}} \cdot \text{proportionality factor} \cdot 2 \cdot \text{dilution factor}}{0.02}$$

Specific activity

$$\text{IU mg}^{-1} = \frac{\Delta\text{Fl min}^{-1} \cdot \text{proportionality factor} \cdot 2 \cdot \text{dilution factor}}{0.02 \text{ mg protein/ml}}$$

C. Oxidation Assay

Assay solutions

0.1 M potassium phosphate pH 7.6

0.1 M NAD ($M_r = 663.4$; 0.663 g in 10 ml 0.1 M potassium phosphate pH 7.5)

0.5 M L-lactate (lithium salt, $M_r = 96$; 480 mg in 10 ml 0.1 M potassium phosphate pH 7.6)

LDH solution (e.g., 10 mg ml^{-1}, 550 IU ml^{-1}; dilute to 1 IU ml^{-1} with 0.1 M potassium phosphate pH 7.0)

Assay mixture

Components	*Concentration (mM)*
8.2 ml 0.1 M potassium phosphate pH 7.6	82
0.2 ml 0.1 M NAD	2.0
1.4 ml 0.5 M sodium lactate	70

Procedure

0.98 ml assay mixture

0.02 ml diluted LDH

The absorption increase at 340 nm is measured at 25 °C; absorption coefficient for NADH: $\varepsilon_{340} = 6.3 \times 10^3 \text{ l mol}^{-1} \text{ cm}^{-1}$.

References

Bergmeyer, H.U. (1983) *Methods of Enzymatic Analysis*, 3rd edn, vol. 2, Verlag Chemie, Weinheim, pp. 232–233.

Everse, J. and Kaplan, N.O. (1973) *Adv. Enzymol.*, **37**, 61–133.

Gravie, E.I. (1980) *Microbiol. Rev.*, **44**, 106–139.

Lee, C.Y., Yuan, J.H., and Goldberg, E. (1982) *Meth. Enzymol.*, **89**, 351–358.

3.3.1.8 Malate Dehydrogenase, EC 1.1.1.37

SN: (*S*)-malate:NAD oxidoreductase, malic dehydrogenase, MDH

$$(S)\text{-malate} + NAD^+ \rightleftharpoons \text{oxalacetate} + NADH + H^+$$

Mammalian enzyme (pig heart, cytosol) M_r 67 000, homodimer (2 × 33 000), K_m (mM): 0.14 (NAD), 0.021 (NADH), 0.0083 (oxalacetate); bacterial enzyme (*Escherichia coli*): homotetramer (4 × 33 500), K_m (mM): 2.5 (L-malate), 0.26 (NAD), 0.061 (NADH), 0.026 (oxalacetate), pH optimum 7.5.

For the assay the back reaction is preferred.

Assay solutions

0.1 M potassium phosphate pH 7.5

0.01 M NADH (disodium salt, $M_r = 709.4$; 71 mg 10 ml^{-1} H_2O)

0.1 M oxaloacetic acid ($M_r = 132.1$; 132 mg in 10 ml 0.1 M potassium phosphate pH 7.5)

MDH solution (dilute to 1 IU ml^{-1} with 0.1 M potassium phosphate pH 7.5)

Assay mixture

Components	*Concentration (mM)*
9.1 ml 0.1 M potassium phosphate pH 7.5	91
0.2 ml 0.01 M NADH	0.2
0.5 ml 0.1 M oxaloacetic acid	5.0

Procedure

0.98 ml assay mixture

0.02 ml diluted MDH

The absorption decrease at 340 nm is measured at 25 °C; absorption coefficient for NADH: $\varepsilon_{340} = 6.3 \times 10^3$ l mol^{-1} cm^{-1}.

References

Banaszak, L.J. and Bradshaw, R.A. (1975) in *The Enzymes*, 3rd edn, vol. 12 (ed. P.D. Boyer), Academic Press, New York, pp. 369–396.

Lin, J.J. *et al.* (2002) *J. Mol. Evol.*, **54**, 107–117.

Muslin, E.H. (1995) *Biophys. J.*, **68**, 2218–2223.

Ochoa, S. (1955) *Meth. Enzymol.*, **1**, 735–739.

Raval, D.N. and Wolfe, R.G. (1962) *Biochemistry*, **1**, 1118–1123.

Yin, Y. and Kirsch, J.F. (2007) *Proc. Natl. Acad. Sci. U.S.A.*, **104**, 17353–17357.

3.3.1.9 Malate Dehydrogenase (Oxaloacetate-Decarboxylating) (NAD^+), EC 1.1.1.38, and Malate Dehydrogenase (Decarboxylating), EC 1.1.1.39

SN: (*S*)-malate:NAD^+ oxidoreductase (oxalacetate-decarboxylating), NAD malic enzyme, ME.

$$(S)\text{-malate} + NAD^+ \rightleftharpoons CO_2 + \text{pyruvate} + NADH + H^+$$

Enzyme from *Escherichia coli*, $M_r = 260\,000$, tetramer (4 × 65 000); K_m (mM): 0.046 (NAD), 0.025 (NADH), 0.19 (L-malate), 2.1 (oxalacetate), pH optimum 7.2; decarboxylates also oxalacetate, in contrast to malate dehydrogenase (decarboxylating) EC 1.1.1.39, SN: (*S*)-malate:NAD^+ oxidoreductase (decarboxylating), mitochondrial enzyme (*Arabidopsis thaliana*): M_r 120 000, homodimer (2 × 58 000), K_m (mM): 3.0 (L-malate), 0.5 (NAD), pH optimum 6.6.

Assay solutions

50 mM Tris/HCl, 20 mM imidazole/HCl pH 6.4 (tris(hydroxymethyl)aminomethane, $M_r = 121.1$; 6.06 g, dissolve in 600 ml H_2O together with 1.36 g imidazole ($M_r = 68.1$), adjust to pH 6.4 with 1 M HCl and bring to 1 l)

0.2 M malate solution (L-malic acid, disodium salt, $M_r = 178.1$; 356 mg in 10 ml H_2O)

0.1 M NAD ($M_r = 663.4$; 3.31 g in 50 ml Tris/imidazole/HCl pH 6.4)

0.1 M $MnCl_2$ ($MnCl_2 \cdot 2H_2O$, $M_r = 161.9$; 162 mg in 10 ml H_2O)

Assay mixture

Components	*Concentration (mM)*
8.95 ml 50 mM Tris/imidazole/HCl pH 6.4	45/18
0.5 ml 0.2 M malate solution	10
1.0 ml 0.1 M NAD	10
0.1 ml 0.1 M $MnCl_2$	1

Procedure

0.98 ml assay mixture

0.02 ml enzyme solution

The absorption increase at 340 nm is measured at 25 °C; absorption coefficient for NADH: $\varepsilon_{340} = 6.3 \times 10^3\ \text{l mol}^{-1}\ \text{cm}^{-1}$.

References

Aktas, D.F. and Cook, P.F. (2008) *Biochemistry*, **47**, 2539–2546.
Tronconi, M. *et al.* (2008) *Plant Physiol.*, **146**, 1540–1552.
Yamaguchi, M. *et al.* (1974) *J. Biochem.*, **76**, 1259–1268.

3.3.1.10 Malate Dehydrogenase (Oxaloacetate-Decarboxylating) (NADP⁺), EC 1.1.1.40

SN: (*S*)-malate:NADP$^+$ oxidoreductase (oxaloacetate-decarboxylating), NADP$^+$ dependent malic enzyme

$$(S)\text{-malate} + \text{NADP}^+ \rightleftharpoons \text{CO}_2 + \text{pyruvate} + \text{NADPH} + \text{H}^+$$

The enzyme exists in the liver extramitochondrial (90%) and mitochondrial (10%); human enzyme: M_r 257 000, homotetramer (4 × 63 000); K_m (mM): 0.0053 (NADPH), 0.0092 (NADP), 0.12 (L-malate), 4.8 (NAD), 5.9 (pyruvate), pH optimum 7.25.

Assay solutions

50 mM Tris/HCl pH 7.8

0.2 M malate solution (L-malic acid, disodium salt, $M_r = 178.1$; 356 mg in 10 ml H_2O)

0.01 M NADP ($M_r = 787.4$; 79 mg in 10 ml H_2O)

0.1 M $MnCl_2$ ($MnCl_2 2\,H_2O$, $M_r = 161.9$; 162 mg in 10 ml H_2O)

Assay mixture

Components	*Concentration (mM)*
8.95 ml 50 mM Tris/HCl pH 7.8	45
0.5 ml 0.2 M malate solution	10
0.25 ml 0.01 M NADP	0.25
0.1 ml 0.1 M $MnCl_2$	1.0

Procedure

0.98 ml assay mixture

0.02 ml enzyme solution

The absorption increase at 340 nm is measured at 25 °C; absorption coefficient for NADPH: $\varepsilon_{340} = 6.3 \times 10^3\ \text{l mol}^{-1}\ \text{cm}^{-1}$.

Reference

Zelewsky, M. and Swierczynski, J. (1991) *Eur. J. Biochem.*, **201**, 339–345.

3.3.1.11 Isocitrate Dehydrogenase (NAD^+), EC 1.1.1.41

SN: isocitrate:NAD^+ oxidoreductase (decarboxylating), IDH

$$\text{isocitrate} + NAD^+ \rightleftharpoons CO_2 + \text{2-oxoglutarate} + NADH + H^+$$

Mitochondrial enzyme from the citric acid cycle, cofactors: Mg^{2+} or Mn^{2+}, oxalosuccinate is not accepted as substrate. Human enzyme: M_r 315 000, heterotetramer ($\alpha_2\beta\gamma$), K_m (mM): 0.32 (isocitrate), 0.004 (NAD), 0.22 (Mn^{2+}), pH optimum 7.2.

Assay solutions

33 mM Tris/acetate pH 7.2

0.1 M NAD (M_r = 663.4; 3.31 g in 50 ml 33 mM Tris/acetate pH 7.2)

0.1 M $MnSO_4$ ($MnSO_4 \cdot H_2O$, M_r = 169; 169 mg in 10 ml H_2O)

0.1 M isocitrate solution (D,L-isocitric acid, trisodium salt, M_r = 258.1; 258 mg in 10 ml H_2O)

isocitrate dehydrogenase solution

Assay mixture

Components	*Concentration (mM)*
9.1 ml 33 mM Tris/acetate pH 7.2	30
0.1 ml 0.1 M NAD	1
0.1 ml 0.1 M $MnSO_4$	1
0.5 ml 0.1 M isocitrate solution	5

Procedure

0.98 ml assay mixture

0.02 ml enzyme solution

The absorption increase at 340 nm is measured at 25 °C; absorption coefficient for NADH: $\varepsilon_{340} = 6.3 \times 10^3\ \text{l mol}^{-1}\ \text{cm}^{-1}$.

Reference

Soundar, S., Park, J.H., Huh, T.L., and Colman, R.F. (2003) *J. Biol. Chem.*, **278**, 52146–52153.

3.3.1.12 Isocitrate Dehydrogenase ($NADP^+$), EC 1.1.1.42

SN: isocitrate:$NADP^+$ oxidoreductase (decarboxylating), ICDH

$$\text{D-isocitrate} + NADP^+ \rightleftharpoons CO_2 + \text{2-oxoglutarate} + NADPH + H^+$$

Mitochondrial enzyme (porcine): M_r 104 000, dimer (2 × 47 000), K_m (mM): 0.00046 (NADP), 0.0075 (isocitrate), 0.001 (Mn^{2+}), pH optimum 7.4.

Assay solutions

0.1 M imidazole/HCl pH 8.0

0.01 M NADP (M_r = 787.4; 79 mg in 10 ml)

0.1 M $MgCl_2$ ($MgCl_2 \cdot 6H_2O$, M_r = 203.3; 203 mg in 10 ml)

0.025 M isocitrate solution (D,L-isocitric acid, trisodium salt, M_r = 258.1; 65 mg in 10 ml)

Isocitrate dehydrogenase solution (10 mg ml^{-1}, dilute 100-fold before usage)

Assay mixture

Components	*Concentration (mM)*
8.8 ml 0.1 M imidazole·HCl pH 8.0	88
0.4 ml 0.01 M NADP	0.4
0.4 ml 0.1 M $MgCl_2$	4.0
0.2 ml 0.025 M isocitrate solution	0.5

Procedure

0.98 ml assay mixture

0.02 ml enzyme solution

The absorption increase at 340 nm is measured at 25 °C; absorption coefficient for NADPH is $\varepsilon_{340} = 6.3 \times 10^3\ l\,mol^{-1}\,cm^{-1}$.

References

Bergmeyer, H.U. (1983) *Methods of Enzymatic Analysis*, 3rd edn, vol. 2, Verlag Chemie, Weinheim, pp. 230–231.
Kim, T.K., Lee, P., and Coleman, R.F. (2003) *J. Biol. Chem.*, **278**, 49323–49331.
Plaut, G.E.W. (1963) in *The Enzymes*, 2nd edn, vol. 7 (eds P.D. Boyer, H. Lardy, and K. Myrbäck), Academic Press, New York, pp. 105–126.

3.3.1.13 Glucose-6-phosphate Dehydrogenase, EC 1.1.1.49

SN: D-glucose 6-phosphate:NADP oxidoreductase, G6P-DH, *Zwischenferment*

$$\text{D-glucose-6-phosphate} + \text{NADP}^+ \rightleftharpoons$$
$$\text{6-phosphogluconate} + \text{NADPH} + \text{H}^+$$

Enzyme from human erythrocytes (human): M_r: 220 000, homotetramer (4 × 57 000); K_m (mM): 0.01 (glucose-6-phosphate), 0.8 (6-phosphogluconate), 0.006 (NADP), 0.025 (NADPH), pH optimum 9. ATP inhibits competitively with glucose-6-phosphate; control enzyme of the pentose phosphate pathway, exists as isoenzyme I and II.

Assay solutions

0.1 M triethanolamine/NaOH pH 7.6 (triethanolamine·HCl, $M_r = 185.7$; 18.6 g adjusted with 1 M NaOH, fill up to 1 l)

0.1 M D-glucose-6-phosphate (sodium salt, $M_r = 282.1$, 282 mg in 10 ml)

0.1 M $MgCl_2$ (hexahydrate, $M_r = 203.3$, 203 mg in 10 ml)

0.04 M NADP (disodium salt, $M_r = 787.4$, 315 mg in 10 ml)

Assay mixture

Components	*Concentration (mM)*
9.1 ml 0.1 M triethanolamine·NaOH pH 7.6	91
0.5 ml 0.1 M $MgCl_2$	5.0
0.1 ml 0.1 M D-glucose-6-phosphate	1.0
0.1 ml 40 mM NADP	0.4

Procedure

0.98 ml assay mixture

0.02 ml enzyme solution

The absorption increase at 340 nm is measured at 25 °C; absorption coefficient for NADPH: $\varepsilon_{340} = 6.3 \times 10^3\ \text{l mol}^{-1}\ \text{cm}^{-1}$.

References

Adediran, S.A. (1996) *Biochimie*, **78**, 165–170.
Bergmeyer, H.U. (1983) *Methods of Enzymatic Analysis*, 3rd edn, vol. 2, Verlag Chemie, Weinheim, pp. 204–205.
Cho, S.W. and Joshi, J.G. (1990) *Neuroscience*, **38**, 819–828.

3.3.1.14 Glucose Oxidase, EC 1.1.3.4

SN: β-D-glucose:oxygen 1-oxidoreductase, GOD.

$$\beta\text{-D-glucose} + H_2O + O_2 \rightleftharpoons \delta\text{-D-glucono-1,5-lactone} + H_2O_2$$

Enzyme from *Aspergillus niger*: $M_r = 150\,000$, homodimer; cofactors: FAD and Fe K_m: 1.5 mM (β-D-glucose), pH optimum 5.5. The assay is coupled with peroxidase(POD), *o*-dianisidine acts as the donor:

$$H_2O_2 + \text{donor} \xrightarrow{\text{POD}} 2H_2O + \text{oxidized donor}$$

Assay solutions

0.1 M potassium phosphate pH 7.0

o-Dianisidine solution (*o*-dianisidine dihydrochloride, $M_r = 317.2$, *carcinogenic* in solid form, prepare 1 ml 25 mM solution (7.9 mg in 1 ml H_2O) and dissolve in 100 ml 0.1 M potassium phosphate pH 7.0, saturate with O_2 for 10 min)

0.5 M D-glucose ($M_r = 180.2$; 9 g in 100 ml 0.1 M potassium phosphate pH 7.0)

Peroxidase (POD) from horseradish (dilute to 120 IU ml^{-1} with 0.1 M potassium phosphate pH 7.0 before use)

Assay mixture

Components	*Concentration*
7.7 ml dianisidine solution in	0.2 mM
0.1 M potassium phosphate pH 7.0	77 mM
2.0 ml 0.5 M D-glucose	0.1 M
0.1 ml POD	1.2 IU ml^{-1}

Procedure

0.98 ml assay mixture

0.02 ml enzyme solution

Record the absorption at 436 nm, 25 °C, $\varepsilon_{436} = 8300\ \text{l mol}^{-1}\ \text{cm}^{-1}$.

Calculation

Volume activity

$$\text{IU ml}^{-1} = \frac{\Delta A\ \text{min}^{-1} \cdot \text{dilution factor}}{8.3 \times 0.02}$$

Specific activity

$$\text{IU mg}^{-1} = \frac{\Delta A\ \text{min}^{-1} \cdot \text{dilution factor}}{8.3 \times 0.02\ \text{mg protein/ml}}$$

References

Bergmeyer, H.U. (1983) *Methods of Enzymatic Analysis*, 3rd edn, vol. 2, Verlag Chemie, Weinheim, pp. 201–202.

Bright, H.B. and Porter, D.J.T. (1975) in *The Enzymes*, 3rd edn, vol. 12B (ed. P.D. Boyer), Academic Press, New York, pp. 421–505.

Keilin, D. and Hartree, E.F. (1952) *Biochem. J.*, **50**, 331–341.

Pazur, J.H. (1966) *Meth. Enzymol*, **9**, 82–87.

3.3.1.15 Formate Dehydrogenase, EC 1.2.1.2

SN: formate:NAD$^+$ oxidoreductase, FDH

$$\text{formate} + \text{NAD}^+ \rightleftharpoons \text{CO}_2 + \text{NADH} + \text{H}^+$$

Reported from various bacteria and plants, for example *Arabidopsis thaliana*. M_r 82 000, dimer, cofactors: FMN and non-heme iron, K_m (mM): 0.011 (formate), 0.075 (NAD), pH optimum 7.5.

Assay solutions

0.05 M sodium phosphate pH 7.5

1 M sodium formate (M_r = 68.0, 0.68 g in 10 ml)

0.1 M NAD (M_r = 663.4; 663 mg in 10 ml)

0.5 mM FMN (flavin mononucleotide, riboflavin 5′-phosphate, sodium salt, M_r = 478.3, 2.4 mg in 10 ml)

Assay mixture

Components	*Concentration*
9.5 ml 0.05 M sodium phosphate pH 7.5	95 mM
0.2 ml 1 M sodium formate	20 mM
0.05 ml 0.1 M NAD	0.5 mM
0.05 ml FMN	2.5 μM

Procedure

0.98 ml assay mixture

0.02 ml enzyme solution

Follow the absorption increase at 340 nm at 25 °C; absorption coefficient for NADH: $\varepsilon_{340} = 6.3 \times 10^3\ \mathrm{l\,mol^{-1}\,cm^{-1}}$.

References

Baack, R.D. (2003) *J. Plant Physiol.*, **160**, 445–450.
Yoch, D.C. *et al.* (1990) *J. Bacteriol.*, **172**, 4456–4463.

3.3.1.16 Glyceraldehyde-3-phosphate Dehydrogenase, EC 1.2.1.12

SN: D-glyceraldehyde-3-phosphate:NAD$^+$ oxidoreductase (phosphorylating), GAPDH, triosephosphate dehydrogenase

$$\text{D-glyceraldehyde-3-phosphate} + \text{NAD}^+ + \text{P}_i \rightleftharpoons \text{1,3-bisphosphoglycerate} + \text{NADH} + \text{H}^+$$

Mammalian enzyme (human): $M_r = 142\,000$, homotetramer, K_m (mM): 0.01 (diphosphoglyceric acid), 0.07 (glyceraldehyde-3-phosphate), 0.01 (NADH), 0.05 (NAD), pH optimum 8. The enzyme can be tested either in the forward reaction or in a coupled test in the reverse reaction. For the forward reaction the unstable D,L-glyceraldehyde-3-phosphate is needed.

A. Oxidation Assay

Assay solutions

0.1 M triethanolamine/NaOH pH 7.6 (triethanolamine·HCl, $M_r = 185.7$; dissolve 18.6 g in 800 ml H_2O, adjust with 1 M NaOH to pH 7.6, fill up to 1 l)

0.3 M D,L-glyceraldehyde-3-phosphate ($M_r = 170.1$, unstable, aqueous solution 50 mg ml^{-1})

0.1 M NAD ($M_r = 663.4$; 663 mg in 10 ml)

0.1 M potassium dihydrogen arsenate ($M_r = 180.0$; 180 mg in 10 ml)

Assay mixture

Components	*Concentration (mM)*
9.37 ml 0.1 M triethanolamine/NaOH, pH 7.6	94
0.03 ml 0.3 M glycerate-3-phosphate	0.9
0.3 ml 0.1 M potassium dihydrogen arsenate	3.0
0.1 ml 0.1 M NAD	1.0

Procedure

0.98 ml assay mixture

0.02 ml enzyme solution

Follow the absorption increase at 340 nm at 25 °C; absorption coefficient for NADH: $\varepsilon_{340} = 6.3 \times 10^3$ l mol^{-1} cm^{-1}.

B. Reduction Assay Coupled with 3-Phosphoglycerate Kinase (PGK)

$$\text{3-phosphoglycerate} + \text{ATP} \overset{\text{PGK}}{\rightleftharpoons} \text{1,3-diphosphoglycerate} + \text{ADP}$$

$$\text{1,3-bisphosphoglycerate} + \text{NADH} + \text{H}^+ \overset{\text{GAPDH}}{\rightleftharpoons} \text{D-glyceraldehyde-3-phosphate} + \text{NAD}^+ + \text{P}_\text{i}$$

Assay solutions

0.1 M triethanolamine/NaOH pH 7.6 (triethanolamine· HCl, $M_r = 185.7$; dissolve 18.6 g in 800 ml H_2O, adjust to pH 7.6 with 1 M NaOH, fill up to 1 l)

0.1 M glycerate-3-phosphate (3-phosphoglyceric acid, disodium salt, $M_r = 230.0$; 230 mg in 10 ml)

0.1 M ATP (disodium salt, trihydrate, $M_r = 605.2$; 605 mg in 10 ml)

0.01 M NADH (disodium salt, $M_r = 709.4$; 71 mg in 10 ml)

0.1 M EDTA (ethylenediaminetetraacetic acid, $M_r = 292.2$; 292 mg in10 ml)

0.1 M magnesium sulfate ($MgSO_4 \cdot 7\,H_2O$, $M_r = 246.5$; 247 mg in 10 ml)

3-phosphoglycerate kinase, from yeast (~2000 IU mg^{-1})

Assay mixture

Components	*Concentration*
8.6 ml 0.1 M triethanolamine/NaOH pH 7.6	86 mM
0.5 ml 0.1 M glycerate-3-phosphate	5.0 mM
0.1 ml 0.1 M ATP	1.0 mM
0.2 ml 0.01 M NADH	0.2 mM
0.1 ml 0.1 M EDTA	1.0 mM
0.2 ml 0.1 M magnesium sulfate	2.0 mM
0.1 ml 3-phosphoglycerate kinase	10 IU

Procedure

0.98 ml assay mixture

0.02 ml enzyme solution

Follow the absorption decrease at 340 nm at 25 °C; absorption coefficient for NADH: $\varepsilon_{340} = 6.3 \times 10^3\ \mathrm{l\,mol^{-1}\,cm^{-1}}$.

References

Beisenherz, G., Boltze, H.J., Bücher, T., Czok, R., Garbade, K.H., Meyer-Arendt, E., and Pfleiderer, G. (1953) *Z Naturforsch.*, **8b**, 555–577.
Heinz, F. and Freimüller, B. (1982) *Meth. Enzymol.*, **89**, 301–305.
Krebs, H. (1955) *Meth. Enzymol.*, **1**, 407–411.
Scheek, R.M. and Slater, E.C. (1982) *Meth. Enzymol.*, **89**, 305–309.

3.3.1.17 Pyruvate Dehydrogenase (Acetyl-Transferring), EC 1.2.4.1

SN: pyruvate:[dihydrolipoyllysine-residue acetyltransferase]-lipoyllysine 2-oxidoreductase (decarboxylating, acceptor acetylating), PDH, E1p

$$\text{pyruvate} + \text{ThDP-E1p} \rightarrow \text{hydroxyethyl-ThDP-E1p} + CO_2$$

$$\text{hydroxyethyl-ThDP-E1p} + \text{electron acceptor} \rightarrow$$

$$\text{acetate} + \text{ThDP-E1p} + \text{reduced electron acceptor}$$

The enzyme exists in the cell only as the E1p component of the pyruvate dehydrogenase complex (see Section 3.3.7.1); cofactors: ThDP and Mg^{2+}; mammalian enzyme (bovine heart) M_r: 154 000 tetramer ($2 \times 41\,000, 2 \times 35\,000$); K_m (mM): 0.027 mM (pyruvate), 0.0034 (thiamin diphosphate); enzyme from *Escherichia coli*, $M_r = 200\,000$, homodimer; K_m: 0.3 mM (pyruvate); strong inhibition by fluoropyruvate. Two photometric assays for the partial reaction of the complex-bound

E1p component are described. Alternatively, the release of CO_2 can be measured manometrically, with a CO_2 electrode or by determining the radioactivity, released as $^{14}CO_2$ when ^{14}C1-pyruvate is the substrate.

A. Ferricyanide as Electron Acceptor The assay is less susceptible against disturbances but not very sensitive; larger amounts of enzyme are required.

Assay solutions

0.05 M Tris/HCl pH 7.6

0.1 M $MgCl_2$ ($MgCl_2 \cdot 6\ H_2O$, $M_r = 203.3$; 203 mg in 10 ml H_2O)

0.1 M pyruvate (sodium salt, $M_r = 110.0$; 110 mg in10 ml H_2O)

0.01 M ThDP (thiamin diphosphate, cocarboxylase, $M_r = 460.8$; 46.1 mg in 10 ml 0.05 M Tris/HCl pH 7.6)

0.1 M $K_3[Fe(CN)_6]$ ($M_r = 329.2$; 330 mg in 10 ml H_2O)

Assay mixture

Components	*Concentration (mM)*
8.9 ml 0.05 M Tris/HCl pH 7.6	45
0.1 ml 0.1 M $MgCl_2$	1.0
0.1 ml 0.01 M ThDP	0.1
0.5 ml 0.1 M pyruvate	5.0
0.2 ml 0.1 M $K_3[Fe(CN)_6]$	2.0

Procedure

0.98 ml assay mixture

0.02 ml enzyme solution

The absorption decrease is measured at 436 nm, 30 °C, $\varepsilon_{436} = 755\ l\,mol^{-1}\,cm^{-1}$.

B. Dichlorophenolindophenol as Electron Acceptor

Assay solutions

0.05 M triethanolamine/HCl pH 7.8

0.1 M $MgCl_2$ ($MgCl_2 \cdot 6\ H_2O$, $M_r = 203.3$; 203 mg in 10 ml)

0.1 M pyruvate (sodium salt, $M_r = 110.0$; 110 mg in 10 ml)

0.01 M thiamin diphosphate (ThDP, cocarboxylase, $M_r = 460.8$; 46.1 mg in 10 ml)

0.01 M 2,6-dichlorophenolindophenol ($M_r = 290.1$; 87 mg in 30 ml *n*-propanol)

Assay mixture

Components	*Concentration (mM)*
8.7 ml 0.05 M triethanolamine/HCl pH 7.8	87
0.2 ml 0.1 M $MgCl_2$	2.0
0.2 ml 0.01 M ThDP	0.2
0.5 ml 0.1 M pyruvate	5.0
0.2 ml 0.01 M 2,6-dichlorophenolindophenol	0.2

Procedure

0.98 ml assay mixture

0.02 ml enzyme solution

The absorption change is pursued at 600 nm, 30 °C.

References

Liu, X. and Bisswanger, H. (2005) *Biol. Chem.*, **386**, 11–18.
Pettit, F.H. and Reed, L.J. (1982) *Meth. Enzymol.*, **89**, 376–386.
Saumweber, H. and Bisswanger, H. (1981) *Eur. J. Biochem.*, **114**, 407–411.
Severin, S.E. and Glemzha, A.A. (1964) *Biokhimiya*, **29**, 1170–1176.
Sümegi, B. and Alkonyi, I. (1983) *Eur. J. Biochem.*, **136**, 347–353.
Schwartz, E.R., Old, L.O., and Reed, L.J. (1968) *Biochem. Biophys. Res. Commun.*, **31**, 495–500.

3.3.1.18 Oxoglutarate Dehydrogenase (Succinyl-Transferring), EC 1.2.4.2

SN: 2-oxoglutarate:[dihydrolipoyllysine-residue succinyltransferase]-lipoyllysine 2-oxidoreductase (decarboxylating, acceptor succinylating), OGDH, E1o

2-oxoglutarate + ThDP-E1o → 2-hydroxy-γ-carboxypropyl-ThDP-E1o + CO_2

2-hydroxy-γ-carboxypropyl-ThDP-E1o + electron acceptor →
succinate + reduced electron acceptor + ThDP-E1o

In the cell, the enzyme is exists only as the E1o component of the α-oxoglutarate dehydrogenase complex (see Section 3.3.7.2); cofactors: thiamin diphosphate (ThDP) and Mg^{2+}; $M_r = 210\,000$, homodimer, K_m: 0.1 mM (α-oxoglutarate), pH optimum 8.0. A photometric assay for the partial reaction of the complex-bound E1o component applying $K_3[Fe(CN)_6]$ as artificial electron acceptor is described. Alternatively, the release of CO_2 can be measured manometrically, or with a CO_2 electrode.

Assay solutions

0.1 M potassium phosphate pH 6.5

0.1 M $MgCl_2$ ($MgCl_2 \cdot 6\,H_2O$, $M_r = 203.3$; 203 mg in 10 ml)

0.5 M 2-oxoglutarate (α-ketoglutarate, 2-oxopentanedioic acid, $M_r = 190.1$, disodium salt, 0.95g in 10 ml)

0.01 M ThDP (thiamin diphosaphte, cocarboxylase, $M_r = 460.8$; 46.1 mg in 10 ml)

0.1 M $K_3[Fe(CN)_6]$ ($M_r = 329.2$; 330 mg in 10 ml H_2O)

1% bovine serum albumine (BSA, 100 mg in 10 ml 0.1 M potassium phosphate pH 6.5)

Assay mixture

Components	*Concentration*
8.1 ml 0.1 M potassium phosphate pH 6.5	81 mM
0.2 ml 0.1 M $MgCl_2$	2.0 mM
0.2 ml 0.01 M ThDP	0.2 mM
1.0 ml 1% BSA	0.1%
0.1 ml 0.5 M α-oxoglutarate	5.0 mM
0.2 ml 0.1 M $K_3[Fe(CN)_6]$	2.0 mM

Procedure

0.98 ml assay mixture

0.02 ml enzyme sample

Absorption decrease at 436 nm, 30 °C, $\varepsilon_{436} = 755\ l \cdot mol^{-1}\ cm^{-1}$

Reference

Hager, L.P. and Gunsalus, I.C. (1953) *J. Am. Chem. Soc.*, **75**, 5767–5768.

3.3.1.19 Pyruvate Ferredoxin Oxidoreductase, EC 1.2.7.1

SN: pyruvate:ferredoxin 2-oxidoreductase (CoA-acetylating), pyruvate synthase

$$\text{pyruvate} + \text{CoA} + 2\ \text{oxidized ferredoxin} \rightleftharpoons \text{acetylCoA} + CO_2 + 2\ \text{reduced ferredoxin} + 2H^+$$

Iron-sulfur and thiamin diphosphate as cofactors, heterodimer, $M_r = 103\,000$ (69 000 and 34 000) K_m: 0.28 mM (pyruvate), pH optimum 7 (*Sulfolobus solfataricus*), K_m: 0.32 mM (pyruvate), 3.7 µM (CoA) pH optimum 7.5 (*Clostridium acetobutylicum*).

Assay with Cytochrome *c* as Electron Acceptor

$$\text{pyruvate} + \text{CoA} + \text{cyt } c_{ox} \rightleftharpoons \text{acetyl CoA} + CO_2 + \text{cyt } c_{red}$$

Assay solutions

0.05 M Tris/HCl pH 7.3

0.01 M cytochrome *c* (from horse heart, $M_r = 12380$, 620 mg in 5 ml 0.05 M Tris/HCl pH 7.3)

0.10 M pyruvate (sodium pyruvate, $M_r = 110.0$; 110 mg in 10 ml H_2O)

0.01 M CoA (free acid: $M_r = 767.5$, CoA· Li_3, $M_r = 785.4$; 23 mg in 3 ml 0.05 M Tris/HCl pH 7.3)

Assay mixture

Components	*Concentration (mM)*
9.1 ml 0.05 M Tris/HCl pH 7.3	93
0.1 ml 0.01 M cytochrome *c*	0.1
0.5 ml 0.10 M pyruvate	5.0
0.1 ml 0.01 M CoA	0.1

Procedure

0.98 ml assay mixture

0.02 ml enzyme solution

Follow the absorption increase at 550 nm, 25 °C. The difference absorption coefficient (reduced–oxidized) for cytochrome *c* is $\varepsilon_{550} = 21 \times 10^3\ \mathrm{l\,mol^{-1}\,cm^{-1}}$.

References

Kerscher, L. and Oesterhelt, D. (1977) *FEBS Lett.*, **83**, 197–201.
Kerscher, L. and Oesterhelt, D. (1982) *Trends Biochem. Sci.*, **7**, 371–374.
Meinecke, B., Bertram, J., and Gottschalk, G. (1989) *Arch. Microbiol.*, **152**, 244–250.
Park, Y.J., Yoo, C.B., Choi, S.Y., and Lee, H.B. (2006) *J. Biochem. Mol. Biol.*, **39**, 46–54.
Rabinowitz, J.C. (1974) *Proc. Natl. Acad. Sci. U.S.A.*, **71**, 1361–1365.
Zhang, Q., Iwasaki, T., Wakagi, T., and Oshima, T. (1996) *J. Biochem.*, **120**, 587–599.

3.3.1.20 Alanine Dehydrogenase, EC 1.4.1.1

SN: L-alanine:NAD^+ oxidoreductase (deaminating)

$$\text{L-alanine} + NAD^+ \rightleftharpoons \text{pyruvate} + NH_4^+ + NADH$$

Bacterial enzyme (*Bacillus subtilis*) homohexamer, M_r 220 000, K_m (mM): 1.7 (L-alanine), 0.65 (pyruvate), 38 (NH_4^+), 0.18 (NAD), 0.023 (NADH), pH optimum 10.0.

A. Oxidation of Alanine

Assay solutions

0.05 M $Na_2CO_3/NaHCO_3$ buffer pH 10.0

0.1 M NAD ($M_r = 663.4$; 0.663 g in 10 ml H_2O)

0.1 M L-alanine ($M_r = 89.1$; 89.1 mg in 10 ml H_2O)

Assay mixture

Components	*Concentration (mM)*
8.7 ml 0.05 M $Na_2CO_3/NaHCO_3$ buffer pH 10.0	43.5
0.1 ml 0.1 M NAD	1
1 ml 0.1 M L-alanine	10

Procedure

0.98 ml assay mixture

0.02 ml enzyme solution

Follow the absorption increase at 340 nm at 25 °C; absorption coefficient for NADH: $\varepsilon_{340} = 6.3 \times 10^3\ \mathrm{l\ mol^{-1}\ cm^{-1}}$.

B. Reduction of Pyruvate

Assay solutions

0.05 M Tris/HCl pH 8.0

0.1 M pyruvate (sodium salt, $M_r = 110$; 110 mg in 10 ml H_2O)

2 M NH_4Cl ($M_r = 53.5$; 10.7 g 100 ml)

0.01 M NADH (disodium salt, $M_r = 709.4$; 71 mg in 10 ml)

Assay mixture

Components	*Concentration*
8 ml 0.05 M Tris/HCl pH 8.0	40 mM
1 ml 2 M NH_4Cl	0.2 M
0.6 ml 0.1 M pyruvate	6 mM
0.2 ml 0.01 M NADH	0.2 mM

Procedure

0.98 ml assay mixture

0.02 ml enzyme solution (~30 μg)

Follow the decrease in the absorption at 340 nm at 25 °C. The absorption coefficient for NADH is $\varepsilon_{340} = 6.3 \times 10^3\ \mathrm{l\ mol^{-1}\ cm^{-1}}$.

References

Vali, Z. *et al.* (1980) *Biochim. Biophys. Acta*, **615**, 34–47.
Yoshida, Y. and Freese, E. (1965) *Biochim. Biophys. Acta*, **96**, 248–262.

3.3.1.21 Glutamate Dehydrogenase, EC 1.4.1.3

SN: L-glutamate:NAD(P)$^+$ oxidoreductase (deaminating); glutamic dehydrogenase

$$\text{L-glutamate} + \text{NAD(P)}^+ + H_2O \rightleftharpoons \text{2-oxoglutarate} + \text{NAD(P)H} + NH_4^+$$

Mammalian enzyme: M_r = 330 000, homohexamer (6 × 56 000), K_m (mM): 0.076 (NADH), 1.25 (2-oxoglutarate), pH optimum 7.5–8.0, ADP activates, ATP inhibits. A similar NAD-dependent glutamate:NAD^+ oxidoreductase (deaminating), EC 1.4.1.2 has been found in bacteria and plants, an NADP-dependent (EC 1.4.1.4) in bacteria.

The back reaction is used for the assay.

Assay solutions

0.1 M imidazole buffer pH 7.9 (imidazole, M_r = 68.1; dilute 0.68 g in 80 ml H_2O, adjust with 1 M HCl to pH 7.9)

0.2 M 2-oxoglutarate (2-oxoglutaric acid, monosodium salt, M_r = 168.1, 336 mg in 10 ml)

10 M ammonium acetate (M_r = 77.1; 0.77 g in 10 ml)

0.01 M NADH (disodium salt, M_r = 709.4; 71 mg 10 ml^{-1})

0.1 M EDTA (M_r = 292.2; 292 mg 10 ml^{-1})

0.1 M ADP (disodium salt, M_r = 471.2; 471 mg 10 ml^{-1})

Assay mixture

Components	*Concentration*
8.4 ml 0.1 M imidazole buffer pH 7.9	84 mM
0.7 ml 0.2 M oxoglutarate	14 mM
0.2 ml 10 M ammonium acetate	0.2 M
0.2 ml 0.01 M NADH	0.2 mM
0.1 ml 0.1 M EDTA	1.0 mM
0.2 ml 0.1 M ADP	2.0 mM

Procedure

0.98 ml assay mixture

0.02 ml enzyme solution

Follow the absorption decrease at 340 nm at 25 °C; absorption coefficient for NADH: $\varepsilon_{340} = 6.3 \times 10^3$ l mol^{-1} cm^{-1}.

References

Goldin, B.R. and Frieden, C. (1971) in *Current Topics in Cellular Regulation* (eds B.L. Horecker and E.R. Stadtman), Academic Press, New York, pp. 4–77.

Lee, E.Y., Huh, J.W., Yang, S.J., Choi, S.Y., Cho, S.W., and Choi, S.J. (2003) *FEBS Lett.*, **540**, 163–166.

Schmidt, E. and Schmidt, F.W. (1983) in *Methods of Enzymatic Analysis*, 3rd edn, vol. 3 (ed. H.U. Bergmeyer), Wiley-VCH Verlag GmbH, Weinheim, pp. 216–217.

3.3.1.22 Leucine Dehydrogenase, EC 1.4.1.9

SN: L-leucine:NAD^+oxidoreductase (deaminating), LeuDH

$$\text{L-leucine} + H_2O + NAD^+ \rightarrow \text{4-methyl-2-oxopentanoate} + NH_4^+ + NADH$$

Bacterial enzyme (*Bacillus* sp.), homohexamer, $M_r = 245\,000$ ($6 \times 41\,000$), reacts also with isoleucine, valine, norvaline, and norleucine; K_m (mM): 1.0 (L-leucine), 1.8 (L-isoleucine), 1.7, (L-valine), 0.31 (4-methyl-2-oxopentanoate), 1.5 (NAD), 0.12 (NADH); inhibitors: Cu^{2+}, Co^{2+}. The reaction amino acid $\rightarrow$ α-oxo-acid is also catalyzed by the branched-chain amino acid transaminase (EC 2.6.1.42).

Assay solutions

0.05 M sodium carbonate/1 mM EDTA, pH 10.0 (Na_2CO_3, $M_r = 106.0$, 5.3 g; EDTA, $M_r = 292.2$, 0.29 g in 1 l and adjust to pH 10.0 with 1 M HCl)

0.2 M L-leucine ($M_r = 131.2$, 0.26 g in 10 ml)

0.1 M NAD ($M_r = 663.4$; 0.663 g in 10 ml)

Enzyme sample (~1 IU ml^{-1})

Assay mixture

Components	*Concentration (mM)*
8.5 ml 0.05 M sodium carbonate/1 mM EDTA, pH 10.0	42.5/0.85
1.0 ml 0.2 M L-leucine	20
0.3 ml 0.1 M NAD	3.0

Procedure

0.98 ml assay mixture

0.02 ml enzyme sample

Follow the absorption increase at 340 nm at 25 °C; absorption coefficient for NADH: $\varepsilon_{340} = 6.3 \times 10^3\ l\ mol^{-1}\ cm^{-1}$.

References

Livesey, G. and Lund, P. (1988) *Meth. Enzymol.*, **166**, 282–288.
Katoh, R. *et al.* (2003) *J. Mol. Catal.*, **B23**, 239–247.
Sekimoto, T. *et al.* (1994) *J. Biochem*, **116**, 176–182.

3.3.1.23 L-Amino Acid Oxidase, EC 1.4.3.2

SN: L-amino acid:oxygen oxidoreductase (deaminating), ophio-amino-acid oxidase

$$\text{L-amino acid} + H_2O + O_2 \rightarrow \text{2-oxo acid} + NH_3 + H_2O_2$$

Lysosomal glycoprotein, involved in the amino acid catabolism; mammalian enzyme (*Mus musculus*): $M_r = 113\ 000$, homodimer; K_m (mM): 6.5 (phenylalanine). pH optimum 4.0; *Rhodococcus opacus*: $M_r = 104\ 000$, homodimer, cofactor: FAD, K_m (mM) 0.3 (L-alanine), 0.028 (L-leucine), 0.022 (L-phenylalanine), pH optimum 8.

Coupled assay with peroxidase (POD):

$$H_2O_2 + \text{donor} \xrightarrow{\text{POD}} 2H_2O + \text{oxidized donor}$$

Assay solutions

0.2 M triethanolamine solution (dissolve 3.7 g triethanolamine hydrochloride, $M_r = 185.7$, and adjust to pH 7.6 in 100 ml, add 0.1 g L-leucine and 6.5 mg *o*-dianisidine)

peroxidase (POD, from horseradish, dilute to 120 IU ml^{-1} with 0.1 M potassium phosphate pH 7.0 before use)

Procedure

0.97 ml triethanolamine solution

0.01 ml peroxidase

Start with 0.02 ml enzyme solution (or H_2O for the blank) and follow the absorption decrease at 436 nm, 25 °C for about 5 min, $\varepsilon_{436} = 8300 l\ mol^{-1}\ cm^{-1}$.

References

Bergmeyer, H.U. (1983) *Methods of Enzymatic Analysis*, 3rd edn., vol. 2, Verlag Chemie, Weinheim, pp. 149–151.
Bright, H.B. and Porter, D.J.T. (1975) in *The Enzymes*, 3rd edn., vol. 12B (ed. P.D. Boyer), Academic Press, New York, pp. 421–505.
Geueke, B. and Hummel, W. (2002) *Enzyme Microb. Technol.*, **31**, 77–87.
Mason, J.M. *et al.* (2004) *J. Immunol.*, **173**, 4561–4567.

Meister, A. and Wellner, D. (1963) in *The Enzymes*, 2nd edn, vol. 7 (eds P.D. Boyer, H. Lardy, and K. Myrbäck), Academic Press, New York, pp. 609–648.
Sun, Y. *et al.* (2002) *J. Biol. Chem.*, **277**, 19080–19086.

3.3.1.24 D-Amino Acid Oxidase, EC 1.4.3.3

SN: D-amino acid:oxygen oxidoreductase (deaminating), ophio-amino acid oxidase, new yellow enzyme.

$$\text{D-amino acid} + H_2O + O_2 \rightarrow \text{2-oxo acid} + NH_3 + H_2O_2$$

Mammalian enzyme (human, porcine) in peroxisomes; M_r 38 000, monomer, cofactor: FAD, K_m (mM) 0.77 (D-alanine), 1.4 (D-phenylalanine), pH optimum 9.0. *Neurospora crassa* in mitochondria, M_r 118000, K_m (mM) 0.28 (D-leucine), 0.24 (D-methionine), pH optimum 9. The coupled assay in Section 3.3.1.23 with peroxidase can be used, replacing L-leucine by D-leucine:

$$H_2O_2 + \text{donor} \xrightarrow{\text{POD}} 2H_2O + \text{oxidized donor}$$

References

Massey, V. *et al.* (1961) *Biochim. Biophys. Acta*, **48**, 1–9.
Molla, G. *et al.* (2006) *FEBS Lett.*, **580**, 2358–2364.
Rosenfeld, M.G. and Leiter, E.H. (1977) *Can. J. Biochem.*, **55**, 66–74.
Setoyama, C. *et al.* (2006) *J. Biochem.*, **139**, 873–879.

3.3.1.25 Monoamine Oxidase, EC 1.4.3.4

SN: amine:oxygen oxidoreductase (deaminating)(flavin-containing), adrenaline oxidase, amine oxidase, tyraminase, tyramine oxidase

$$RCH_2NH_2 + H_2O + O_2 \rightarrow RCHO + NH_3 + H_2O_2$$

Catalyzes oxidative deamination of neurotransmitter and biogenic amines

Mammalian liver enzyme (bovine); K_m (mM) 0.19 (phenylethylamine), 0.51 (benzylamine), 2.0 (dimethylaminobenzylamine), pH optimum 8.0, *Aspergillus niger*: K_m (mM): 0.56 (benzylamine), 20 (ethylamine).

Assay solutions

0.05 M sodium phosphate buffer pH 7.2

Triton X-100

0.02 M kynuramine solution (3-(2-aminophenyl-3-oxopropanamine) dihydrobromide, $M_r = 326$; 65.2 mg in 10 ml H_2O)

Assay mixture

Components	*Concentration*
9.28 ml 0.05 M sodium phosphate buffer pH 7.2	46 mM
0.02 mM Triton X-100	0.2%
0.5 ml kynuramine solution	1 mM

For provision of oxygen the mixture must be air-equilibrated.

Procedure

0.98 ml assay mixture

0.02 ml enzyme solution

Follow the absorption change at 314 nm, 25 °C, $\varepsilon_{314} = 12.3 \times 10^3$ l mol^{-1} cm^{-1}.

Modification

Kynuramine in the assay mixture can be replaced by 0.5 ml 60 mM D-dimethylaminobenzylamine 2HCl (M_r = 223.1, 134 mg in 10 ml, assay concentration 3 mM). The absorption is measured at 355 nm, $\varepsilon_{355} = 27.7 \times 10^3$ l mol^{-1} cm^{-1}.

References

Deitrich, R.A. and Erwin, V.G. (1969) *Anal. Biochem.*, **30**, 396–402.
Husain, M., Edmondson, D.E. and Singer, T.B. (1982) *Biochemistry*, **21**, 595–600.
Ramsey, R.R. (1991) *Biochemistry*, **30**, 4624–4629.
Salach, I. and Weyler, W. (1987) *Meth. Enzymol.*, **142**, 627–637.
Suzuki, H., Ogura, Y., and Yamada, H. (1972) *J. Biochem.*, **72**, 703–712.

3.3.1.26 Primary Amine Oxidase, EC 1.4.3.21

SN: primary amine:oxygen oxidoreductase (deaminating), amine oxidase (copper-containing), benzymamine oxidase

$$RCH_2NH_2 + H_2O + O_2 \rightarrow RCHO + NH_3 + H_2O_2$$

Closely related to 1.4.3.22, but not to 1.4.3.4, preferentially oxidizes primary monoamines, but neither diamines such as histamine nor secondary and tertiary amines. In some mammalian tissues also functions as vascular-adhesion protein (VAP-1). Microbial enzyme (yeast): M_r 136 000, dimer; *Candida boidinii*): K_m (mM): 0.01 (O_2), 0.2 (methylamide), pH optimum 7.0.

A. Spectrophotometric Assay H_2O_2 formed with methylamine is coupled to the oxidation of ABTS with peroxidase:

$$ABTS + H_2O_2 \rightarrow \text{oxidized ABTS} + 2H_2O$$

Assay solutions

0.05 M potassium phosphate pH 7.0

0.1 M methylamine (hydrochloride, $M_r = 67.5$; 0.67 g in 100 ml 0.05 M potassium phosphate pH 7.0)

0.02 M ABTS (2,2′-azino-bis-3-ethylbenzothiazoline-6-sulfonic acid, diammonium salt, $M_r = 548.7$; 1.1 g in 0.1 M potassium phosphate pH 7.0)

peroxidase (POD) from horseradish (prepare a solution of 150 IU ml^{-1} in 0.1 M potassium phosphate pH 7.0 before usage)

Assay mixture

Components	*Concentration*
8.8 ml 0.05 M potassium phosphate pH 7.0	50 mM
0.3 ml 0.1 M methylamine	3 mM
0.5 ml 0.02 M ABTS	1 mM
0.2 ml peroxidase	3 IU ml^{-1}

Procedure

0.98 ml assay mixture

0.02 ml enzyme sample

Follow absorption increase at 414 nm, 25 °C, $\varepsilon_{414} = 24.6 \times 10^3$ l mol^{-1} cm^{-1}; 1 mol H_2O_2 oxidizes 2 mol ABTS.

B. Polarographic Assay of O_2 Uptake with O_2 Electrode

Asaay Solutions

0.05 M potassium phosphate pH 7.0

0.1 M methylamine (hydrochloride, $M_r = 67.5$; 0.67 g in 100 ml 0.05 M potassium phosphate pH 7.0)

Assay mixture

Components	*Concentration (mM)*
9.5 ml 0.05 M potassium phosphate pH 7.0	50
0.3 ml 0.1 M methylamine	3

Procedure

The O_2 consumption is measured at 25 °C with an O_2 electrode. One enzyme unit is defined as the consumption of 0.5 μmol min^{-1} (corresponding to the spectrophotometric assay)

C. Assays for Benzylamine Oxidase Activity The assays A and B can be similarly used, replacing methylamine by

- 0.01 M benzylamine (hydrochloride, $M_r = 67.5$; 0.67 g in 100 ml 0.05 M potassium phosphate pH 7.0; assay concentration = 0.3 mM)

References

Haywood, G.W. and Large, P.J. (1981) *Biochem. J.*, **199**, 187–201.
Andrés, M. *et al.* (2001) *J. Histochem. Cytochem.*, **49**, 209–217.
Lyles, G.A. (1996) *Int. J. Biochem. Cell Biol.*, **28**, 259–274.
Saysell, C.G. *et al.* (2002) *Biochem. J.*, **365**, 809–816.

3.3.1.27 Diamine Oxidase, EC 1.4.3.22

SN: histamine:oxygen oxidoreductase (deaminating), amine oxidase (copper containing), copper amine oxidase (CAO), histamine deaminase, histaminase, semicarbazide-sensitive amine oxidase (SSAO)

$$RCH_2NH_2 + H_2O + O_2 \rightarrow RCHO + NH_3 + H_2O_2$$

$$\text{histamine} + H_2O + O_2 \rightarrow \text{(imidazol4-yl)acetaldehyde} + NH_3 + H_2O_2$$

Copper quinoproteins (2,4,5-trihydroxyphenylalanine quinone and copper as cofactors), oxidizes diamines such as histamine and primary monoamines, but not secondary and tertiary amines; sensitive to inhibition by carbonyl group reagents such as semicarbazide. Mammalian (human, porcine): M_r 200 000, homodimer, K_m (mM) 0.0028 (histamine), 0.03 (cadaverine), 0.02 (putrescine), pH optimum 7.0.

Assay solutions

0.1 M potassium phosphate pH 7.6

[β ^{3}H]-Putrescine ($M_r = 88.2$, 1.5×10^6 cpm, ~3 nmol ml^{-1} in 0.1 M potassium phosphate pH 7.6)

Scintillation liquid (Instabray)

Procedure

0.1 ml [β ^{3}H]-putrescine

0.1 ml enzyme solution

Incubate together for 1 h at 37 °C in sealed cups

Extract the deaminated product in two successive steps with 1 ml toluene, each

Add 0.2 ml of the toluene phase to 10 ml scintillation liquid and determine the radioactivity

Histaminase activity can be determined in a similar procedure, replacing [β ^{3}H]-putrescine by [β ^{3}H]-histamine (0.5×10^6 cpm, ~0.15 nmol ml^{-1} in 0.1 M potassium phosphate pH 6.8).

References

Baylin, S.B. and Margolis, S. (1975) *Biochim. Biophys. Acta*, **397**, 294–306.
Elmore, B.O., Bollinger, J.A. and Dodey, D.M. (2002) *J. Biol. Inorg. Chem.*, **7**, 565–571.
Frebort, I. (1996) *Eur. J. Biochem.*, **237**, 255–265.
Costa, M.T., Rotilio, G., Agrò, A.F., Vallogini, M.P., and Mondovi, B. (1971) *Arch. Biochem. Biophys.*, **147**, 8–13.
Stevanoto, R., Porgia, M., Befani, O., Mondovi, B., and Rigo, A. (1989) *Biotechnol. Appl. Biochem.*, **11**, 266–272.

3.3.1.28 Urate Oxidase, EC 1.7.3.3

SN: urate:oxygen oxidoreductase, uricase

$$\text{urate} + H_2O + O_2 \rightarrow \text{5-hydroxyisourate} + H_2O_2$$
$$(\text{urate} + 2H_2O + O_2 \rightarrow \text{allantoin} + H_2O_2 + CO_2)$$

Enzyme from *Bacillus subtilis*: M_r 150 000, dimer, Cu^{2+} as cofactor, K_m (mM) 0.0034 (urate), inhibitor: nitroxanthine K_i (mM) 0.0021 (competitive with urate)

Assay solutions

0.1 M sodium borate pH 8.5 ($Na_2B_4O_7 \cdot 10H_2O$, $M_r = 381.4$; 3.8 g in 100 ml, adjust to pH 8.5 with 0.1 M boric acid, saturated with gaseous oxygen for 15 min prior to use)

1.0 mM urate ($M_r = 168.1$, dissolve 17 mg in 100 ml borate buffer pH 8.5, heat to ~50 °C if necessary)

Assay mixture

Components	*Concentration (mM)*
9.4 ml 0.1 M sodium borate pH 8.5	94
0.4 ml 1.0 mM urate	0.04

Procedure

0.98 assay mixture

0.02 ml enzyme solution

Follow the absorption change at 295 nm, 25 °C; $\varepsilon_{295} = 12.6 \times 10^3$ l mol^{-1} cm^{-1}.

References

Bergmeyer, H.U. (1983) *Methods of Enzymatic Analysis*, 3rd edn, vol. 2, Verlag Chemie, Weinheim, pp. 322–323.

Colloch, N. *et al.* (2006) *Biochim. Biophys. Acta*, **1764**, 391–397.

Doll, C., Bell, A.F., Power, N., Tonge, P.J., and Tipton, P.A. (2005) *Biochemistry*, **44**, 11440–11446.

Leone, E. (1955) *Meth. Enzymol.*, **2**, 485–489.

Mahler, H.R. (1963) in *The Enzymes*, 2nd. edn., vol. 8 (eds P.D. Boyer, H.A. Lardy, and K. Myrbäck), Academic Press, New York, pp. 285–296.

3.3.1.29 Dihydrolipoamide Dehydrogenase, EC 1.8.1.4

SN: NADH:lipoamide oxidoreductase, diaphorase, E3

$$\text{lipoamide} + \text{NADH} + \text{H}^+ \overset{\text{E3}}{\rightleftharpoons} \text{dihydrolipoamide} + \text{NAD}^+$$

In the cell, the enzyme exists in a free form as a homodimer (M_r = 112 000), and as the E3 component bound to the pyruvate dehydrogenase complex, the α-oxoglutarate dehydrogenase complex and the branched chain α-oxoacid dehydrogenase complex (see Section 3.3.7). The designation *diaphorase* refers to enzymes that can catalyze the oxidation of NAD(P)H by artificial electron acceptors such as dichlorophenolindophenol, ferrricyanide, and quinones.

The reaction can be tested from both sides. Lipoic acid and its amide, which is about five times more active, are commercially available in the oxidized form and can be reduced by a simple procedure (see Section 3.3.2.1).

A. Oxidation of Dihydrolipoamide

Assay solutions

0.1 M potassium phosphate pH 7.5

0.1 M NAD (free acid, $M_r = 663.4$; 663 mg in 10 ml)

0.1 M DTE (dithioerythritol, $M_r = 154.2$; 154 mg in10 ml)

0.2 M D,L-dihydrolipoamide (M_r 207.3; 52 mg in 1.25 ml ethanol; alcohol dehydrogenase if present, for example in crude extracts, forms NADH from ethanol and NAD. In this case, acetone must be taken as solvent)

Assay mixture

Components	*Concentration (mM)*
9.46 ml 0.1 M potassium phosphate pH 7.5	95
0.1 ml 0.1 M NAD	1.0
0.04 ml 0.1 M DTE	0.4

Procedure

0.96 ml assay mixture	–
0.02 ml 0.2 M D,L-dihydrolipoamide	4.0
0.02 ml enzyme solution	–

The absorption increase is measured at 340 nm, 30 °C, $\varepsilon_{340} = 6.3 \times 10^3$ l mol^{-1} cm^{-1}.

B. Reduction of Lipoamide To avoid overreduction of the enzyme into the inactive state, NAD must be present in addition to NADH.

Assay solutions

0.1 M potassium phosphate pH 7.5

0.1 M NAD (free acid, $M_r = 663.4$; 663 mg in10 ml)

0.01 M NADH (disodium salt, $M_r = 709.4$, 71 mg in10 ml)

0.1 M DTE (dithioerythritol, $M_r = 154.2$; 154 mg in 10 ml)

0.1 M EDTA (ethylenediaminetetraacetic acid, $M_r = 292.2$; 292 mg in 10 ml)

0.2 M D,L-lipoamide (6,8-thioctic acid amide, M_r 205.3; 51 mg in 1.25 ml ethanol or acetone, see Section 3.3.1.29.A)

Assay mixture

Components	*Concentration (mM)*
9.06 ml 0.1 M potassium phosphate pH 7.5	91
0.1 ml 0.1 M NAD	1.0
0.02 ml 0.01 M NADH	0.02
0.1 ml 0.1 M EDTA	1.0
0.02 ml 0.1 M DTE	0.2

Procedure

0.93 ml assay mixture	–
0.05 ml 0.2 M D,L-lipoamide	10 mM
0.02 ml enzyme solution	–

The absorption decrease is measured at 340 nm, 30 °C, $\varepsilon_{340} = 6.3 \times 10^3$ l mol^{-1} cm^{-1}.

References

Reed, L.J., Koike, M., Levitch, M.E., and Leach, F.R. (1958) *J. Biol. Chem.*, **232**, 143–158.
Reed, L.J. and Willms, C.R. (1966) *Meth. Enzymol.*, **9**, 247–269.
Schmincke-Ott, E. and Bisswanger, H. (1981) *Eur. J. Biochem.*, **114**, 413–420.
Schwartz, E.R. and Reed, L.J. (1970) *Biochemistry*, **9**, 1434–1439.
Straub, F.B. (1939) *Biochem. J.*, **33**, 787–792.
Willms, C.R., Oliver, R.M., Henney, H.R., Mukherjee, B.B., and Reed, L.J. (1967) *J. Biol. Chem.*, **242**, 889–897.

3.3.1.30 Glutathione Disulfide Reductase, EC 1.8.1.7

SN: glutathione:$NADP^+$ oxidoreductase, glutathione reductase

$$2\ \text{glutathione} + NADP^+ \rightarrow 2\ \text{glutathione disulfide} + NADPH + H^+$$

Mammalian enzyme (bovine), M_r 118 000, homodimer, coenzyme FAD, K_m (mM): 0.008 (NADPH), 0.04 (glutathione), 0.065 (glutathione disulfide), pH optimum 7.3.

Assay solutions

0.1 M potassium phosphate, 1 mM EDTA pH 7.0 (dissolve 0.29 g EDTA in 1 l 0.1 M potassium phosphate pH 7.0)

0.01 M NADPH (M_r = 833.4, Na_4 salt, 83 mg in 10 ml 0.1 M potassium phosphate, 1 mM EDTA pH 7.0, freshly prepared)

0.01 M GSSG (oxidized glutathione, M_r = 612.6; 61 mg in 10 ml 0.1 M potassium phosphate, 1 mM EDTA pH 7.0)

Assay mixture

Components	*Concentration (mM)*
8.6 ml 0.1 M potassium phosphate, 1 mM EDTA pH 7.0	86/0.86
0.2 ml 0.01 M NADPH	0.2
1.0 ml 0.01 M GSSG	1.0

Procedure

0.98 ml assay mixture

0.02 ml enzyme sample

The absorption decrease at 340 nm is measured at 25 °C, $\varepsilon_{340} = 6.3 \times 10^3$ l mol^{-1} cm^{-1}.

References

Böhme, C.C. *et al.* (2000) *J. Biol. Chem.*, **275**, 37317–37323.
Dolphin, D., Poulson, R., and Avromonic, O. (1989) *Glutathione: Chemical, Biochemical and Metabolic Aspects*, Part A, John Wiley & Sons, Inc., New York.
Dringen, R. and Gutterer, J.M. (2002) *Meth. Enzymol.*, **348**, 281–288.
Sakirolu, E.M. and Ciftci, M. (2003) *Prep. Biochem. Biotechnol.*, **33**, 283–300.
Smith, L.K. *et al.* (1988) *Anal. Biochem.*, **175**, 408–413.

3.3.1.31 Catalase, EC 1.11.1.6

SN: hydrogenperoxide: hydrogenperoxide oxidoreductase

$$2H_2O_2 \rightarrow O_2 + 2H_2O$$

The green enzyme (M_r = 240 000, bovine liver) consists of four identical subunits each carrying a protohematin IX group in a high-spin state with a central trivalent iron. Four coordination sites of the iron ion are occupied by the porphyrin ring structure, the fifth by a histidine residue of the protein. The sixth position remains free and may be occupied by anions like cyanide or fluoride, and block the enzyme reaction. Binding of cyanide induces the iron to move into the plane of the porphyrin ring, the high spin changes to the low spin state, accompanied by a remarkable spectral shift. Catalase possesses a strong tendency to crystallize, so the crystal suspension is the preferred storage form. At higher concentrations hydrogen peroxide damages the enzyme, so it cannot be tested at a saturating substrate. The enzyme has an absorption coefficient of 380×10^3 l mol^{-1} cm^{-1} at 405 nm. *Escherichia coli*: K_m (mM): 3.5 (H_2O_2), pH optimum 10.5; human, K_m (mM): 80 (H_2O_2), pH optimum 6.8.

Assay solutions

0.05 M potassium phosphate pH 7.0

H_2O_2 solution (~10 mM), dilute 0.06 ml 30% H_2O_2 in 50 ml 0.05 M potassium phosphate pH 7.0, the absorption against buffer at 240 nm should be 0.50 ± 0.01; otherwise, adjust to this value with buffer or H_2O_2

Catalase solution (20 mg ml^{-1}, dilute in 0.05 M potassium phosphate, pH 7.0 before use, according to the specific activity of the sample, for example 2000-fold)

Procedure

0.98 ml H_2O_2 solution

0.02 ml catalase solution

Follow the absorption decrease at 240 nm, 25 °C in quartz cuvettes; $\varepsilon_{240} = 40 \times 10^3$ l mol^{-1} cm^{-1}. An absorption change of 0.04/min corresponds to a decomposition of 1 μmol H_2O_2/min ~1 IU (16.67 nkat).

References

Aebi, H.E. (1984) *Meth. Enzymol.*, **105**, 121–126.
Bergmeyer, H.U. (1955) *Biochem. Z.*, **327**, 255–258.
Chance, B. and Maehly, A.C. (1955) *Meth. Enzymol.*, **2**, 764–775.
Moore, R.L. (2008) *J. Inorg. Biochem.*, **102**, 1819–1824.
Nicholls, P. and Schonbaum, G.R. (1963) in *The Enzymes*, vol. 8 (eds P.D. Boyer, H.A. Lardy, and K. Myrbäck), Academic Press, New York, pp. 147–225.
Sumner, J.B. and Dounce, A.L. (1937) *J. Biol. Chem.*, **121**, 417–424.
Switala, J. and Loewen, P.C. (2002) *Arch. Biochem. Biophys.*, **401**, 145–154.

3.3.1.32 **Peroxidase, EC 1.11.1.7**

SN: donor:hydrogenperoxide oxidoreductase, lactoperoxidase, myeloperoxidase

$$\text{donor} + H_2O_2 \rightarrow \text{oxidized donor} + 2H_2O$$

Glycoprotein, M_r = 44 000 (horseradish), related to catalase with protohematin IX as prosthetic group. It is specific for the acceptor H_2O_2, but reacts with various donor substrates, all applicable for the enzyme assay: 5-amino-2,3-dihydro-1,4-phthalazinedion (luminol), 3-amino-9-ethylcarbazole (AEC), 5-aminosalicylic acid, 2,2′-azino-bis-(3-ethylbenzothiazoline-6-sulfonic acid (ABTS), 4-chloro-1-naphthol, 3,3′-diaminobenzidine, *o*-dianisidine, guaiacol, *o*-phenylenediamine, ascorbic acid, resorcinol, pyrogallol, hydroquinon, 3,3′,5,5′-tetramethylbenzidine. Peroxidase is applied in various conjugates; for example, with antibodies in Western blots and ELISA, or colloidal gold in cell biology. K_m (mM): 10.5 (guajacol), 0.48 (dianisidin), 0.21 (ABTS); inhibitors: azide, 1,10-phenathrolin, pH optimum 4.5.

A. Assay with 2,2′-Azino-bis-3-ethylbenzothiazoline-6-sulfonic Acid (ABTS)

Assay solutions

0.1 M potassium phosphate pH 6.0

0.02 M ABTS (2,2′-azino-bis-3-ethylbenzothiazoline-6-sulfonic acid, diammonium salt, M_r = 548.7; 1.1 g in 0.1 M potassium phosphate pH 6.0)

H_2O_2 solution (~ 10 mM), dilute 0.06 ml 30% H_2O_2 in 50 ml 0.05 M potassium phosphate pH 7.0, the absorption against buffer at 240 nm should be 0.50 ± 0.01. Otherwise, adjust to this value with buffer or H_2O_2

Assay mixture and procedure

	Concentration (mM)
0.78 ml 0.1 M potassium phosphate pH 6.0	98
0.1 ml 0.02 M ABTS	2.0
0.1 ml H_2O_2 solution	1.0
0.02 ml enzyme solution	–

Follow absorption change at 414 nm, 25 °C, ε_{414}= 24.6 × 10^3 l mol^{-1} cm^{-1}; 1 mol H_2O_2 oxidizes 2 mol ABTS.

B. Assay with Guaiacol

Assay solutions

0.1 M potassium phosphate pH 7.0

0.02 M guaiacol (2-methoxyphenol, catechol monomethylether, $M_r = 124.1$; 24.8 mg in 10 ml)

H_2O_2 solution (~10 mM), dilute 0.06 ml 30% H_2O_2 in 50 ml 0.05 M potassium phosphate pH 7.0; the absorption against buffer at 240 nm should be 0.50 ± 0.01, otherwise adjust to this value with buffer or H_2O_2

peroxidase solution (dilute concentrated peroxidase sample about 2000-fold in 0.1 M potassium phosphate pH 7.0)

Assay mixture

Components	*Concentration (mM)*
9.55 ml 0.1 M potassium phosphate pH 7.0	98
0.15 ml 0.02 M guaiacol	0.3

Procedure

0.97 ml assay mixture

0.02 ml peroxidase solution

Start by addition of

0.01 ml H_2O_2 solution	0.1 mM

Follow the absorption change by 436 nm, 25 °C, $\varepsilon_{436} = 25501 \cdot mol^{-1}\ mm^{-1}$.

C. Assay with Dianisidine

Assay solutions

0.1 M potassium phosphate pH 6.0

0.03 M *o*-dianisidine (3,3′-dimethoxybenzidine dihydrochloride, $M_r = 317.2$; 95 mg in 10 ml methanol, *carcinogenic!*)

H_2O_2 solution (~10 mM), dilute 0.06 ml 30% H_2O_2 in 50 ml 0.05 M potassium phosphate pH 7.0; the absorption against buffer at 240 nm should be 0.50 ± 0.01, otherwise adjust to this value with buffer or H_2O_2

peroxidase solution (dilute concentrated peroxidase sample about 2000-fold in 0.1 M potassium phosphate pH 7.0)

Assay mixture

Components	*Concentration (mM)*
9.6 ml 0.1 M potassium phosphate pH 6.0	98
0.1 ml 0.03 M *o*-dianisidine	0.3

Procedure

0.97 ml assay mixture

0.02 ml peroxidase solution

Start by addition of

0.01 ml H_2O_2 solution 0.1 mM

Record the absorption at 436 nm, 25 °C, $\varepsilon_{436} = 8.3 \times 10^3\ l\ mol^{-1}\ cm^{-1}$.

References

Basbaum, A.I. (1989) *J. Histochem. Cytochem.*, **37**, 1811–1815.
Chance, B. and Maehly, A.C. (1955) *Meth. Enzymol.*, **2**, 764–775.
Childs, R.E. and Bardsley, W.G. (1975) *Biochem. J.*, **145**, 93–103.
Engvall, E. (1980) *Meth. Enzymol.*, **70**, 419–439.
Mäkinen, K.K. and Tenovuo, J. (1982) *Anal. Biochem.*, **126**, 100–108.
Pütter, J. and Becker, R. (1983) in *Methods of Enzymatic Analysis*, vol. 3 (ed. H.U. Bergmeyer), Verlag Chemie, Weinheim, pp. 286–293.
Shindler, J.S., Childs, R.E., and Bardsley, W.G. (1976) *Eur. J. Biochem.*, **65**, 325–331.
Szutowicz, A., Kobes, R.D., and Orsulak, P.J. (1984) *Anal. Biochem.*, **138**, 86–94.

3.3.1.33 **Luciferase, EC 1.13.12.7**

SN: Photinus-luciferin:oxygen oxidoreductase (decarboxylating, ATP-hydrolyzing), Photinus-luciferin 4-monooxygenase, firefly luciferase

$$\text{Photinus-luciferin} + \text{ATP} + O_2 \rightarrow \text{oxidized photinus-luciferin} + CO_2 + \text{AMP} + PP_i + H_2O + \text{light}$$

Enzyme from *Photinus pyralis*: M_r 100 000, dimer; K_m (mM): 0.004 (ATP), 0.008 (luciferin), pH optimum 7.8.

The light intensity I is related to the Michaelis–Menten equation:

$$I = \frac{d(h\nu)}{dt} = \frac{V[\text{ATP}]}{K_m + [\text{ATP}]}$$

h is Planck's constant and ν the frequency of emitted light. At very low concentrations of ATP ($[\text{ATP}] \ll K_m$) the Michaelis–Menten equation reduces to

$$I = \text{const} \times [\text{ATP}];$$

const $= V\,K_m{}^{-1}$, the light intensity becomes proportional to [ATP].

Assay solutions

0.1 M Tris/acetate pH 7.75

10 μM ATP (disodium salt, trihydrate, $M_r = 605.2$; prepare a 10 mM solution with 60.5 mg in 10 ml, and dilute 10 μl in 10 ml 0.1 M Tris/acetate pH 7.75)

0.1 M $MgSO_4$ ($MgSO_4 \cdot 7H_2O$, $M_r = 246.5$; 246 mg in 10 ml H_2O)

1 mM D-luciferin (4,5-dihydro-[6-hydroxy-2-benzothiazolyl]-4-thiazole carboxylic acid, $M_r = 280.3$; sodium salt, $M_r = 302.3$; 3 mg in 10 ml 0.1 M Tris/acetate pH 7.75, store at −20 °C)

Luciferase (1 μg ml^{-1} in 0.1 M Tris/acetate pH 7.75)

Assay mixture and procedure

	Concentration
0.34 ml 0.1 M Tris/acetate pH 7.75	68 mM
0.05 ml 0.1 M $MgSO_4$	10 mM
0.05 ml 1 mM D-luciferin	0.1 mM
0.05 ml luciferase	50 ng
Start with addition of	
0.01 ml 10 μM ATP	0.2 μM

Read total light intensity (emission maximum at 562 nm) after 10 s (25 °C). Prepare a standard curve with ATP between 0.05 and 0.5 μM.

One enzyme unit is defined as the light output from a 50-μl assay mixture containing 5 pmol ATP and 7.5 nmol luciferin in glycine-Tris pH 7.6 at 25 °C. One light unit produces a biometer peak height equivalent to 0.02 μCi of ^{14}C in PPO/POPOP (2,5-diphenyloxazole/2,2′-p-phenylene-bis[5-phenyloxazole]) cocktail (before 1991, 1 unit was the amount of luciferase producing 1 nmol pyrophosphate/min at pH 7.7, 25 °C in the presence of 0.6 mM ATP and 0.1 mM D-luciferin).

References

DeLuca, M. and McElroy, W.D. (1978) *Meth. Enzymol.*, **57**, 3–15.
Wulff, K. (1982) in *Methods of Enzymatic Analysis*, 3rd edn, vol. 1 (ed. H.U. Bergmeyer), Verlag Chemie, Weinheim, pp. 340–368.
Yousefi-Nejad, M. *et al.* (2007) *Enzyme Microb. Technol.*, **40**, 740–746.

3.3.2 Transferases, EC 2

3.3.2.1 Dihydrolipoamide Acetyltransferase, EC 2.3.1.12

SN: acetyl-CoA :enzyme N6-(dihydrolipoyl)lysine *S*-acetyltransferase, E2p

$$\text{hydroxyethyl ThDP-E1p} + \text{lipoyl E2p} \rightarrow \text{acetyl dihydrolipoyl-E2p} + \text{ThDP-E1p}$$

$$\text{acetyl dihydrolipoyl-E2p} + \text{CoA} \rightarrow \text{acetyl CoA} + \text{dihydrolipoyl-E2p}$$

In the cell, the enzyme exists only as E2p component of the pyruvate dehydrogenase complex (E1p, pyruvate dehydrogenase, EC 1.2.4.1, see Section 3.3.7.1); lipoic acid is a covalent bound cofactor; enzyme from *Escherichia coli*: $M_r = 86\,400$, *Azotobacter vinelandii*: K_m (mM) 0.022 (acetyl-CoA), 1.1 (dihydrolipoamide), pH optimum 8.

A. Spectrophotometric Assay

$$\text{CoA} + \text{acetylphosphate} \overset{\text{PTA}}{\rightleftharpoons} \text{acetyl-CoA} + P_i$$

$$\text{acetyl CoA} + \text{D,L-dihydrolipoamide} \overset{\text{E1p}}{\rightleftharpoons} \text{acetyl dihydrolipoamide} + \text{CoA}$$

Assay solutions

0.1 M Tris/HCl pH 7.6

0.01 M CoA (trilithium salt, $M_r = 785.4$; 24 mg in 3 ml)

1.0 M acetylphosphate (LiK salt, $M_r = 184.1$; 1.84 g in 10 ml)

phosphotransacetylase (acetyl-CoA:orthophosphate acetyltransferase, EC 2.3.1.8, 4000 IU ml^{-1})

0.2 M D,L-dihydrolipoamide (50 mg in 1.25 ml ethanol). The available oxidized form of D,L-lipoamide (D,L-6.8-thioctic acid amide, $M_r = 205.3$) can be reduced according to Reed, Leach, and Koike (1958)

- Suspend 2 g D,L-lipoamide at 0 °C in 40 ml methanol and 10 ml H_2O
- Add 2 g $NaBH_4$ successively during 2 h; the solution must become colorless, if necessary add more $NaBH_4$
- Acidify the solution is with 5 N HCl
- Extract the reagent three times with 20 ml chloroform
- Wash the collected chloroform phases three times with water
- Remove the solvent in the vacuum
- Dissolve the remaining dry powder in 5 ml hot benzene (or toluene)
- Add petroleum ether dropwise until a faint turbidity remains, crystals will be formed overnight at 4 °C. The crystallization procedure should be repeated two times, and the final product must be completely white

Assay mixture

Components	*Concentration*
9.38 ml 0.1 M Tris/HCl pH 7.6	94 mM
0.2 ml 0.2 M D,L-dihydrolipoamide	4.0 mM
0.1 ml 0.01 M CoA	0.1 mM
0.1 ml 1 M acetylphosphate	10 mM
0.02 ml phosphotransacetylase	4 IU ml^{-1}

Procedure

0.98 ml assay mixture

0.02 ml enzyme sample

Follow the absorption change at 240 nm, 30 °C, in quartz cuvettes.

B. Stopped Assay

Assay solutions

0.1 M Tris/HCl pH 7.6

0.01 M CoA (trilithium salt, $M_r = 785.4$; 24 mg in 3 ml)

1.0 M acetylphosphate (LiK salt, $M_r = 184.1$; 1.84 g in 10 ml)

phosphotransacetylase (acetyl-CoA:orthophosphate acetyltransferase, EC 2.3.1.8, 4000 IU ml^{-1})

0.2 M D,L-dihydrolipoamide

1 N HCl

2 M hydroxylamine (mix equal volumes of 4 M hydroxylamine hydrochloride, $M_r = 69.49$, and 4 M KOH, $M_r = 56.11$)

$FeCl_3$ reagent:
- 100 ml 5% $FeCl_3$ (8.33 g $FeCl_3 \cdot 6H_2O$, $M_r = 270.3$, bring to 100 ml with 0.1 N HCl)
- 100 ml 12% TCA (trichloroacetic acid, $M_r = 163.4$)
- 100 ml 3 M HCl (mix 1 part 37% HCl with 3 parts H_2O)

Mix the three solutions together

Assay mixture

Components	*Concentration*
8.98 ml 0.1 M Tris/HCl pH 7.6	90 mM
0.2 ml 0.01 M CoA	0.2 mM
0.1 ml 1 M acetylphosphate	10 mM
0.5 ml 0.2 M D,L-dihydrolipoamide	10 mM
0.02 ml phosphotransacetylase	4 IU ml^{-1}

Procedure

0.98 ml assay mixture

0.02 ml enzyme solution

Incubate for 15 min at 30 °C

0.1 ml 1 N HCl

5 min in a boiling water bath

Remove 0.4 ml and mix with

0.2 ml 2 M hydroxylamine

0.6 ml $FeCl_3$ reagent.

10 min at room temperature

Centrifuge for 5 min

Measure the absorption at 546 nm.

Calculation

The specific activity is expressed in special units, referring to 1 h of reaction time:

$$\text{specificactivity} = \frac{A_{546} \times 20 \times \text{assay volume (1.1 ml)} \times \text{dilution factor}}{\text{enzyme volume (ml)} \times \text{protein (mg ml}^{-1}\text{)}}$$

References

Fuller, C.C., Reed, L.J., Oliver, R.M., and Hackert, M.L. (1979) *Biochem. Biophys. Res. Commun.*, **90**, 431–438.
Hager, L.P. and Gunsalus, I.C. (1953) *J. Am. Chem. Soc.*, **75**, 5767–5768.
Hendle, J. *et al.* (1995) *Biochemistry*, **34**, 4287–4298.
Reed, L.J., Leach, F.R., and Koike, M. (1958) *J. Biol. Chem.*, **232**, 123–142.
Schwartz, E.R. and Reed, L.J. (1969) *J. Biol. Chem.*, **244**, 6074–6079.
Willms, C.R., Oliver, R.M., Henney, H.R., Mukherjee, B.B., and Reed, L.J. (1967) *J. Biol. Chem.*, **242**, 889–897.

3.3.2.2 Fatty Acid Synthase, EC 2.3.1.85

SN: acyl-CoA:malonyl-CoA C-acyltransferase (decarboxylating, oxoacyl- and enoyl-reducing and thioester-hydrolyzing)

$$\text{acetyl CoA} + n\ \text{malonyl-CoA} + 2n\ \text{NADPH} + 2n\ H^+ \rightarrow$$
$$CH_3\text{-}(CH_2\text{-}CH_2)_n\text{-CO-CoA} + n\ \text{CoA} + nCO_2 + 2n\ \text{NADP}^+ + n\ H_2O$$
$$n = 6\text{–}8$$

Assay solutions

0.1 M potassium phosphate pH 6.5

0.1 M EDTA (ethylenediamine tetraacetic acid, $M_r = 292.2$; 292 mg in 10 ml)

0.1 M DTE (dithioerythritol, $M_r = 154.2$; 154 mg in 10 ml)

0.01 M acetyl-CoA (Li_3 salt, $M_r = 827.4$; 8.3 mg in 1 ml)

7 mM malonyl-CoA ($M_r = 877.3$, Li_4 salt; 6.1 mg ml^{-1})

0.01 M NADPH (Na_4 salt, $M_r = 833.4$, 83 mg in 10 ml)

2.4% BSA (240 mg in 10 ml)

Assay mixture

Components	*Concentration*
8.64 ml 0.1 M potassium phosphate pH 6.5	86 mM
0.25 ml 0.1 M EDTA	2.5 mM
0.25 ml 0.1 M DTE	2.5 mM
0.25 ml 2.4% BSA	0.06%
0.06 ml 0.01 M acetyl-CoA	0.06 mM
0.15 ml 0.01 NADPH	0.15 mM

Procedure

0.96 ml assay mixture

0.02 ml enzyme solution

Observe absorption decrease at 340 nm, 25 °C, for 3–4 min
Start the reaction by addition of

0.02 ml 7 mM malonyl-CoA	0.14 mM

Follow the absorption decrease at 340 nm, 25 °C, $\varepsilon_{340} = 6.3 \times 10^3$ l mol^{-1} cm^{-1}.

Reference

Lynen, F. (1969) *Meth. Enzymol.*, **14**, 17–33.

3.3.2.3 Phosphorylase a, EC 2.4.1.1

SN: 1.4-α-D-glucan:orthophosphate-α-D-glucosytransferase, glycogenphosphorylase

The enzyme splits α-1,4-glucosidic bonds in polysaccharides:

$$(\text{glucose})_n + P_i \rightarrow (\text{glucose})_{n-1} + \text{glucose-1-P}$$

Enzyme from *Zea mays*, homodimer, M_r 2 × 112 000, K_m (mM): 0.25 (glycogen), 0.028 (amylopectine) 1.5 (glucose 1-phosphate), pH optimum 5.5–6.5.

For the assay, the reaction is coupled to phosphoglucomutase (EC 5.4.2.2)

glucose-1-P $\rightarrow$ glucose-6-P

and glucose-6-phosphate dehydrogenase (EC 1.1.1.49)

glucose-6-P + $NADP^+$ $\rightarrow$ gluconate-6-P + NADPH + H^+

Assay solutions

0.05 M potassium phosphate pH 6.8

0.01 M EDTA (ethylenediaminetetraacetic acid, $M_r = 292.3$, 29.2 mg in 10 ml)

glycogen (60 mg in 10 ml)

1 mM glucose 1,6-bisphosphate (tetracyclohexylammonium salt, $M_r = 736.8$; 7.4 mg in 10 ml)

0.01 M NADP ($M_r = 787.4$; 79 mg in 10 ml)

1 M $MgCl_2$ (hexahydrate, $M_r = 203.3$; 2g in 10 ml)

phosphoglucomutase 100 IU mg^{-1} (solution of 100 IU ml^{-1})

Glucose 6-phosphate dehydrogenase, ~400 IU mg^{-1} (solution of 300 IU ml^{-1})

Both enzyme preparations must be free of ammonium salt. If they are present as ammonium sulfate suspensions, they must be dialyzed against 0.05 M potassium phosphate pH 6.8 before use.

Assay mixture

Components	*Concentration*
8.5 ml 0.05 M potassium phosphate pH 6.8	42.5 mM
0.1 ml 0.01 M EDTA	0.1 mM
0.3 ml glycogen	5.0 mM
0.05 ml 1 mM glucose-1,6-bisphosphate	5 µM
0.4 ml 0.01 M NADP	0.4 mM
0.15 ml 1 M $MgCl_2$	15 mM
0.1 ml phosphoglucomutase	1 U/ml
0.2 ml glucose-6-phosphate dehydrogenase	6 U/ml

Procedure

0.98 ml assay mixture

0.02 ml enzyme solution

The absorption increase at 340 nm is measured at 25 °C, $\varepsilon_{340} = 6.3 \times 10^3$ l mol^{-1} cm^{-1}.

References

Bergmeyer, H.U. (1983) *Methods of Enzymatic Analysis*, 3rd edn, vol. 2, Verlag Chemie, Weinheim, pp. 293–295.

Helmreich, E., Michaelides, M.C., and Cori, C.F. (1967) *Biochemistry*, **6**, 3695–3710.

Mu, H.H. *et al.* (2001) *Arch. Biochem. Biophys.*, **388**, 155–164.

3.3.2.4 Aspartate Transaminase, EC 2.6.1.1

SN: L-aspartate:2-oxoglutarate aminotransferase, AAT, aspartate amidotransferase, glutamic-aspartic transaminase, glutamate-oxalacetate transaminase, transaminase A, 2-oxoglutarate-glutamate aminotransferase.

$$\text{L-aspartate} + \text{2-oxoglutarate} \rightleftharpoons \text{oxaloacetate} + \text{L-glutamate}$$

Mammalian enzyme (rat, cytoplasm,) M_r 44 600, monomer; cofactor pyridoxal 5-phosphate; K_m (mM): 0.35 (L-aspartate), 1.25 (2-oxoglutarate); enzyme from *Escherichia coli*: M_r 82 000, dimer, K_m (mM): 4.4 (L-aspartate), 0.07 (2-oxoglutarate), 0.37 (2-oxaloacetate), 0.00025 (pyridoxalphosphate), pH optimum 8.0.

Assay reaction coupled with malate dehydrogenase (EC 1.1.1.37):

$$\text{oxaloactetate} + \text{NADH} + \text{H}^+ \rightleftharpoons \text{malate} + \text{NAD}^+$$

Assay solutions

0.1 M Tris/HCl pH 7.4

0.2 M 2-oxoglutarate (sodium salt, $M_r = 168.1$; 3.4 g in 10 ml H_2O)

0.2 M L-aspartate (sodium salt, hydrate, $M_r = 155.1$, 3.1 g in 10 ml H_2O)

0.01 M NADH (disodium salt, $M_r = 709.4$; 71 mg in 10 ml H_2O)

Malate dehydrogenase (1000 IU mg^{-1})

Assay mixture

Components	*Concentration*
9.0 ml 0.1 M Tris/HCl pH 7.4	90 mM
0.35 ml 0.2 M 2-oxoglutarate	7 mM
0.35 ml 0.2 M L-aspartate	7 mM
0.03 ml 0.01 M NADH	0.03 mM
0.07 ml MDH	7 IU

Procedure

0.98 ml assay mixture

0.02 ml enzyme sample

The absorption decrease at 340 nm is measured at 25 °C, $\varepsilon_{340} = 6.3 \times 10^3$ l mol^{-1} cm^{-1}.

References

Krista, M.L. and Fonda, M.L. (1973) *Biochim. Biophys. Acta*, **309**, 83–96.
Mavrides, C. and Orr, W. (1975) *J. Biol. Chem.*, **250**, 4128–4133.
Rakhmanova, T.I. and Popova, T.N. (2006) *Biochemistry*, **91**, 211–217.

3.3.2.5 Alanine Transaminase, EC 2.6.1.2

SN: L-alanine:2-oxoglutarate aminotransferase, alanine aminotransferase, glutamic:alanine transaminase, glutamine:pyruvate transaminase

$$\text{L-alanine} + \text{2-oxoglutarate} \rightleftharpoons \text{pyruvate} + \text{L-glutamate}$$

Mammalian enzyme (rat liver): M_r 18 300, monomer, cofactor pyridoxal 5-phosphate, K_m (mM): 0.12 (2-oxoglutarate), 0.5 (L-alanine), 0.4 (pyruvate), pH optimum 8.5. *Clamydomonas reinhardtii*: M_r 105 000, homodimer; K_m (mM) 0.05 (2-oxoglutarate), 0.52 (L-glutamate), 0.24 (pyruvate), 2.7 (L-alanine), pH optimum 7.3. The assay is coupled with the lactate dehydrogenase reaction:

$$\text{pyruvate} + \text{NADH} + \text{H}^+ \rightleftharpoons \text{L-lactate} + \text{NAD}^+$$

Assay solutions

0.1 M Tris/HCl, pH 7.3

0.5 M L-alanine (M_r 89.1, 4.46 g in 100 ml 0.1 M Tris/HCl, pH 7.3)

0.01 M NADH (disodium salt, M_r = 709.4; 71 mg in 10 ml H_2O)

4 mM PLP (pyridoxal 5-phosphate, $M_r = 247.1$; 10 mg in 10 ml H_2O)

0.2 M 2-oxoglutarate (sodium salt, $M_r = 168.1$; 3.4 g in 10 ml H_2O)

LDH (lactate dehydrogenase, commercial products contain about 500 IU mg^{-1}, prepare a solution with 120 IU ml^{-1} in 0.1 M Tris/HCl, pH 7.3)

Assay mixture

Components	*Concentration*
8.5 ml 0.5 M L-alanine	0.425 M
0.2 ml 0.01 M NADH	0.2 mM
0.25 ml 4 mM PLP	0.1 mM
0.75 ml 0.2 M 2-oxoglutarate	15 mM
0.1 ml LDH	1.2 IU

Procedure

0.98 ml assay mixture

0.02 ml enzyme sample

The absorption decrease at 340 nm is measured at 25 °C, $\varepsilon_{340} = 6.3 \times 10^3$ l mol^{-1} cm^{-1}.

References

Lain-Guelbenzu, B. *et al.* (1991) *Eur. J. Biochem.*, **202**, 881–887.
Ruscak, M. *et al.* (1982) *J. Neurochem.*, **39**, 210–216.
Vedavathi, M. *et al.* (2006) *Biochemistry*, **91**, S105–S112.

3.3.2.6 **Tyrosine Transaminase, EC 2.6.1.5, Tryptophan Transaminase, EC 2.6.1.27, Phenylalanine Transaminase, EC 2.6.1.58**

The following assay is similar for the three transaminases, only the respective amino acid must be exchanged

Tyrosine transaminase, SN: L-tyrosine:2-oxoglutarate aminotransferase; TAT, tyrosine aminotransferase

L-tyrosine + 2-oxoglutarate ⇌ 4-hydroxyphenylpyruvate + L-glutamate

Mammalian enzyme (rat): M_r 110 500, dimer (2 × 50 000), cofactor: pyridoxal 5-phosphate; K_m (mM): 0.4 (2-oxoglutarate), 0.5 (pyruvate), 1.2 (L-tyrosine), 2.2 (L-glutamate); pH optimum 7.7.

Tryptophan transaminase, SN: L-tryptophan:2-oxoglutarate aminotransferase, Tam 1

$$\text{L-tryptophan} + \text{2-oxoglutarate} \rightleftharpoons \text{(indol-3-yl)pyruvate} + \text{L-glutamate}$$

Mammalian enzyme (porcine): M_r 55 000, cofactor: pyridoxal 5-phosphate; K_m (mM): 0.74 (2-oxoglutarate), 15 (tryptophan): pH optimum 8.

Phenylalanine (histidine) transaminase, SN: L-phenylalanine:pyruvate aminotransferase, histidine:pyruvate transaminase

$$\text{L-phenylalanine} + \text{pyruvate} \rightleftharpoons \text{phenylpyruvate} + \text{L-alanine}$$

Mammalian enzyme (mouse): M_r 75 000, dimer, cofactor: pyridoxal 5-phosphate; K_m (mM): 6.5 (histidine), 21 (pyruvate); pH optimum 9.

Assay solutions

0.2 M potassium phosphate pH 7.3

40 mM L-tyrosine (M_r = 181.2, 72.4 mg in 10 ml 0.1 N HCl)

40 mM tryptophan (M_r = 204.2, 81.6 mg in 10 ml H_2O)

40 mM L-phenylalanine (M_r = 165.2; 66 mg in 10 ml H_2O)

0.2 M 2-oxoglutarate (sodium salt, M_r = 168.1; 3.4 g in 10 ml 0.2 M potassium phosphate pH 7.3)

4 mM PLP (pyridoxal 5-phosphate, M_r = 247.1; 10 mg in 10 ml H_2O)

10 N NaOH

Assay mixture

Components	*Concentration (mM)*
7.7 ml 0.2 M potassium phosphate pH 7.3	164
1.5 ml 40 mM L-tyrosine (respectively L-phenylalanine L-tryptophan)	6
0.5 ml 0.2 M 2-oxoglutarate	10
0.1 ml 5 mM PLP	0.05

Procedure

1 ml assay mixture

0.02 ml enzyme sample

10 min, 37 °C

0.2 ml 10 N NaOH

30 min, room temperature

Measure the absorption at the respective wavelength:

Tyrosine transaminase: 331 nm, $\varepsilon_{331} = 19.9 \times 10^3\ \text{l mol}^{-1}\ \text{cm}^{-1}$

Tryptophan transaminase: 335 nm, $\varepsilon_{335} = 10 \times 10^3\ \text{l mol}^{-1}\ \text{cm}^{-1}$

Tyrosine transaminase: 315 nm, $\varepsilon_{315} = 17.5 \times 10^3\ \text{l mol}^{-1}\ \text{cm}^{-1}$

References

Diamondstone, T.I. (1966) *Anal. Biochem.*, **16**, 395–401.
George, H. and Gabay, S. (1968) *Biochim. Biophys. Acta*, **167**, 555–565.
Hargrove, J.L. and Granner, D.K. (1981) *J. Biol. Chem.*, **256**, 8012–8017.
Minatogawa, I. *et al.* (1976) *J. Neurochem.*, **27**, 1097–1101.
Minatogawa, I. *et al.* (1977) *Hoppe-Seyler's Z. Physiol. Chem.*, **358**, 59–67.
Sobrado, V.R. (2003) *Protein Sci.*, **12**, 1039–1050.

3.3.2.7 Hexokinase, EC 2.7.1.1

SN: ATP:D-hexose-6-phosphotransferase, HK

$$\text{D-hexose} + \text{ATP} \rightarrow \text{D-hexose-6-phosphate} + \text{ADP}$$

Hexokinase exists in two forms: in most cells it shows high affinity for glucose (K_m 0.057 mM), while the liver enzyme (*glucokinase*) has low affinity (K_m 5.5 mM); human enzyme: M_r 100 000, homodimer, Mg^{2+} as cofactor, K_m (mM) 0.56 (ATP), pH optimum 7.8.

There exists no simple photometric assay for this reaction. Instead, the phosphorylation of glucose may be followed by different methods, like phosphate determination or labeling with ATP^{32}. For these assays, glucose-6-phosphate (G6P) must be completely separated from ATP and ADP by chromatographic methods, especially the HPLC technique. Photometric detection can be achieved by coupling with glucose-6-phosphate dehydrogenase (G6PDH):

$$\text{D-glucose} + \text{ATP} \xrightarrow{\text{HK}} \text{glucose-6-phosphate} + \text{ADP}$$
$$\text{glucose-6-phosphate} + \text{NADP}^+ \xrightarrow{\text{G6PDH}} \text{6-phosphogluconate} + \text{NADPH} + \text{H}^+$$
$$\text{D-glucose} + \text{ATP} + \text{NADP}^+ \rightarrow \text{6-phosphogluconate} + \text{ADP} + \text{NADPH} + \text{H}^+$$

When first performing this assay the G6PDH reaction should be tested separately (cf. assay in Section 3.3.1.13) and thereafter, G6P should be replaced by glucose and ATP to follow the complete reaction sequence in the presence of hexokinase.

Assay solutions

100 mM triethanolamine · HCl/NaOH pH 7.6 (triethanolamine · HCl, M_r = 185.7; 18.6 g adjusted with 1 M NaOH, fill up to 1 l)

1.0 M D-glucose (M_r = 180.2; 18 g in 100 ml)

0.1 M ATP (disodium salt, trihydrate, M_r = 605.2; 605 mg in 10 ml)

0.1 M $MgCl_2$ (hexahydrate, M_r = 203.3; 203 mg in 10 ml)

0.01 M NADP (disodium salt, M_r = 787.4; 79 mg in 10 ml)

Glucose-6-phosphate dehydrogenase (commercial products have 100–500 IU mg^{-1}, prepare a dilution ~5 IU 0.1 ml^{-1})

Assay mixture

Components	*Concentration*
5.8 ml 0.1 mM triethanolamine · HCl/NaOH pH 7.6	58 mM
0.6 ml 0.1 mM $MgCl_2$	6.0 mM
2.0 ml 1.0 M D-glucose	200 mM
1.0 ml 10 mM NADP	1.0 mM
0.3 ml 0.1 M ATP	3.0 mM
0.1 ml glucose-6-phosphate dehydrogenase	0.5 IU ml^{-1}

Procedure

0.98 ml assay mixture

0.02 ml enzyme sample

The absorption increase at 340 nm is measured at 25 °C, $\varepsilon_{340} = 6.3 \times 10^3$ l mol^{-1} cm^{-1}.

References

Aleshin, A.E. *et al.* (1999) *Biochemistry*, **38**, 8359–8366.

Bergmeyer, H.U. (1983) *Methods of Enzymatic Analysis*, vol. 2, Verlag Chemie, Weinheim, pp. 222–223.

Darrow, R.A. and Colowick, S.P. (1962) *Meth. Enzymol.*, **5**, 226–235.

Rose, J.A. and Rose, Z.B. (1969) *Compr. Biochem.*, **17**, 93–161.

Vandercammen, A. and Van Schaftingen, E. (1991) *Eur. J. Biochem*, **200**, 545–551.

3.3.2.8 Pyruvate Kinase, EC 2.7.1.40

SN: ATP:pyruvate 2-O-phosphotransferase, PK, phosphoenolpyruvate kinase

$$\text{phosphoenolpyruvate} + \text{ADP} \overset{\text{PK}}{\rightleftharpoons} \text{pyruvate} + \text{ATP}$$

Mammalian enzyme (porcine), M_r 230 000, homotetramer (4 × 60 000), cofactor: Mg^{2+}, K_m (mM): 0.4 (ADP), 0.3 (phosphoenolpyruvate), pH optimum 7.4; exists in different isoenzymes: L (liver), M (muscle), A (other tissues). According to other kinases, the reaction can directly be tested with P^{32}; to avoid radioactivity, a coupled photometric assay with the LDH is described:

$$\text{pyruvate} + \text{NADH} + \text{H}^+ \overset{\text{LDH}}{\rightleftharpoons} \text{L-lactate} + \text{NAD}^+$$

Assay solutions

0.1 M triethanolamine · HCl/KOH pH 7.6 (triethanolamine · HCl, M_r = 185.7; dissolve 18.6 g in 800 ml H_2O, adjust to pH 7.6 with 1 N KOH and fill up to1 l)

0.5 M KCl (M_r = 74.6; 373 mg in 10 ml)

0.25 M $MgCl_2$ (hexahydrate, M_r = 203.3; 508 mg in 10 ml)

0.01 M phosphoenolpyruvate (PEP, tricyclohexylammonium salt, M_r = 465.6; 46.6 mg in 10 ml)

0.1 M ADP (disodium salt, M_r = 471.2; 471 mg in 10 ml)

0.01 M NADH (disodium salt, M_r = 709.4; 71 mg in 10 ml)

LDH (commercial products contain about 500 IU mg^{-1}, prepare a solution with 150 IU 0.1 ml^{-1})

Assay mixture

Components	*Concentration*
8.2 ml 0.1 M triethanolamine/KOH pH 7.6	82 mM
0.2 ml 0.5 M KOH	10 mM
0.1 ml 0.25 M $MgCl_2$	2.5 mM
0.5 ml 0.01 M PEP	0.5 mM
0.5 ml 0.1 M ADP	5.0 mM
0.2 ml 0.01 M NADH	0.2 mM
0.1 ml LDH	15 IU ml^{-1}

Procedure

0.98 ml assay mixture

0.02 ml enzyme sample

The absorption decrease at 340 nm is measured at 25 °C, $\varepsilon_{340} = 6.3 \times 10^3$ l mol^{-1} cm^{-1}.

References

Beisenherz, G., Boltze, H.J., Bücher, T., Czok, R., Garbade, K.H., Meyer-Arendt, E., and Pfleiderer, G. (1953) *Z. Naturforsch.*, **8b**, 555–577.
Bergmeyer, H.U. (1983) *Methods of Enzymatic Analysis*, vol. **2**, Verlag Chemie, Weinheim, pp. 303–303.
Bücher, T. and Pfleiderer, G. (1955) *Meth Enzymol.*, **1**, 435–440.
Farrar, G. and Farrar, W.W. (1995) *Int. J. Biochem. Cell Biol.*, **27**, 1145–1151.
Imamura, K. and Tanaka, T. (1982) *Meth. Enzymol.*, **96**, 150–165.
Rose, I.A. and Rose, Z.B. (1969) *Compr. Biochem.*, **17**, 93–161.

3.3.2.9 Acetate Kinase, EC 2.7.2.1

SN: ATP:acetate phosphotransferase, AK

$$\text{acetate} + \text{ATP} \rightarrow \text{acetyl phosphate} + \text{ADP}$$

Enzyme from *Escherichia coli*: dimer, M_r 70 000, K_m (mM): 7 (acetate), 0.16 (acetyl-phosphate), 0.07 (ATP), 0.05 (ADP), pH optimum 7.5. For the assay, the reaction is coupled with pyruvate kinase (PK) and lactate dehydrogenase (LDH):

$$\text{phosphoenolpyruvate} + \text{ADP} \overset{\text{PK}}{\rightleftharpoons} \text{pyruvate} + \text{ATP}$$

$$\text{pyruvate} + \text{NADH} + \text{H}^+ \overset{\text{LDH}}{\rightleftharpoons} \text{L-lactate} + \text{NAD}^+$$

Assay solutions

0.1 M triethanolamine · HCl/NaOH pH 7.6 (triethanolamine · HCl, $M_r = 185.7$; 18.6 g adjusted with 1 M NaOH, fill up to 1 l)

2 M sodium acetate ($M_r = 82.0$; 16.4 g in 100 ml)

0.1 M ATP (disodium salt, trihydrate, $M_r = 605.2$; 605 mg in 10 ml)

0.1 M $MgCl_2$ (hexahydrate, $M_r = 203.3$; 203 mg in 10 ml)

0.01 M NADH (disodium salt, $M_r = 709.4$; 71 mg in 10 ml)

0.01 M phosphoenol pyruvate (PEP, tricyclohexylammonium salt, $M_r = 465.6$; 46.6 mg in 10 ml)

LDH, $20\,mg\,ml^{-1}$, about $500\,IU\,mg^{-1}$

PK, $10\,mg\,ml^{-1}$, $500\,IU\,mg^{-1}$

Assay mixture

Components	*Concentration*
6.3 ml 0.1 M triethanolamine/NaOH pH 7.6	63 mM
1.5 ml 2.0 M sodium acetate	0.3 M
0.5 ml 0.1 M ATP	5.0 mM
0.15 ml 0.1 M $MgCl_2$	1.5 mM
1.0 ml 0.01 M PEP	1.0 mM
0.3 ml 0.01 M NADP	0.3 mM
0.025 ml LDH	$25\,IU\,ml^{-1}$
0.015 ml PK	$7.5\,IU\,ml^{-1}$

Procedure

0.98 ml assay mixture

0.02 ml enzyme sample

The absorption decrease at 340 nm is measured at 25 °C, $\varepsilon_{340} = 6.3 \times 10^3\,l\,mol^{-1}\,cm^{-1}$.

References

Bergmeyer, H.U. (1983) *Methods of Enzymatic Analysis*, vol. 2, Verlag Chemie, Weinheim, pp. 127–128.

Fox, D.K. and Roseman, S. (1986) *J. Biol. Chem.*, **261**, 13487–13497.

Rose, I.A., Grunberg-Manago, M., and Korey, S.R. (1954) *J. Biol. Chem.*, **211**, 737–756.

3.3.2.10 Phosphoglycerate Kinase, EC 2.7.2.3

ATP:3-phospho-D-glycerate 1-phosphotransferase, PGK

$$\text{3-phospho-D-glycerate} + \text{ATP} \overset{\text{PGK}}{\rightleftharpoons} \text{1,3-diphosphoglycerate} + \text{ADP}$$

Mammalian enzyme (human): M_r 50 000, Mg^{2+} as cofactor, K_m (mM): 0.1 (3-phospho-D-glycerate), 0.005 (1,3-diphosphoglycerate), 1.2 (ADP), 1.1 (ATP), pH optimum 7.5. For the assay the reaction is coupled with glyceraldehyde-3-phosphate dehydrogenase (GAPDH)

$$1,3\text{-diphoshoglycerate} + \text{NADH} + \text{H}^+ \overset{\text{GAPDH}}{\rightleftharpoons} \text{D-glyceraldehyde-3-phosphate} + \text{NAD}^+$$

Assay solutions

0.1 M triethanolamine/NaOH pH 7.6 (triethanolamine · HCl, $M_r = 185.7$; dissolve 18.6 g in 800 ml H_2O, adjust to pH 7.6 with 1 M NaOH, replenish to 1 l)

0.1 M glycerate-3-phosphate (3-phosphoglyceric acid, disodium salt, $M_r = 230.0$; 230 mg in 10 ml)

0.1 M ATP (disodium salt, trihydrate, $M_r = 605.2$; 605 mg in 10 ml)

0.01 M NADH (disodium salt, $M_r = 709.4$; 71 mg in 10 ml)

0.1 M EDTA (ethylenediaminetetraacetic acid, $M_r = 292.2$; 292 mg in 10 ml)

0.1 M magnesium sulfate ($MgSO_4\ 7H_2O$, $M_r = 246.5$; 247 mg in10 ml)

3-phosphoglycerate kinase, from yeast (~2 000 IU ml^{-1})

GAPDH (glyceraldehyde-3-phosphate dehydrogenase from rabbit muscle, 800 IU ml^{-1})

Assay mixture

Components	*Concentration*
8.67 ml 0.1 M triethanolamine/NaOH pH 7.6	87 mM
0.5 ml 0.1 M glycerate-3-phosphate	5.0 mM
0.1 ml 0.1 M ATP	1.0 mM
0.2 ml 0.01 M NADH	0.2 mM
0.1 ml 0.1 M EDTA	1.0 mM
0.2 ml 0.1 M magnesium sulfate	2.0 mM
0.03 ml GAPDH	2.4 IU ml^{-1}

Procedure

0.98 ml assay mixture

0.02 ml enzyme solution

Follow the absorption decrease at 340 nm at 25 °C, $\varepsilon_{340} = 6.3 \times 10^3$ l mol^{-1} cm^{-1}.

References

Bergmeyer, H.U. (1983) *Methods of Enzymatic Analysis*, vol. 2, Verlag Chemie, Weinheim, pp. 280–281.
Bücher, T. (1955) *Meth. Enzymol.*, **1**, 415–422.
Rose, I.A. and Rose, Z.B. (1969) *Compr. Biochem.*, **17**, 93–161.
Scopes, R.K. (1973) in *The Enzymes*, 3rd edn, vol. 8 (ed. P. Boyer), Academic Press, New York, pp. 335–351.
Szabo, J. (2008) *Biochemistry*, **47**, 6735–6744.
Yoshida, A. (1975) *Meth. Enzymol.*, **42C**, 144–148.

3.3.2.11 Aspartokinase, EC 2.7.2.4

SN : ATP: aspartate 4-phosphotransferase, AK

$$\text{L-aspartate} + \text{ATP} \rightarrow \text{4-phospho-L-aspartate} + \text{ADP}$$

Allosteric enzyme in *Escherichia coli*, fused mitochondrial homoserine dehydrogenase (cf. assay in Section 3.3.1.5) to aspartokinase–homoserine dehydrogenase I, M_r 360 000, homotetramer (4 × 84 000, dimer if separate from homoserine dehydrogenase), requires Mg^{2+}, K_m (mM): 0.18 (ATP), 0.51 (L-aspartate), 16 (L-asparagine), pH optimum 7. For the assay the reaction is coupled to pyruvate kinase (PK) and lactate dehydrogenase (LDH)

$$\text{ADP} + \text{phosphoenolpyruvate} \overset{\text{PK}}{\rightleftharpoons} \text{ATP} + \text{pyruvate}$$

$$\text{pyruvate} + \text{NADH} + \text{H}^+ \overset{\text{LDH}}{\rightleftharpoons} \text{lactate} + \text{NAD}^+$$

Assay solutions

0.1 M HEPES/NaOH, 0.1 M KCl pH 8.0 (*N*-(2-hydroxyethyl)piperazine-*N*′-ethanesulfonic acid, HEPES, $M_r = 238.3$; dissolve 23.8 g and 7.6 g KCl ($M_r = 74.6$) in 600 ml H_2O, adjust to pH 8.0 with 0.1 M NaOH and bring to 1 l with H_2O)

0.36 M magnesium acetate ($Mg^{2+}C_4H_6O_4 \cdot 4H_2O$, $M_r = 214.5$, 772.2 mg in 10 ml H_2O)

0.01 M PEP (phosphoenolpyruvate, tricyclohexylammonium salt, $M_r = 465.6$; 46.6 mg in 10 ml)

0.1 M ATP (disodium salt, trihydrate, $M_r = 605.2$; 605 mg in 10 ml H_2O)

0.01 M NADH (disodium salt, $M_r = 709.4$; 71 mg in 10 ml H_2O)

LDH (commercial products contain about 500 IU mg^{-1}, prepare a solution with 400 IU in 0.2 ml buffer)

pyruvate kinase (commercially available, prepare a solution with 200 IU in 0.2 ml buffer)

0.1 M sodium aspartate (L-aspartic acid, sodium salt, $NaC_4H_5NO_4 \cdot H_2O$, $M_r = 173.1$, 173 mg in 10 ml H_2O)

Assay mixture

Components	*Concentration*
9 ml 0.1 M HEPES/NaOH, 0.1 M KCl pH 8.0	90 mM
0.1 ml 0.36 M magnesium acetate	3.6 mM
0.7 ml 0.01 M PEP	0.7 mM
0.18 ml 0.1 M ATP	1.8 mM
0.1 ml 0.01 M NADH	0.1 mM
1.0 ml 0.1 M sodium aspartate	10 mM
0.2 ml LDH	40 IU ml^{-1}
0.2 ml PK	20 IU ml^{-1}

Procedure

0.98 ml assay mixture

0.02 ml enzyme sample (about 10 μg)

Follow the absorption decrease at 340 nm at 25 °C, $\varepsilon_{340} = 6.3 \times 10^3$ l mol^{-1} cm^{-1}.

References

Angeles, T.S. *et al.* (1989) *Biochemistry*, **28**, 8771–8777.
Truffa-Bachi, P. (1973) in *The Enzymes*, 3rd edn, vol. 8 (ed. P.D. Boyer), Academic Press, New York, pp. 509–553.

3.3.3 Hydrolases, EC 3

3.3.3.1 Lipase, EC 3.1.1.3

SN: triacylglycerol acyl-hydrolase, triacylglycerol lipase

$$\text{triacylglycerol} + H_2O \rightarrow \text{diacylglycerol} + \text{long-chain fatty acid}$$

Pancreatic enzyme (human, porcine): M_r = 44 000, monomer, K_m (mM): 0.86 (1-olein), 0.36 (triolein), 1,2 (1-caprin); pH optimum 8–9. Two different assays are described. The first one is based on the formation of free carboxyl groups due to the cleavage of the triglyceride, which causes a decrease in the pH. This is detected by a pH-stat connected to an automatic burette (see Section 2.3.2.2), which keeps the pH constant. If a pH-stat is not available, a pH meter may be used but care must be taken that the pH remains constant during the assay. The fluorometric assay observes the release of the fluorescent residue from a synthetic substrate.

A. Assay with pH Stat (Autotitrator)

Assay solutions

3.0 M sodium chloride (NaCl, $M_r = 58.4$; 17.5 g in 100 ml)

75 mM calcium chloride ($CaCl_2 \cdot 6H_2O$; $M_r = 219.1$; 1.62 g in 100 ml)

5 mM calcium chloride (dilute the 75 mM calcium chloride solution 15-fold)

0.5% BSA (bovine serum albumine, 0.5 g in 100 ml H_2O)

27 mM sodium taurocholate ($M_r = 537.7$; 1.45 g in 100 ml)

Olive oil–gum arabic emulsion: Dissolve 16.5 g gum arabic in 130 ml H_2O and fill up to 165 ml. Add 20 ml olive oil and 15 g crushed ice, treat in a blender at low speed until an emulsion is formed. Filter through glass wool

Lipase (dissolve to 1 mg ml^{-1} in 5 mM calcium chloride, dilute further for assay in 5 mM calcium chloride)

Use 0.01 M NaOH standard solution as titrant for the autotitrator

Assay mixture and procedure

5 ml olive oil–gum arabic emulsion

5 ml H_2O

2 ml 3.0 M NaCl

1 ml 75 mM calcium chloride

2 ml 0.5% BSA for yeast lipase or 27 mM sodium taurocholate for porcine lipase

---

15 ml, adjust to pH 8.0

Blank: Record the volume per unit time of the titrant added by the auto-titrator for about 3 min

Sample: Add an appropriate amount of diluted enzyme (according to the activity) to the assay mixture and record the volume per unit time of the titrant added by the autotitrator for about 5 min

Calculation

One unit releases 1 μmol fatty acid/min from emulsified olive oil at 25 °C and pH 8.0.

B. Fluorimetric Assay This assay is also applicable for chymotrypsin, cholinesterase, and acylase. In case of acetylcholinesterase, diacetylfluorescein instead of dibutyrylfluorescein should be used as substrate.

$$\underset{\textit{nonfluorescent}}{\text{dibutyrylfluorescein}} + 2H_2O \rightarrow \underset{\textit{fluorescent}}{\text{fluorescein}} + 2\,\text{butyric acid}$$

Assay solutions

0.1 M Tris/HCl pH 8.0

0.05 mM substrate solution (dibutyrylfluorescein, $M_r = 472.5$; dissolved 2.36 g in 5 ml ethyleneglycol–monomethylether (methylcellosolve) and 95 ml 0.1 M Tris/HCl pH 8.0)

Procedure

To 2.9 ml of the substrate solution, 0.1 ml of the enzyme sample is added and the fluorescence, excited at 470 nm, is measured at 510 nm.

References

Bernbäck, S., Hernell, O., and Bläckberg, C. (1985) *Eur. J. Biochem.*, **148**, 233–238.

Desnuelle, P. (1972) in *The Enzymes*, 3rd edn, vol. 7 (ed. P.D. Boyer), Academic Press, New York, p. 575.

Guilbault, G.G. and Kramer, D.N. (1964) *Anal. Chem.*, **36**, 409–412.

Lagocki, J.W., Law, J.H., and Kezdy, F.J. (1973) *J. Biol. Chem.*, **248**, 580–587.

Oosterbaum, R.A. and Jansz, H.S. (1965) *Compr. Biochem.*, **16**, 1–54.

Sarda, L., Marchis-Mouren, G., Constantin, M.J., and Denuelle, P. (1957) *Biochim. Biophys. Acta*, **23**, 264–227.

Tsuzuki, W. *et al.* (2004) *Biochim. Biophys. Acta*, **1684**, 1–7.

3.3.3.2 Phospholipase A_2, EC 3.1.1.4

SN: phosphatidylcholine 2-acylhydrolase, lecithinase A

$$\text{phosphatidylcholine} + H_2O \rightarrow \text{1-acylglycerophosphocholine} + \text{carboxylate}$$

Mammalian enzyme (human, porcine), cofactor Ca^{2+}, monomer M_r 14 000, K_m (mM): 8.3 (lecithin), 0.1 (1,2-dioctanoyl-*sn*-glycero-3-phosphocholine), 2(1,2-dipalmitoyl-phosphatidylcholine), EDTA inhibits, pH optimum 8. For the following pH-stat assay preparation of vesicles is required.

Assay solutions and reagents

1,2-dipalmitoylphosphatitylcholine (DPPC, 1,2-dihexadecanoyl-*sn*-glycero-3-phosphocholine, $M_r = 734.0$)

50 mM KCl ($M_r = 74.5$; 0.37 g in 100 ml H_2O)

0.1 M $CaCl_2$ ($M_r = 111$; 1.11 g in 100 ml H_2O)

Enzyme solution (~0.2 mg ml^{-1}, pH 8.0)

Preparation of vesicles

Dilute individual phospholipids or mixtures in chloroform

Lyophilize to dryness

Suspend in 2–8 ml 50 mM KCl

Sonicate under nitrogen in a glass tube with flat bottom (the clearance to the tube walls should not exceed 0.5 cm) together for 30 min, after every 3 min of sonication interrupt for 1 min for cooling

Centrifuge at 100 000 × *g* for 30 min in an ultracetrifuge

Centrifuge the supernatant again at 159 000 × *g* for 3 h to obtain a homogeneous vesicle preparation

Procedure

1.75 ml of the vesicle preparation

0.2 ml 0.1 M $CaCl_2$ solution

Adjust to pH 8.0

0.05 ml enzyme solution

Record the volume per unit time of the titrant added by the autotitrator.

References

Barenholz, Y. *et al.* (1977) *Biochemistry*, **16**, 2806–2810.
Dennis, E.A. (1983) in *The Enzymes*, 3rd edn, vol. 16 (ed. P.D. Boyer), Academic Press, New York, pp. 307–353.
Franken, P.A. *et al.* (1992) *Eur. J. Biochem.*, **203**, 89–93.
Menashe, M. *et al.* (1986) *J. Biol. Chem.*, **261**, 5328–5333.

3.3.3.3 Acetylcholine Esterase, EC 3.1.1.7
SN: acetylcholine acetylhydrolase, AChE

$$\text{acetylcholine} + H_2O \rightarrow \text{choline} + \text{acetate}$$

Mammalian enzyme (mouse): M_r 66 000, monomer, K_m (mM): 0.046 (acetylcholine), pH optimum 8.0. For assays, see also choline esterase (assay in Section 3.3.3.2) and lipase (assay in Section 3.3.3.1). An assay for a pH-stat is described, the amount of base required to neutralize the acetic acid should be determined. If a pH-stat is not available, a pH meter may be used but care must be taken that the pH remains constant during the assay.

Assay solutions

0.02 M potassium phosphate pH 7.0

5 mM acetylcholine chloride ($M_r = 181.7$; 91 mg in 100 ml)

0.01% gelatin, containing 0.2 M sodium chloride and 0.04 M magnesium chloride (0.1 g gelatin, 11.7 g NaCl, 8.1 g $MgCl_2 \cdot 6H_2O$ in 1 l H_2O)

0.01 M sodium hydroxide, standard solution as titrant for the autotitrator

Acetylcholinesterase, dilute in 0.02 M potassium phosphate pH 7.0 to about 20 IU ml^{-1}

Assay mixture and procedure

	Concentration
7.5 ml 5 mM acetylcholine	2.5 mM
7.5 ml 0.01% gelatin-salt solution	0.005%
15 ml, adjust to pH 7.0; 25 °C	

Blank: record for some minutes the volume of 0.01 M sodium hydroxide needed per unit time to maintain a pH of 7.0

Sample: add enzyme solution (0.1–1 ml) and record for about five min, the volume of 0.01 M sodium hydroxide needed per unit time to maintain a pH of 7.0

Calculation

One unit hydrolyzes 1 μmol acetylcholine/min at pH 7.0, 25 °C.

$$\text{unit mg}^{-1} = \frac{\text{mL base min}^{-1} \times \text{base normality} \times 1000}{\text{mg enzyme in reaction mixture}}$$

References

Boyd, A.E. *et al.* (2000) *J. Biol. Chem.*, **275**, 22401–22408.

Froede, H.C. and Wilson, I.B. (1971) in *The Enzymes*, 3rd edn, vol. 5 (ed. P.D. Boyer), Academic Press, New York, pp. 87–114.

Gomez, J.L. (2003) *Int. J. Biochem. Cell Biol.*, **35**, 1109–1118.

Oosterbaum, R.A. and Jansz, H.S. (1965) *Compr. Biochem.*, **16**, 1–54.

Reed, D.L., Goto, K., and Wang, C.H. (1966) *Anal. Biochem.*, **16**, 59–64.

3.3.3.4 Choline Esterase, EC 3.1.1.8

SN: acylcholine acylhydrolase, pseudocholine esterase, butyrylcholine esterase, ButChE

$$\text{acylcholine} + H_2O \rightarrow \text{choline} + \text{carboxylate}$$

Mammalian enzyme (human): 85 000, monomer; K_m (mM) 0.018 (butyrylthiocholine), 0.05 (acetylthiocholine), 0.003 (benzoylcholine), 0.08 (acetylcholine), pH optimum 8.0. The enzyme hydrolyzes butyrylcholine 2.5 times faster than acetylcholine. The assay is performed preferentially with a pH-stat or a pH meter determining the amount of base required to neutralize the acetic acid. Besides, a colorimetric assay is described. Assays for acetylcholine esterase (assay in Section 3.3.3.3) and lipase (assay in Section 3.3.3.1) are principally applicable.

A. pH-Stat Assay

Assay solutions

0.02 M Tris/HCl pH 7.4

2.2 M acetylcholine chloride ($M_r = 181.7$; 40 g in 100 ml^{-1}, or butyrylcholine iodide, $M_r = 301.2$; 66 g in 100 ml^{-1})

0.2 M magnesium chloride ($MgCl \cdot 6H_2O$; $M_r = 203.3$; 4.1 g in 100 ml H_2O)

0.01 M sodium hydroxide, standard solution as titrant to fill the burette of the autotitrator

Cholinesterase, dissolve 1 mg in 1 ml^{-1} H_2O

Assay mixture and procedure

	Concentration
7.0 ml H_2O	–
3.0 ml 0.2 M magnesium chloride	43 mM
3.0 ml 0.02 M Tris/HCl pH 7.4	4.3 mM
1 ml 2.2 M acetylcholine chloride	157 mM
14 ml, adjust to pH 7.4; 25 °C	–

Blank: record for some minutes the volume of 0.01 M sodium hydroxide needed per unit time to maintain a pH of 7.4
Sample: add enzyme solution (e.g., 0.1–1 ml) and record for about 5 min the volume of 0.01 M sodium hydroxide needed per unit time to maintain a pH of 7.4

Calculation

One unit hydrolyzes 1 μmol acetylcholine/min at pH 7.0, 25 °C.

$$\text{unit mg}^{-1} = \frac{\text{ml base min}^{-1} \times \text{base normality} \times 1000}{\text{mg enzyme in reaction mixture}}$$

B. Colorimetric Assay

indophenyl acetate → indophenol + acetic acid

The color changes from red to blue.

Assay solutions

0.1 M potassium phosphate pH 8.0

0.62 mM indophenyl acetate (M_r = 241.2; 0.375 g indophenyl acetate in 25 ml ethanol)

Procedure

	Concentration
0.68 ml 0.1 M potassium phosphate pH 8.0	68 mM
0.3 ml indophenyl acetate solution	0.19 mM
0.02 ml enzyme solution	–

Determine the absorption at 625 nm in comparison to a calibration curve at 25 °C.

References

Augustinson, K.-B. and Olsson, B. (1959) *Biochem. J.*, **71**, 477–484.
Ellmann, G.L. (1961) *Biochem. Pharmacol.*, **7**, 88–95.
Glick, D. (1937) *Biochem. J.*, **31**, 521–525.
Goodson, L.H. and Jacobs, W.B. (1973) *Anal. Biochem.*, **51**, 362–367.
Koelle, G. (1953) *Biochem. J.*, **53**, 217–226.
Kramer, D.N. and Gamson, R.M. (1958) *Anal. Chem.*, **30**, 251–254.
Main, R., Soucie, W.C., Buxton, I.L., and Arnic, E. (1974) *Biochem. J.*, **143**, 733–744.
Masson, P. *et al.* (2004) *Eur. J. Biochem.*, **271**, 220–234.
Mehrani, H. (2004) *Proc. Biochem.*, **39**, 877–882.
Nachmanson, D.W. and Wilson, J.B. (1951) *Adv. Enzymol.*, **12**, 259–339.

3.3.3.5 *S*-Formylglutathione Hydrolase, EC 3.1.2.12

SN: *S*-Formylglutathione hydrolase

S-formylglutathione + H_2O → glutathione + formate

Enzyme from human liver: M_r 58 200, dimer K_m (mM): 0.29 (*S*-formylglutathione). 0.12 (*S*-acetylglutathione), pH optimum 6.9

Assay solutions

0.1 M Tris/HCl 7.4

10 mM DTNB (5,5′-dithio-bis(2-nitrobenzoic acid), Ellman's reagent, M_r = 396.3; 39.6 mg in 10 ml)

10 mM *S*-formylglutathione, (M_r = 335.3, 33.5 mg in 10 ml)

Assay mixture

Components	*Concentration (mM)*
9.45 ml 0.1 M Tris/HCl 7.4	95
0.15 ml 10 mM DTNB	0.15
0.2 ml *S*-formylglutathione	0.2

Procedure

0.98 assay mixture

0.02 enzyme sample

Follow the absorption increase at 412 nm, 25 °C, $\varepsilon_{412} = 14\,150\,l\,mol^{-1}\,cm^{-1}$.

References

Baylin, S.B. and Margolis, S. (1975) *Biochim. Biophys. Acta*, **397**, 294–306.
Uolila, L. and Koivusola, M. (1974) *J. Biol. Chem.*, **249**, 7664–7672.

3.3.3.6 Alkaline Phosphatase, EC 3.1.3.1

SN: phosphate monoester phosphohydrolase (alkaline optimum)

$$\text{phosphate monoester} + H_2O \rightarrow \text{alcohol} + P_i$$

Mammalian enzyme (human) dimer, M_r 158 000 (2 × 75 000), cofactor: Zn^{2+}, Mg^{2+}, K_m (mM): 2.5 (2′-AMP), 0.8 (*p*-nitrophenylphosphate, human), 0.0078 (*Escherichia coli*), phosphate inhibits, pH optimum 10.5. Alkaline phosphatase is frequently applied as a conjugate to antibodies, for example in Western blots and ELISA. Apart from the assays presented below, the cleavage of *o*-carboxyphenyl phosphate described for the acid phosphatase (assay in Section 3.3.3.7) can be observed in an alkaline medium.

A. Mammalian Alkaline Phosphatase

$$p\text{-nitrophenylphosphate} + H_2O \rightarrow p\text{-nitrophenol} + P_i$$

Assay solutions

0.1 M glycine/KOH pH 10.5, 1 mM Mg^{2+}, 0.1 mM Zn^{2+} (dissolve 7.5 g glycine, 203 mg $MgCl_2$ $6H_2O$, and 14 mg $ZnCl_2$ in 750 ml H_2O; adjust to pH 10.5 with 1 N KOH and fill up to 1 l with H_2O)

0.5 M *p*-nitrophenylphosphate (disodium salt, hexahydrate, $M_r = 371.1$; 1.86 g in 10 ml)

Procedure

	Concentration
0.97 ml 0.1 M glycine/KOH pH 10.5	97 mM
0.01 ml 0.5 M nitrophenylphosphate	5.0 mM
0.02 ml enzyme solution	–

Follow the absorption increase at 405 nm, 25 °C, $\varepsilon_{405} = 18\,500\,\mathrm{l\,mol^{-1}\,cm^{-1}}$.

B. Bacterial Alkaline Phosphatase

Assay solutions

0.5 M Tris/HCl, pH 8.2

6.6 mM nitrophenylphosphate (disodium salt, hexahydrate, $M_r = 371.1$; dissolve 49 mg in 20 ml 0.5 M Tris/HCl pH 8.2)

Procedure

	Concentration
0.98 ml 6.6 mM nitrophenylphosphate	6.5 mM
0.02 ml enzyme solution	–

Follow the absorption increase at 405 nm, 25 °C, $\varepsilon_{405} = 18.5\,\mathrm{l\,mmol^{-1}\,cm^{-1}}$.

References

Bessey, O.A., Lowry, H.O., and Broch, M.J. (1946) *J. Biol. Chem.*, **164**, 321–329.
Fernley, K.N. (1971) in *The Enzymes*, vol. 4 (ed. P.D. Boyer), Academic Press, New York, pp. 417–447.
Garattini, E. *et al.* (1986) *Arch. Biochem. Biophys.*, **245**, 331–337.
Garen, A. and Levinthal, C. (1960) *Biochim. Biophys. Acta*, **38**, 470–483.
Harkness, D.R. (1968) *Arch. Biochem. Biophys.*, **126**, 513–523.
Morton, R.K. (1965) *Compr. Biochem.*, **16**, 55–84.
Orhanovic, S. *et al.* (2006) *Int. J. Biol. Macromol.*, **40**, 54–58.
Stadtman, T.C. (1961) in *The Enzymes*, 2nd edn, vol. 5 (ed. P.D. Boyer), Academic Press, New York, pp. 55–71.
Torriani, A. (1968) *Meth. Enzymol.*, **12B**, 212–218.

3.3.3.7 Acid Phosphatase, EC 3.1.3.2

SN: phosphate monoester phosphohydrolase (acid optimum)

$$\text{phosphate monoester} + H_2O \rightarrow \text{alcohol} + P_i$$

Mammalian enzyme (human); M_r 30 000, monomer, cofactor Fe^{2+}, K_m (mM):1 (*p*-nitrophenylphosphate), 0.2 (*o*-carboxyphenyl phosphate), 0.27 (ATP), 0.33 (ADP), pH optimum 5.7. The assay system described for alkaline phosphates with *o*-nitrophenylphosphate (assay in Section 3.3.3.6) can be used, applying acid conditions.

Assay Reaction

$$o\text{-carboxyphenyl phosphate} + H_2O \rightarrow \text{salicylic acid} + P_i$$

Assay solutions

0.15 M acetate buffer pH 5.0 (8.55 ml acetic acid, 100%, M_r = 60, in 800 ml H_2O, adjust to pH 4.0 with 1 N NaOH, fill up to 1 l)

3.65 mM *o*-carboxyphenyl phosphate (salicylic acid phosphate, fosfosal, M_r = 218.1; 80 mg in 100 ml)

Procedure

	Concentration
0.7 ml 0.15 M acetate buffer pH 5.0	105 mM
0.2 ml 3.65 mM *o*-carboxyphenyl phosphate	0.73 mM
Incubate for 3 min at 25 °C	
0.1 ml enzyme solution	

Record the absorbance at 300 nm, 25 °C, molar absorption coefficient for salicylic acid $\varepsilon_{300} = 3500\ l\ mol^{-1}\ cm^{-1}$.

References

Belfield, A. *et al.* (1972) *Enzymologia*, **42**, 91–106.
Bessey, O.A., Lowry, H.O., and Broch, M.J. (1946) *J. Biol. Chem.*, **164**, 321–329.
Brouillard, J. and Quellet, L. (1965) *Can. J. Biochem.*, **43**, 1899–1905.
Hayman, A.R. *et al.* (1989) *Biochem. J.*, **261**, 601–609.
Hofstee, B.H.J. (1954) *Arch. Biochem. Biophys.*, **51**, 139–146.
Hollander, V.P. (1971) in *The Enzymes*, vol. 4 (ed. P.D. Boyer), Academic Press, New York, pp. 449–498.

3.3.3.8 Ribonuclease (Pancreatic), EC 3.1.27.5

RNase I Cleaves RNA to 3′-phospho-mononucleotides and 3′-phospho-oligonucleotides ending in Cp or Up forming 2′,3′-cyclic phosphate intermediates. Human enzyme: M_r 14 000 monomer, K_m (mM) 0.063 (poly(A)poly(U)), 0.1 (poly(C)), pH optimum 7.7.

Assay solutions

0.1 M acetate buffer pH 5.0 (5.7 ml acetic acid, 100%, $M_r = 60$, in 800 ml H_2O, adjust to pH 4.0 with 1 N NaOH, fill up to 1 l)

Uranylacetate-perchloric solution (0.75% uranyl acetate in 25% perchloric acid)

RNA (yeast) solution (50 mg 10 ml^{-1} in 0.1 M sodium acetate buffer pH 5.0)

RNase (1 mg ml^{-1}, dilute 100-fold in 0.1 M sodium acetate buffer pH 5.0 prior to use)

Procedure

0.6 ml 1% RNA solution

0.1 ml enzyme (1.0 ml 0.1 M respectively of acetate buffer pH 5.0 for the blank).

Incubate for 4 min at 37 °C and stop the reaction by adding

0.3 ml uranylacetate-perchloric acid solution.

Five min on ice, centrifuge for five min, dilute

0.1 ml supernatant with 0.9 ml H_2O

Measure the absorption increase at 260 nm.

Calculation

One unit is defined as the amount of enzyme causing a change in absorbance of 1.0 at 37 °C and pH 5.0 under the described conditions.

References

Dalaly, B.K. *et al.* (1980) *Biochim. Biophys. Acta*, **615**, 381–391.
Dey, P. *et al.* (2007) *Biochem. Biophys. Res. Commun.*, **350**, 809–814.
Kalnitsky, G., Hummel, J.P., and Dierks, C. (1959) *J. Biol. Chem.*, **234**, 1512–1516.
Kunitz, M. (1946) *J. Biol. Chem.*, **164**, 563–568.

3.3.3.9 α-Amylase, EC 3.2.1.1

SN: 1,4-α-D-glucan glucanohydrolase

Cleaves starch into reducing fragments and maltose; bacterial enzyme (*Bacillus subtilis*) M_r 41 000, monomer, K_m 1.2 mM (soluble starch), pH optimum 6.9.

Assay solutions

0.1 M potassium phosphate pH 7.0

10 mM maltose solution (maltose monohydrate, $M_r = 360.3$; 36 mg in 10 ml)

Starch solution (dissolve 0.5 g soluble starch and 17.5 mg NaCl in 50 ml 0.1 M potassium phosphate pH 7.0, boil for dissolving)

Dinitrosalicylate reagent (3,5-dinitrosalicylic acid, 2-hydroxy-3,5 dinitrobenzoic acid, $M_r = 228.1$; dissolve 1 g in 20 ml 2 N NaOH and 50 ml H_2O, dissolve 30 g K-Na-tartrate and fill up to 100 ml with H_2O)

Procedure

0.05 ml starch solution

0.05 ml enzyme sample (diluted in 0.1 M potassium phosphate pH 7.0).

Incubate for 5 min at 25 °C

0.1 ml dinitrosalicylate reagent

Mix and incubate 10 min a 100 °C

1 ml H_2O

Measure the absorption at 546 nm
The enzyme activity is determined with the aid of a calibration curve:

Fill up different amounts of the maltose solution in the range between 4 and 30 μl with H_2O to 0.1 ml

Add 0.1 ml of the dinitrosalicylate reagent

10 min at 100 °C

Add 1 ml H_2O

Measure the absorption at 546 nm.

Calculation

One unit is defined as the enzyme amount releasing 1 μmol reducing groups (calculated as maltose)/min from soluble starch at 25 °C, pH 7.0.

References

Bergmeyer, H.U. (1983) *Methods of Enzymatic Analysis*, 3rd edn, vol. 2, Verlag Chemie, Weinheim, pp. 151–152.
Bernfeld, P. (1951) *Adv. Enzymol.*, **12**, 379–428.
Das, K., Doley, R., and Mukherjee, A.K. (2004) *Biotechnol. Appl. Biochem.*, **40**, 291–298.
Fischer, E.H. and Stein, E.A. (1960) in *The Enzymes*, 2nd edn, vol. 4, Academic Press, New York, pp. 313–343.
Takagi, T., Toda, H., and Isemura, T. (1971) in *The Enzymes*, 3rd edn, vol. 5 (ed. P.D. Boyer), Academic Press, New York, pp. 235–271.

3.3.3.10 Amyloglucosidase, EC 3.2.1.3

SN: 1,4-α-D-glucan glucohydrolase, glucan 1,4-α-glucosidase, glucoamylase.

$$\text{starch} + H_2O \rightarrow \beta\text{-D-glucose}$$

Releases β-D-glucose by hydrolyzing terminal 1,4-and 1,6-linked α-1,6-glucoside bonds successively from the nonreducing ends of the chains. Human enzyme: M_r 105 360, K_m (mM) 7.7 (maltose), 0.05 (maltotetraose), inhibition by salanicol (K_i 0.19 mM), pH optimum 6.5, *Aspergillus awamori*, K_m (mM): 3.4 (4-nitrophenyl-α-glucoside), 3 (maltose), 0.18 (maltotetraose), pH optimum 5.5; shrimps (*Penaeus japonicus*), M_r 105 000, monomer, K_m (mM) 0.125 (4-methylumbelliferyl-α-D-glucoside), 0.2 (soluble starch, 2.3 (amylose), 4.2 (amylopectin), pH optimum 5.

A. Coupled Assay with HK and G6PDH The assay is similar to that described for α-glucosidase (assay in Section 3.3.3.14), except that glycogen is used instead of maltose as substrate. After a certain time (e.g., 5 min) the reaction is stopped and glucose determined by the coupled hexokinase/glucose-6-phosphate dehydrogenase assay:

$$\text{glycogen} + n - 1\text{H}_2\text{O} \xrightarrow{\text{Amyloglucosidase}} n\,\text{D-glucose}$$

$$n\,\text{D-glucose} + n\,\text{ATP} \xrightarrow{\text{HK}} n\,\text{glucose-6-phosphate} + n\,\text{ADP}$$

$$n\,\text{glucose-6-phosphate} + n\,\text{NADP}^+ \xrightarrow{\text{G6PDH}} n\,\text{6-phosphogluconat} + n\,\text{NADPH} + n\,\text{H}^+$$

Assay solutions

0.1 M acetate buffer pH 4.0 (5.7 ml acetic acid, 100%, $M_r = 60$, in 800 ml H_2O, adjust to pH 4.0 with 1 N NaOH, fill up to 1 l)

Glycogen solution (80 mg glycogen from oysters in 10 ml)

0.3 M tris(hydroxymethyl)aminomethan ($M_r = 121.1$; 363 mg in 10 ml)

Procedure

0.54 ml 0.1 M acetate buffer pH 4.0

0.25 ml glycogen solution

0.01 ml enzyme sample

5 min at 25 °C

0.2 ml 0.3 M tris(hydroxymethyl)aminomethane

Determine the glucose content (against a blank without enzyme sample) as described in Section 3.3.3.14 for α-glucosidase.

B. Photometric Assay with 4-Nitrophenyl-D-Glucose

$$\text{4-nitrophenyl-}\alpha\text{-D-glucopyranoside} \rightarrow \text{4-nitrophenol} + \text{D-glucose}$$

The assay is described for 96 well microplates for serial measurements, but can be performed in a conventional spectrophotometer, thereby increasing the reaction volume (RV) tenfold. The reaction can be followed continuously at 405 nm.

Assay solutions

100 mM MES buffer pH 6.5 (2-[*N*-morpholino]ethanosulfonic acid, $M_r = 213.2$, $21.3\ g\ l^{-1}$, adjust to pH 6.5 with 1 M NaOH)

33.3 mM *p*NP-glucose (4-nitrophenyl-α-D-glucopyranoside, $M_r = 301.3$; 100 mg in 10 ml MES buffer pH 6.5)

0.5 M Na_2CO_3 solution ($M_r = 106$; 5.3 g in 100 ml H_2O)

Procedure

	Concentration (mM)
45 µl 33.3 mM *p*NP-glucose	30
5 µl enzyme sample	–
Incubate for 30 min at 37 °C	
50 µl 0.5 M Na_2CO_3 solution	–

Measure the absorption at 405 nm with a microplate reader. $\varepsilon_{405} = 18\,500\ l\ mol^{-1}\ cm^{-1}$.

C. Fluorimetric Assay with 4-Methylumbelliferyl-α-D-Glucoside

Assay solutions

0.1 M sodium citrate buffer pH 5.5

1 mM substrate solution (4-methylumbelliferyl-α-D-glucoside, $M_r = 338.3$; 33.8 mg in 100 ml 0.1 M sodium citrate buffer pH 5.5)

50 mM glycine/NaOH pH 9.0

10 mM 4-methylumbelliferone (7-hydroxy-4-methylcoumarin, $M_r = 176.2$, 17.6 mg in 10 ml), for the standard curve

Procedure

0.2 ml 1 mM substrate solution

0.02 ml enzyme solution

30 min, 37 °C

1.78 ml 50 mM glycine/NaOH pH 9.0

Measure the fluorescence, excitation at 358 nm, emission at 448 nm. For quantification, prepare a standard curve with various concentrations of 4-methylumbelliferone.

References

Bergmeyer, H.U. (1983) *Methods of Enzymatic Analysis*, 3rd edn, vol. 2, Verlag Chemie, Weinheim, pp. 154–155.
Chuand, N.-N., Lin, K.-S., and Yang, B.-C. (1992) *Comp. Biochem. Physiol.*, **B102**, 273–277.
Liu, H. *et al.* (2007) *J. Org. Chem.*, **72**, 6562–6572.
Norouziam, D. *et al.* (2006) *Biotechnol. Adv.*, **24**, 80–85.

3.3.3.11 Cellulases, β-1,4-Glucanase, EC 3.2.1.4, and β-Glucosidase, EC 3.2.1.21

The designation "cellulase" is used for a series of enzymes degradating cellulose from its ends (exocellulases) or within the chain (endocellulases); two examples are mentioned.

β-1,4-glucanase, EC 3.2.1.4, SN: 4-β-D-glucan 4-glucano-hydrolase, cellulase, avicelase.

The enzyme catalyzes the endohydrolysis of 1,4-β-D-glucosidic linkages in cellulose, lichenin, and cereal β-D-glucans. Bacterial enzyme (*Trichoderma viride*): M_r 64 000, monomer; K_m (mM): 0.04 (cellotetraose), 0.15 (cellopentaose), 0.33 (cellohexaose), 1.63 (cellotriose), pH optimum 5.5.

β-glucosidase, EC 3.2.1.21, SN: 4-β-D-glucoside glucano-hydrolase, β-1,6-glucosidase, cellobiase.

The enzyme catalyzes the hydrolysis of terminal nonreducing β-D-glucose residues with release of β-D-glucose. Mammalian enzyme (human): M_r 53 000, monomer, K_m (mM): 4.1 (methylumbelliferyl-β-D-glucoside), pH optimum 5–6.5.

An orcinol assay determining the release of reducing units (glucose) from cellulose and an activity staining to detect active enzyme bands in electrophoresis gels are described.

A. Oricinol Assay

Assay solutions

0.1 M sodium phosphate/citric acid pH 6.1

Cellulose suspension, (microcrystalline cellulose, Avicel®, 2.5% (w/v) suspended in 0.1 M sodium phosphate/citric acid pH 6.1)

0.2% orcinol (5-methylresorcinol, 3,5-dihydroxytoluene, M_r = 142.2) in 70% sulfuric acid

1 M glucose solution (M_r = 180.2, 1.8 g in 10 ml) for a standard curve

Assay mixture and procedure

0.4 ml cellulose suspension

0.1 ml enzyme solution.
- 60 min 37 °C with permanent shaking (temperature depends on the source of the enzyme)
- Chill on ice
- Centrifuge 3 min, 5000 rpm

0.1 ml supernatant

0.9 ml 0.2% orcinol in 70% sulfuric acid.
- 20 min at 100 °C
- Cool to room temperature

Measure absorption at 550 nm, prepare a standard curve with glucose solution for quantification.

B. Activity Staining This staining can be performed in SDS polyacrylamide gels. To avoid inactivation, the enzyme sample should not be heated before applying it onto the electrophoresis gel. It is recommended to run two identical gels: one for activity staining, another one for staining with Coomassie Blue according to the standard procedure of SDS gel electrophoresis (cf. Laemmli, 1970).

Assay solutions

50 mM Na_2HPO_4/12.5 mM citric acid, pH 6.3

50 mM Na_2HPO_4/12.5 mM citric acid, pH 6.3, 25% isopropanol

Agar (agar-agar)

Carboxymethylcellulose, sodium salt

0.1% Congo red solution ($M_r = 696.7$)

1 M NaCl ($M_r = 58.44$, 5.84 g in 100 ml H_2O)

5% acetic acid

2 glass plates (~6 × 10 cm, ~2 mm thickness)

Preparation of the substrate gel

Heat 2% agar, 0.1% carboxymethylcellulose in 50 mM Na_2HPO_4/12.5 mM citric acid, pH 6.3 in a boiling water bath

Warm up two glass plates (size comparable to the polyacrylamide gel) in an incubator for 10 min at 60 °C

Immediately pour the hot agar solution on one glass plate and cover with the second one, maintaining a distant of 2 mm between both; this is achieved by mounting stoppers between both (avoid enclosure of air)

Store the gel in the refrigerator before usage

Preparation of the polyacrylamide gel

After running, wash the electrophoresis gel two times for 30 min with 50 mM Na_2HPO_4/12.5 mM citric acid, pH 6.3, 25% isopropanol

Wash two times for 30 min with 50 mM Na_2HPO_4/12.5 mM citric acid, pH 6.3 (omitting isopropanol)

Remove remaining fluid by rolling a glass tube over the gel surface

Performing the activity staining

Warm both, the substrate gel on the glass plate and the polyacrylamide gel, for 10 min at 60 °C in an incubator and remove eventually remaining fluid by rolling a glass tube over the gel surface

Layer the polyacrylamide gel onto the substrate gel, avoiding enclosure of air bubbles

Wrap the gels together with a parafilm foil and incubate for 60 °C at the assay temperature (e.g., 37 °C) in an incubator

Cool for 15 min in the refrigerator

Remove the polyacrylamide gel, stain with Coomassie Brilliant Blue according the to standard electrophoresis procedure (cf. Laemmli, 1970)

Immerse the substrate gel in 0.1% Congo red solution for 30 min at room temperature

Wash the substrate gel with 1 M NaCl until unbound dye is completely removed

Wash with 5% acetic acid; the hydrolyzed zones indicating enzyme activity should become visible

References

Beguin, P. (1983) *Anal. Biochem.*, **131**, 333–336.
Daniels, L.B. *et al.* (1981) *J. Biol. Chem.*, **256**, 13004–13013.
Laemmli, U.K. (1970) *Nature*, **227**, 680–685.
Michelin, K. *et al.* (2004) *Clin. Chim. Acta*, **343**, 145–153.
Okado, G. and Nisizawa, K. (1975) *J. Biochem.*, **78**, 297–306.

3.3.3.12 Lysozyme, EC 3.2.1.17

SN : peptidoglycan *N*-acetylmuramoylhydrolase, muramidase

Hydrolysis of 1,4-β-linkages between *N*-acetylmuramic acid and *N*-acetyl-D-glucosamine residues in peptidoglycans and between *N*-acetyl-D-glucosamine residues in chitodextrins. Enzyme from *Micrococcus lysodeikticus*: M_r 12 000, K_m (mM): 3.0 (4-nitrophenyl N,N',N'',N''',N'''' pentaacetylchitopentanose), pH optimum 5.2. The enzyme is tested following the turbidity decrease due to the lysis of the bacterium *Micrococcus lysodeikticus (M. luteus)*.

Assay solutions

0.1 M potassium phosphate pH 7.0

Suspension of 10 mg of lyophilized cells of *Micrococcus lysodeikticus* in 50 ml 0.1 M potassium phosphate pH 7.0

Procedure

0.98 ml *Micrococcus lysodeikticus* suspension

0.02 ml enzyme sample (about 500 IU ml^{-1})

Record the absorption (turbidity) decrease at 450 nm, 25 °C, for 5 min.

Calculation

$$\text{units mg}^{-1} = \frac{\text{absorbance change min}^{-1} \times 1000}{\text{mg enzyme in reaction mixture}}$$

References

Bergmeyer, H.U. (1984) *Methods of Enzymatic Analysis*, 3rd edn, vol. 4, Verlag Chemie, Weinheim, pp. 189–195.

Jollès, P. (1960) in *The Enzymes*, 2nd edn, vol. 4 (eds P.D. Boyer, H. Lardy, and K., Myrbäck), Academic Press, New York, pp. 431–445.

Miyauchi, K. *et al.* (2007) *Fish Sci.*, **73**, 1404–1406.

Shugar, D. (1952) *Biochim. Biophys. Acta*, **8**, 302–309.

3.3.3.13 Sialidase, EC 3.2.1.18

SN: acetylneuraminyl hydrolase, exo-α-sialidase, α-neuraminidase

Hydrolyzes $\alpha(2\to3)$-, $\alpha(2\to6)$-, $\alpha(2\to8)$-glycosidic linkages of terminal sialic acid residues in oligosaccharides, glycoproteins, and glycolipids. Mammalian enzyme (human): monomer, M_r 48 000, requires Ca^{2+}; K_m (mM): 5.6 (*N*-acetyl-neuraminic acid-α-2,6-lactose, 0.028 (2-(4-methylumbelliferyl)-α-D-*N*-acetylneuraminic acid); pH optimum 4.6.

A. Fluorimetric Assay

2-(4-methylumbelliferyl)-α-D-*N*-acetylneuraminic acid + H_2O →
4-methylumbelliferone + α-D-*N*-acetylneuraminic acid

Assay solutions

0.05 M acetic acid/sodium acetate pH 4.6

1 M $CaCl_2$ ($M_r = 111$, 1.11 g in 10 ml H_2O)

BSA (bovine serum albumine,200 mg in 10 ml H_2O)

10 mM Mumana (2-(4-methylumbelliferyl)-α-D-*N*-acetylneuraminic acid, sodium salt, $M_r = 489.4$; 4.89 mg ml^{-1} H_2O)

10 mM 4-methylumbelliferone (7-hydroxy-4-methylcoumarin, $M_r = 176.2$, 17.6 mg in 10 ml), for the standard curve

Assay mixture

Components	*Concentration*
9.7 ml 0.05 M acetic acid/sodium acetate pH 4.6	48.5 mM
0.05 ml BSA	0.1 mg/ml
0.05 ml 10 mM Mumana	0.05 mM

Procedure

0.98 ml assay mixture

0.02 ml enzyme sample

Assay volume 1 ml for reduced fluorescence cuvettes, for standard cuvettes take double the amount (2 ml assay volume).
Determine the fluorescence intensity (excitation 358 nm, emission 448 nm); for quantification, prepare a standard curve with various concentrations of 4-methylumbelliferone.

B. Activity Staining Specific method for staining active enzyme bands in electrophoresis gels.

Solutions and substances

0.1 M potassium phosphate buffer pH 6.0.

MPN (2-(3′-methoxyphenyl)-*N*-acetyl-α-neuramininc acid, $M_r = 415.4$)

Black K salt (diazonium salt of 4-amino-2,5-dimethoxy-4-nitroazobenzene, $M_r = 835.8$)

7% acetic acid

Staining solution

1 mg MPN

1 mg Black K salt.

Dissolve in 1 ml 0.1 M potassium phosphate buffer pH 6.0

Procedure

Cover the gel with sufficient volume (usually 1–2 ml) of the staining solution

Incubate at 37 °C for about 1 h (the staining solution can be exchanged every 20 min); active enzyme bands develop a deep red color

Wash with 7% acetic acid to intensify the bands

References

Groome, N.P. and Belyavin, G. (1975) *Anal. Biochem.*, **63**, 249–254.

Schauer, R., Wember, M., and Tschesche, H. (1984) *Hoppe-Seyler's Z. Physiol. Chem.*, **370**, 419–426.

Tuppy, H. and Palese, P. (1969) *FEBS Lett.*, **3**, 72–75.

Wang, Z.M. *et al.* (2001) *J. Virol. Methods*, **98**, 53–61.

3.3.3.14 α-Glucosidase, EC 3.2.1.20

SN: α-D-glucoside glucohydrolase, maltase.

Cleaves α-D-glucose from terminal, nonreducing 1,4-linked α-D-glucose residues. Human enzyme: M_r 110 000, K_m (mM) 2.1 (4-methylumbelliferyl-α-D-glucoside) pH optimum 4. *Aspergillus niger*, glycoprotein, M_r 120 000, K_m (mM): 0.62 (4-nitrophenyl-α-D-glucoside), 1.26 (maltose), pH optimum 4.5.

In the assay maltose is split into two glucoses, which are determined separately in the hexokinase reaction (HK), coupled with glucose-6-phosphate dehydrogenase (G6PDH).

$$\text{maltose} + H_2O \xrightarrow{\alpha\text{-glucosidase}} 2\ \text{D-glucose}$$

$$2\ \text{D-glucose} + 2\ \text{ATP} \xrightarrow{\text{HK}} 2\ \text{glucose-6-phosphate} + 2\ \text{ADP}$$

$$2\ \text{glucose-6-phosphate} + 2\ \text{NADP}^+ \xrightarrow{\text{G6P-DH}} 2,6\text{-phosphogluconate} + 2\ \text{NADPH} + 2\ \text{H}^+$$

A. α-Glucosidase Assay

Assay solutions

0.1 M acetate buffer pH 6.0, containing 1.5 mM EDTA (5.7 ml acetic acid, 100%, $M_r = 60$, in 800 ml H_2O, add 560 mg $EDTA \cdot Na_2 \cdot 2H_2O$, $M_r = 372.2$, adjust to pH 6.0 with 1 N NaOH, fill up to 1 l)

0.5 M maltose solution (monohydrate, $M_r = 360$, 18 g in 100 ml H_2O)

Procedure

	Concentration (mM)
0.74 ml 0.1 M acetate buffer pH 6.0	74
0.25 ml 0.5 M maltose solution	125
0.01 ml enzyme solution	

5 min at 25 °C

Transfer the reaction tube into a 100 °C water bath or heating block

Glucose is determined in 0.2 ml aliquots.

B. Glucose Determination

Assay solutions

100 mM triethanolamine · HCl/NaOH pH 7.6 (triethanolamine · HCl, adjusted with NaOH)

1.0 M D-glucose ($M_r = 180.2$; 18 g in 100 ml)

0.1 M ATP (disodium salt, trihydrate, $M_r = 605.2$; 605 mg in 10 ml)

0.1 M $MgCl_2$ (hexahydrate; $M_r = 203.3$; 203 mg in 10 ml)

0.01 M NADP (disodium salt, $M_r = 787.4$; 79 mg in 10 ml)

Glucose-6-phosphate dehydrogenase (G6PDH, commercial products are 100–500 IU mg^{-1}, dilute to ~5 IU 0.1 ml^{-1})

Hexokinase (HK, ~200 IU ml^{-1})

Assay mixture

Components	*Concentration*
5.8 ml 0.1 mM triethanolamine/NaOH pH 7.6	58 mM
0.6 ml 0.1 mM $MgCl_2$	6.0 mM
1.0 ml 10 mM NADP	1.0 mM
0.3 ml 0.1 M ATP	3.0 mM
0.1 ml G6PDH	0.5 IU ml^{-1}

Procedure

0.78 ml assay mixture

0.2 ml D-glucose sample (1 M for the reference 200 mM)

0.01 ml HK

The absorption increase at 340 nm is measured at 25 °C, $\varepsilon_{340} = 6.3 \times 10^3\ l \cdot mol^{-1}\ cm^{-1}$.

Calculation

$$\text{glucose concentration} = \frac{\Delta A \times \text{RV glucosidase} \times \text{RV glucose}}{\varepsilon_{340} \times 1 \times 0.1} = \frac{\Delta A \times 1 \times 1}{6.3}$$

= μmol glucose in 1 ml incubation mix (the blank must be subtracted), RV = reaction volume (ml).

C. Assay with 4-Nitrophenylglucopyranoside

Assay solutions

50 mM acetate buffer pH 4.5

1.25 mM 4-nitrophenyl-α-D-glucopyranoside solution ($M_r = 301.3$; 37.7 mg in 100 ml 50 mM acetate buffer pH 4.5)

0.5 M Na_2CO_3 solution ($M_r = 106$, dissolve 5.3 g in 100 ml H_2O)

Procedure

0.8 ml 1.25 mM 4-nitrophenyl-α-D-glucopyranoside solution

0.2 ml enzyme sample

10 min 37 °C

0.1 ml 0.5 M Na_2CO_3 solution

Measure absorption at 405 nm; $\varepsilon_{405} = 18\,500\,l\,mol^{-1}cm^{-1}$.

References

Bergmeyer, H.U. (1983) *Methods of Enzymatic Analysis*, 3rd edn, vol. 2, Verlag Chemie, Weinheim, pp. 205–206.
Bruni, C.B., Sica, V., Auricchio, F., and Covelli, I. (1970) *Biochim. Biophys. Acta*, **212**, 470–477.
Dahlquist, A. (1968) *Anal. Biochem.*, **22**, 99–107.
Flanagan, P.R. and Forstner, G.G. (1978) *Biochem. J.*, **173**, 553–563.
Fuller, M. *et al.* (1995) *Eur. J. Biochem.*, **234**, 903–909.
Larner, J. (1960) in *The Enzymes*, 2nd edn, vol. 4 (eds P.D. Boyer, H. Lardy, and K. Myrbäck), Academic Press, New York, pp. 369–378.
Ogawa, M. (2006) *J. Appl. Glycosci.*, **53**, 13–16.
Sorensen, S.H., Nòren, O., Sjöström, H., and Danielsen, E.M. (1982) *Eur. J. Biochem.*, **126**, 559–568.

3.3.3.15 β-Galactosidase, EC 3.2.1.23

SN: β-D-Galactoside galactohydrolase

Hydrolysis of terminal nonreducing β-D-galactose residues in β-D-galactosides:

$$\text{lactose} + H_2O \rightarrow \beta\text{-D-galactose} + \text{D-glucose}$$

Bacterial enzyme (*Escherichia coli*): M_r 516 000, homotetramer (4 × 116 000), K_m (mM): 0.03 (4-nitrophenyl-β-D-galactose).

Assay solutions

0.1 M potassium phosphate pH 7.6

0.05 M *o*-nitrophenyl β-D-galactopyranoside (ONPG, M_r = 301.3; 150 mg in 10 ml 0.1 M potassium phosphate pH 7.6)

β-Galactosidase from *E. coli* (~300 IU mg^{-1})

Procedure

	Concentration
0.88 ml 0.1 M potassium phosphate pH 7.6	88 mM
0.1 ml 0.05 M ONPG	5.0 mM
0.02 ml enzyme solution	–

Follow the absorption increase at 405 nm at 25 °C, $\varepsilon_{405} = 18\,500\,l\,mol^{-1}\,cm^{-1}$.

References

Bergmeyer, H.U. (1983) *Methods of Enzymatic Analysis*, 3rd edn, vol. 2, Verlag Chemie, Weinheim, pp. 197–198.
Craven, R.G., Steers, E., and Anfinsen, C.B. (1965) *J. Biol. Chem.*, **240**, 2468–2477.
Rotman, B., Zderic, J.A., and Edelstein, M. (1963) *Proc. Natl. Acad. Sci.*, **50**, 1–6.
Wallenfels, K. and Well., R. (1972) in *The Enzymes* 3rd edn, vol. 7 (ed. P.D. Boyer), Academic Press, New York, pp. 617–663.
Woolen, J.W. and Turner, P. (1965) *Clin Chim. Acta*, **12**, 647–658.

3.3.3.16 **β-Fructosidase, EC 3.2.1.26**

SN: β-D-fructosefuranoside fructohydrolase, β-fructofuranosidase, invertase, invertin, saccharase, sucrase.

The enzyme hydrolyzes the terminal nonreducing β-D-fructofuranoside residues in β-D-fructofuranosides, especially saccharose. This is one of the most intensely studied enzymes in the early epoch of biochemical research; the Michaelis–Menten equation was originally derived from the invertase reaction. The assay utilizes the fact that the deflection of the plane of polarized light becomes inversed ("invertase") when saccharose is cleaved into an equimolar mixture of glucose and fructose. The reaction can be continuously followed with a polarimeter (ORD spectrometer). The assay is described for 10 ml samples for a 10 cm cuvette and may be performed accordingly if smaller cuvettes and a more sensitive instrument are used. Enzyme from *Aspergillus niger*: M_r 95 000, homodimer (2 × 47 000), K_m (mM): 0.06625 (saccharose), pH optimum 5.

Assay solutions

0.05 M acetate buffer pH 4.62 (2.85 ml acetic acid, 100%, $M_r = 60$, in 800 ml H_2O, adjust to pH 4.62 with 1 N NaOH, fill up to 1 l)

0.3 M D-(+)-saccharose, ($M_r = 342.3$; 10.3 g in 100 ml H_2O)

1 M sodium carbonate ($Na_2CO_3 \cdot 10H_2O$, $M_r = 286.1$; 28.6 g in 100 ml H_2O)

Invertase from yeast (~300 IU mg^{-1})

Procedure

	Concentration (mM)
1.9 ml 0.05 M sodium acetate buffer pH 4.62	24
2.0 ml 0.3 M saccharose	150

Bring to 30 °C

0.1 ml invertase.

Incubate for different times (1–10 min) and add

6 ml 1 M Na_2CO_3

Measure the polarization in a 10 cm cell of a polarimeter. 1 unit of the enzyme splits 1 μmol substrate/min.

References

Myrbäck, K. (1960) *The Enzymes*, 2nd edn, vol. 4 (eds P.D. Boyer, H. Lardy, and K. Myrbäck), Academic Press, New York, pp. 379–396.
Rubio, M.C. and Maldonado, M.C. (1995) *Curr. Microbiol.*, **31**, 80–83.
Sumner, J.B. and Howell, S.F. (1935) *J. Biol. Chem.*, **108**, 51–54.

3.3.3.17 β-Glucuronidase EC 3.2.1.31

SN: β-D-glucuronoside glucuronosohydrolase

$$\beta\text{-D-glucuronoside} + H_2O \rightarrow \text{D-glucuronate} + \text{alcohol}$$

Glycoprotein, human enzyme: M_r 390 000, tetramer (4 × 83 000), K_m (mM): 1.3 (4-umbelliferyl-β-D-glucuronide), 0.13 (*p*-nitrophenyl-β-D-glucuronide), 0.018 (phenolphthalein-β-D-glucuronide), pH optimum 4.5.

A. Fluorimeteric Assay

Assay solutions

0.2 M sodium acetate pH 4.8

10 mM 4-umbelliferyl-β-D-glucuronide (7-hydroxycoumarin-β-D-glucuronide, sodium salt, M_r = 361.3; 36.1 mg in 10 ml 0.2 M sodium acetate pH 4.8)

Procedure

0.1 ml 10 mM 4-umbelliferyl-β-D-glucuronide

0.01 ml enzyme sample.

Incubate for 30 min at 37 °C

Measure the fluorescence, excitation at 358 nm, emission at 448 nm. For quantification, prepare a standard curve with various concentrations of 4-methylumbelliferone.

References

Brot, F.E., Bell, E., and Sly, W.S. (1978) *Biochemistry*, **17**, 385–391.
Islam, M.R. *et al.* (1999) *J. Biol. Chem.*, **274**, 23451–23455.

3.3.3.18 Proteases, EC 3.4, General Assays

Due to the different kinds and specificities of the proteases, many different protease assays have been described. Proteases are classified according to distinct criteria, with respect to (i) the catalysis mechanism as serine proteases, cysteine proteases, and metalloproteases; (ii) the mode of attacking the substrate as exo- and endoproteases; (iii) the substrate specificities, for example as trypsin- and chymotrypsin-like proteases. These criteria must be considered for performing a protease assay. Principally, two types of protease assays exist: (i) general assays demonstrating only the peptide-cleaving capacity and (ii) specific protease assays. Usually, in the former case, proteins are used and their digestion into fragments is detected, while in the latter case, synthetic substrates or defined peptides are applied.

Reference

Cunningham, L. (1965) *Compr. Biochem.*, **16**, 85–188.

A. Anson Assay Proteins in their native structure often resist proteolytic attack. Therefore, hemoglobin, denatured by urea, is applied for this assay. The digested fragments are separated from nonhydrolyzed protein and detected in the supernatant with the Folin–Ciocalteau reagent, which preferentially recognizes tryptophan and tyrosine and to a lesser extent, cysteine and histidine. According to the respective protease assay, temperatures from room temperature (25 °C) to 50 °C are chosen.

Assay solutions

1 N NaOH (M_r = 40.0; 4 g NaOH in 100 ml H_2O, store in PE flasks)

0.5 N NaOH (1 part 1 N NaOH, 1 part H_2O)

1 M KH_2PO_4 (1 M_r = 136.1; 13.61 g in 100 ml H_2O)

0.3 M trichloroacetic acid (M_r = 163.4; 4.9 g in H_2O)

Folin–Ciocalteau phenol reagent (commercially available, keep at 4 °C; dilute 1 volume with 2 volumes H_2O before use)

0.2 M HCl (bring 3.31 ml HCl (37%) to 200 ml with H_2O)

1 mM L-tyrosine (M_r = 181; 18.1 mg in 100 ml 0.2 M HCl)

Hemoglobin substrate: place a 100 ml beaker, containing 10 ml in a water bath and keep at 25 °C. The components listed below are added one after the other and dissolved by stirring
- 11.54 g urea
- 2.4 g 1 N NaOH
- 0.635 g lyophilized hemoglobin

Stir for 45 min

- Dissolve 3.16 g 1 M KH_2PO_4 and add. H_2O to give a total weight of 32.6 g
- Adjust the pH to 7.5 with 1 N HCl

The solution should be stored in the cold (5 °C) and used within 14 days. For every new preparation of the hemoglobin substrate, a distinct standard curve is needed.

Procedure

	Sample	**Blank**
Enzyme solution	250 μl	250 μl
Hemoglobin substrate	250 μl	–
15 min incubation at assay temperature (25 °C)		
0.3 M trichloroacetic acid	1000 μl	1000 μl
Hemoglobin substrate	–	250 μl
5 min centrifugation		
Supernatant	500 μl	500 μl
0.5 N NaOH	1000 μl	1000 μl
Folin–Ciocalteau phenol reagent (1 : 2 diluted)	250 μl	250 μl
Measure absorption difference sample – blank at 750 nm		

Calibration curve

A calibration curve between 0 and 250 nmol L-tyrosine is prepared. Increasing amounts (~12 different samples) from 0 to 250 μl 1 mM L-tyrosine (1 μl to 1 nmol) are filled up with 0.2 M HCl to a final volume of 0.5 ml and 1 ml 0.5 N NaOH and 0.250 ml Folin–Ciocalteau phenol reagent (1 : 2 diluted) are added to each sample. Absorption difference at 750 nm against the blank (without tyrosine) is determined. A linear relationship of absorbance with tyrosine concentration should be obtained, the slope of the resulting line is absorbance per nmol Tyr.

Calculation of Enzyme Activity

Since determination of molarity of product formation from hemoglobin is difficult, special Anson Units (AU) are defined: 1 AU is the amount of enzyme producing a color intensity corresponding to 1 mEq of tyrosine (1 nmol tyrosine ~1 μAU)/min under assay conditions.

The absorbance at 750 nm divided by the slope of the calibration curve gives the amount of product formed, equivalent to nanomole Tyr. To obtain μAU, the value must be divided by the incubation time (min) and multiplied by a factor of 3 (since only 0.5 ml out of the total assay volume of 1.5 ml is taken for the color reaction) and by the dilution factor.

Enzyme activity in the assay:

$$\mu AU = \frac{A_{750} \times 3 \times \text{dilution factor}}{\text{slope} \times \text{time(min)}}$$

With reference to the total enzyme solution, the value obtained must be multiplied by the total volume of the enzyme solution and divided by the enzyme amount used for the assay and a factor for converting to AU:

Total enzyme activity:

$$AU = \frac{A_{750} \times 3 \times \text{dilution factor} \times \text{total enzyme volume (ml)}}{\text{slope} \times \text{time (min)} \times \text{test enzyme volume (ml)} \times 10^6}$$

References

Anson, M.L. (1939) *J. Gen. Physiol*, **22**, 79–89.
Kunitz, M. (1939) *J. Gen. Physiol.*, **22**, 447–450.
Peterson, G.L. (1979) *Anal. Biochem.*, **100**, 201–220.

B. Casein Assay This is a simple, but not very sensitive assay. The absorption of the amino acids cleaved from casein by protease action is measured at 280 nm in the supernatant after separation from the remaining protein by TCA precipitation. For quantification, a tyrosine calibration curve must be prepared.

Assay solutions

0.01 M Tris/HCl pH 8.0, containing 10 mM $CaCl_2$ (1.47 g $CaCl_2 \cdot 2H_2O$ in 1 l 0.01 M Tris/HCl pH 8.0)

0.2 M NaOH ($M_r = 40.0$; 0.8 g in 100 ml H_2O)

0.2 M acetic acid ($M_r = 60.05$; 12 g in 1 l)

1.2 M trichloroacetic acid (TCA, $M_r = 163.4$; 19.6 g in 100 ml H_2O)

1 mM L-tyrosine ($M_r = 181$; 18.1 mg in 100 ml 0.2 M HCl)

Casein substrate: dissolve 2 g casein in 90 ml 0.01 M Tris/HCl pH 8.0, 10 mM $CaCl_2$ under stirring, add 0.2 M NaOH until a clear solution is obtained, adjust the pH to 8.0 with 0.2 M acetic acid and bring the solution to 100 ml with H_2O. Divide it in aliquots and store at −20 °C

Procedure

Warm up 0.4 ml casein substrate to 35 °C

Add 0.2 ml protease solution in 0.01 M Tris/HCl pH 8.0, 10 mM $CaCl_2$

Incubate at 35 °C for 10 min

Add 1 ml 1.2 M TCA to stop the reaction

Treat the blank accordingly, except that the protease solution is given to the casein substrate after addition of TCA

Centrifuge the samples for 5 min

Measure the absorption of the supernatant at 275 nm

One unit is defined as the amount of protease, producing an absorption corresponding to 1 μmol tyrosine of the calibration curve after 1 min at 35 °C. The calibration curve is prepared in the micromolar range with the 1 mM tyrosine solution.

Reference

Kunitz, M. (1947) *J. Gen. Physiol.*, **30**, 291–310.

C. Azocasein Assay The sensitivity of the casein assay is essentially improved by azo groups covalently bound to casein. Upon liberation by proteolytic digestion, an intense color is developed in the supernatant.

Assay solutions

0.1 M potassium phosphate pH 8.0

0.2% azocasein (sulfanilamide azocasein) in 0.1 M potassium phosphate pH 8.0

10% trichloroacetic acid (TCA, w/v)

Procedure

0.5 ml azocasein solution

0.2 ml protease solution

Incubate at 25 °C for 30 min

Add 0.2 ml 10% trichloroacetic acid

Centrifugate for 5 min

Measure the absorption in the supernatant at 340 nm, 1 unit is defined as the protease amount producing an absorption change of 1 within 1 h

References

Brock, F.M., Frosberg, C.W., and Buchanan-Smith, J.G. (1982) *Appl. Environ. Microbiol.*, **44**, 561–569.

Peek, K., Daniel, R.M., Monk, C., Parker, L., and Coolbear, T. (1992) *Eur. J. Biochem.*, **207**, 1035–1044.

Tomarelli, R.M., Charney, J., and Harding, M.L. (1949) *J. Lab. Clin. Med.*, **34**, 428–433.

D. Ninhydrin Assay Ninhydrin develops a very intense color with free amino acids and the high sensitivity compensates for the relatively laborious method. Casein is used as protease substrate. After precipitation of nonhydrolyzed protein with trichloroacetic acid the ninhydrin reaction is performed in the supernatant.

Assay solutions

0.01 M Tris/HCl pH 8.0, containing 10 mM $CaCl_2$ (1.47 g $CaCl_2$ $2H_2O$ in 1 l 0.01 M Tris/HCl pH 8.0)

0.2 M NaOH ($M_r = 40.0$; 0.8 g in 100 ml H_2O)

0.2 M acetic acid ($M_r = 60.05$; 12 g in 1 l)

1.2 M trichloroacetic acid (TCA, $M_r = 163.4$; 19.6 g in 100 ml H_2O)

Casein substrate: dissolve 2 g casein in 90 ml 0.01 M Tris/HCl pH 8.0, 10 mM $CaCl_2$ under stirring, add 0.2 M NaOH until a clear solution is obtained, adjust the pH to 8.0 with 0.2 M acetic acid and bring the solution to 100 ml with H_2O. Divide it in aliquots and store at −20 °C

Acetate buffer pH 5.4 (dissolve 270 g sodium acetate trihydrate, $C_2H_3O_2Na \cdot 3H_2O$, $M_r = 136.1$, in 200 ml H_2O, add 50 ml glacial acetic acid and fill up to 750 ml)

10 mM KCN ($M_r = 65.1$; 65 mg in 100 ml H_2O, *volatile, even the vapor is extremely poisonous!*)

0.2 mM KCN in acetate buffer (0.2 ml 10 mM KCN + 9.8 ml acetate buffer, prepare fresh)

Ninhydrin reagent (2,2-dihydroxy-1,3-indanedione, $M_r = 178.1$, commercially available)

50% ethanol

Standard BSA solution, 40 μg ml^{-1}, prepared by 100-fold dilution of a stock solution, 100 mg in 25 ml

10 mM leucine (D,L-leucine, $M_r = 131.2$; 131 mg in 100 ml H_2O)

Protease solution

Procedure for casein degradation

Warm up 0.4 ml casein substrate to 35 °C

Add 0.2 ml protease solution in 0.01 M Tris/HCl pH 8.0, 10 mM $CaCl_2$

Incubate at 35 °C for 10 min

Stop the reaction by addition of 1 ml 1.2 M TCA

Treat the blank accordingly, except that the protease solution is given to the casein substrate after addition of TCA

Centrifuge the samples for 5 min to get the TCA supernatant

Ninhydrin reaction

0.2 ml TCA supernatant

0.1 ml 0.2 mM KCN in acetate buffer

0.1 ml ninhydrin reagent

10 min in a boiling water bath or a heating block at 100 °C

Chill in ice for 1 min and add 0.5 ml 50% ethanol

Measure the absorbance at 570 nm against a blank where the hydrolysate is displaced by water

Calibration curve

For quantification, a calibration curve is generated with 12 aliquots (1–50 µl, fill up to 200 µl) of the 10 mM leucine solution.

Reference

Rosen, H. (1957) *Arch. Biochem. Biophys.*, **67**, 10–15.

3.3.3.19 Leucine Aminopeptidase, EC 3.4.11.1

Leucyl aminopeptidase, LAP, peptidase S, cathepsin III; bacterial leucyl aminopeptidase, EC 3.4.11.10.

Zn-containing exopeptidase, splits *N*-terminal of an amino acid Xaa-/-Yaa, in which Xaa is preferably leucine, but also other amino acids, including proline, but not arginine and lysine; Yaa may be proline. Amides and methylesters from amino acids are hydrolyzed with considerably lower rates. Human enzyme: M_r 120 000, K_m (mM) 1.28 (L-leucine-4-nitroanilide), pH optimum 7.5.

A. Assay with Leucineamide

Assay solutions

0.5 M Tris/HCl pH 8.5

0.025 manganese chloride ($MnCl_2 \cdot 2H_2O$, $M_r = 161.9$; 40.5 mg in 10 ml)

0.1 M magnesium chloride ($MgCl_2 \cdot 6H_2O$, $M_r = 203.3$; 203 mg in 10 ml)

0.0625 M L-leucine ($M_r = 131.2$; 820 mg in 100 ml, adjust to pH 8.5)

0.125 M L-leucinamide (hydrochloride, $M_r = 166.7$; 2.08 g in 100 ml, adjust to pH 8.5)

LAP solution (0.1 ml, corresponding to 0.4 mg, activate prior to use by incubating at 37 °C for 2 h in a mixture of 0.05 ml 0.025 M $MnCl_2$, 0.05 ml 0.5 M Tris/HCl pH 8.5, 0.8 ml water)

Assay mixture and procedure

	Concentration (mM)
0.05 ml 0.5 M Tris/HCl pH 8.5	25
0.4 ml 0.125 M leucinamide	50
0.05 ml 0.1 M magnesium chloride	5
0.4 ml H_2O	–
0.1 ml LAP solution	–

Follow the decrease of absorption at 238 nm, 25 °C.

For the blank, replace leucinamide and H_2O by 0.8 ml 0.0625 M leucine.

Calculation

Prepare a calibration curve with different concentrations of leucinamide at 238 nm (α_m = absorption/[leucinamide]):

$$\text{unit mg}^{-1} = \frac{A_{238}\text{min}^{-1} \times 1000 \times \text{ml reaction mixture}}{\alpha_m \times \text{mg enzyme}}$$

B. Assay with Leucine-*p*-nitroanilide

L-leucine-4-nitroanilide → L-leucine + *p*-nitroaniline

Assay solutions

0.05 M Tricine pH 8.0 (*N*-tris[hydroxymethyl]methylglycine, $M_r = 179.2$, dissolve 8.96 g in 0.8 l H_2O, adjust with 1 N NaOH to pH 8.0 and bring to 1 l)

1 mM L-leucine-*p*-nitroanilide (hydrochloride, $M_r = 287.7$; 28.8 mg in 100 ml 0.05 M tricine, pH 8.0)

Assay mixture

Components	*Concentration (mM)*
7.8 ml 0.05 M tricine pH 8.0	50
2 ml 0.01 M L-leucine-*p*-nitroanilide	0.2

Procedure

0.98 ml assay mixture

0.02 ml enzyme sample

Measure the absorption increase at 405 nm ($\varepsilon_{405,\,p\text{-nitroaniline}} = 10\,800$ $l \cdot mol^{-1}\,cm^{-1}$), 25 °C.

References

Delange, J.R. and Smith, E.L. (1971) in *The Enzymes*, 3rd edn, vol. 3 (ed. P.D. Boyer), Academic Press, New York, pp. 81–118.
Hattori, A. *et al.* (2000) *J. Biochem.*, **128**, 755–762.
Himmelhoch, R. (1970) *Meth. Enzymol.*, **19**, 508–513.
Mitz, M.A. and Schlueter, R.J. (1958) *Biochim. Biophys. Acta*, **27**, 168–172.
Spungnin, A. and Blumberg, S. (1989) *Eur. J. Biochem.*, **183**, 471–477.

3.3.3.20 α-Chymotrypsin, EC 3.4.21.1

Protease preferentially cleaving peptides behind tyrosine, tryptophan, phenylalanine, and leucine. Mammalian enzyme (bovine): M_r 25 210, K_m (mM): 0.32 (*N*-glutaryl-L-phenylalanine-4-nitroanilide, GLUPHEPA), pH optimum 8. A fluorimetric assay is described under lipase (assay in Section 3.3.3.1.B).

A. Assay with SUPHEPA

Assay solutions

Buffer I: 0.2 M TEA/NaOH pH 7.8 (triethanolamine · HCl, $M_r = 185.7$, 37.13 g, 2.2 g calcium chloride in 800 ml H_2O, adjust with 2 N NaOH to pH 7.8, fill up to 1 l)

Buffer II: 0.2 M TEA/NaOH pH 7.8 (37.13 g triethanolamine · HCl in 800 ml H_2O, adjust with 2 N NaOH to pH 7.8, fill up to 1 l)

60 μM α-chymotrypsin (15.2 mg in 10 ml 1 mM HCl)

0.1 M SUPHEPA (*N*-succinyl-L-phenylalanine-*p*-nitroanilide, M_r = 385.4; 385 mg in 10 ml buffer II)

Assay mixture and procedure

	Concentration (mM)
0.95 ml buffer I	190
0.04 ml 0.1 M SUPHEPA	4.0
0.01 ml α-chymotrypsin solution	–

Follow the increase of absorbance at 405 nm, 25 °C. Absorption coefficient of *p*-nitroanilide $\varepsilon_{405} = 10.2 \times 10^3$ l mol^{-1} cm^{-1}.

B. Assay with GLUPHEPA

Assay solutions

0.1 M potassium phosphate pH 7.6

10 mM GLUPHEPA (*N*-glutaryl-L-phenylalanin-4-nitroanilide, M_r = 399.4; 40 mg in 10 ml 0.1 M potassium phosphate pH 7.6)

Assay mixture and procedure

	Concentration (mM)
0.98 ml GLUPHEPA	10
0.02 ml α-chymotrypsin solution	–

Measure the absorption increase at 405 nm, 25 °C. Absorption coefficient of *p*-nitroanilide $\varepsilon_{405} = 10.2 \times 10^3$ l mol^{-1} cm^{-1}.

References

Bergmeyer, H.U. (1984) *Methods of Enzymatic Analysis*, 3rd edn, vol. 5, Verlag Chemie, Weinheim, pp. 99–104.

Erlanger, B.F., Cooper, A.G., and Bendich, H.J. (1964) *Biochemistry*, **3**, 1880–1883.

Fiedler, F., Geiger, R., Hirschauer, C., and Leysath, G. (1978) *Hoppe Seyler's Z. Physiol. Chem.*, **259**, 1667–1673.
Nagel, W., Willig, F., Peschke, W., and Schmid, F.H. (1965) *Hoppe Seyler's Z. Physiol. Chem.*, **340**, 1–10.
Rick, W. (1974) *Methods of Enzymatic Analysis*, 3rd edn, vol. 1, Verlag Chemie, Weinheim, pp. 1045–1051.
Schwert, G.W. and Takenaka, Y. (1955) *Biochim. Biophys. Acta*, **16**, 570–575.
Spreti, N. *et al.* (2001) *Eur. J. Biochem.*, **268**, 6491–6497.

3.3.3.21 Pancreatic Elastase, EC 3.4.21.35 (Previous EC 3.4.4.7)

Hydrolysis of proteins, including elastin, preferential cleavage at Ala.

Mammalian pancreatic enzyme (porcine): M_r 25 900, monomer; enzyme from salmon: K_m (mM): 1.47 (succinyl-Ala-Ala-Ala-*p*-nitroanilide); 0.3 (NBA, *p*-nitrophenyl-*N*-*tetr*-butyloxycarbonyl-L-alaninate), pH optimum 8.7.

A. Assay with Succinyl-Ala-Ala-Ala-*p*-Nitroanilide

Assay solutions

0.1 M Tris/HCl pH 8.5

2.5 mM succinyl-Ala-Ala-Ala-*p*-nitroanilide (M_r = 451.4, 11.3 mg in 10 ml)

Procedure

	Concentration (mM)
0.88 ml 0.1 M Tris/HCl pH 8.5	88
0.1 ml 2.5 mM succinyl-Ala-Ala-Ala-*p*-nitroanilide	0.25
0.02 enzyme sample	–

Measure the absorption increase at 405 nm ($\varepsilon_{405,\ p\text{-nitroaniline}}$ = 10 200 l · mol^{-1} cm^{-1}), 25 °C.

B. Esterase Activity of Elastase

Assay solutions

0.05 M potassium phosphate pH 6.5

0.01 M NBA (*p*-nitrophenyl-*N*-*tetr*-butyloxycarbonyl-L-alaninate, *N*-tert-BOC-L-alanine *p*-nitrophenylester, M_r = 310.3, 31 mg in 10 ml acetonitrile or methanol)

Assay mixture

Components	*Concentration (mM)*
9.5 ml 0.05 M potassium phosphate pH 6.5	47.5
0.3 ml 0.01 M NBA	0.3

Procedure

0.98 ml assay mixture

0.02 ml enzyme sample (~5 μg)

Measure the absorption increase at 347.5 nm ($\varepsilon_{374.5} = 5500\,M^{-1}\,cm^{-1}$, the absorption of *p*-nitrophenol at this wavelength is independent of pH), 25 °C.

References

Berglund, G.I. *et al.* (1998) *Mol. Marine Biol. Biotechnol.*, **7**, 105–114.
Hilpert, K. *et al.* (2003) *J. Biol. Chem.*, **278**, 24986–24993.
Visser, L. and Blout, E.R. (1972) *Biochim. Biophys. Acta*, **268**, 257–266.

3.3.3.22 Pepsin, EC 3.4.23.1

Pepsin A Endopeptidase in the gastric juice; cleaves hydrophobic, preferentially aromatic amino acid residues. Mammalian enzyme (human): M_r 35 000, monomer, activated by Ca^{2+}; K_m (mM, *Gallus gallus*): 0.1 (hemoglobin), 1.05 (casein), pH optimum 2.

Assay solutions

0.3 M HCl (24.8 ml 37% HCl, density 1.18, fill up with H_2O to 1 l)

2% hemoglobin (dissolve 2.5 g bovine hemoglobin in 100 ml H_2O, mix for 5 min in a Waring blender, filter through glass wool and add to 80 ml filtrate 20 ml 0.3 M HCl)

0.3 M trichloroacetic acid ($M_r = 163.4$, 4.9 g in 100 ml H_2O)

0.01 M HCl (dilute 0.3 M HCl 30-fold)

Pepsin solution (5 mg in 10 ml 0.01 M HCl, dilute for the assay 50-fold in 0.01 M HCl)

Procedure

	Sample	Blank
2% hemoglobin substrate	0.3 ml	0.3 ml
Incubation at 37 °C (water bath)		
0.3 M trichloroacetic acid	–	0.6 ml
Pepsin solution	0.1 ml	0.1 ml
10 min incubation at 37 °C		
0.3 M trichloroacetic acid	0.6 ml	–
5 min, 37 °C		
5 min centrifugation		
Record absorption at 280 nm		

Calculation

$$\text{units mg}^{-1} = \frac{A_{280,\text{sample}} - A_{280,\text{blank}} \times 1000}{10\ \text{min} \times \text{mg enzyme}}$$

References

Althauda, S.B.P. *et al.* (1989) *J. Biochem.*, **106**, 920–927.

Bergmeyer, H.U. (1984) *Methods in Enzymatic Analysis*, 3rd edn, vol. 5, Verlag Chemie, Weinheim, pp. 232–233.

Fruton, J.S. (1971) in *The Enzymes*, 3rd edn, vol. 3 (ed. P.D. Boyer), Academic Press, New York, pp. 119–164.

Kassell, B. and Meitner, P.A. (1970) *Meth. Enzymol.*, **19**, 337–347.

Klomklao, S. *et al.* (2007) *Comp. Biochem. Physiol., B*, **147B**, 682–689.

3.3.3.23 Trypsin, EC 3.4.21.4

Cleaves preferentially behind Arg- and Lys-; pancreatic enzyme (*Engraulis japonicus*, human): M_r 24 000, monomer, strong inhibition by the soybean inhibitor ($IC50$ = 0.009 µM), K_m (mM): 0.033 (benzoylarginine-4-nitroanilide), 0.014 (benzoylarginine-ethylester), 0.1 (leupeptin); pH optimum 9.5.

Assay solutions

0.3 M potassium phosphate pH 8.0

20 mM *N′*-benzoyl-L-arginine-*p*-nitroanilide (BAPNA, M_r= 434.9; 174 mg in 20 ml DMSO)

1 mM HCl (83 µl 37% HCl in 1 l H_2O)

1 mM trypsin, 23.4 mg in 1 ml 1 mM HCl (dilute 50-fold in 0.05 M potassium phosphate pH 8.0 immediately before testing)

Assay mixture and procedure

	Concentration
0.70 ml H_2O	–
0.18 ml 20 mM BAPNA	3.0 mM
0.10 ml 0.3 M potassium phosphate pH 8.0	30 mM
0.02 ml trypsin solution	9 μg ml^{-1}

Record the absorbance at 405 nm 25 °C, $\varepsilon_{405} = 9.62 \times 10^3$ l mol^{-1} cm^{-1}.

References

Ahsan, M.N. and Watabe, S. (2001) *J. Protein Chem.*, **20**, 49–58.
Gaertner, H.F. and Puigserver, A.J. (1992) *Enzyme Microb. Technol.*, **14**, 150–155.
Gravett, P.S., Viljoen, C.C., and Oosthuizen, M.M.J. (1991) *Int. J. Biochem.*, **23**, 1085–1099.
Knecht, W. *et al.* (2007) *J. Biol. Chem.*, **282**, 26089–26100.
Kunitz, M. (1947) *J. Gen. Physiol.*, **30**, 291–310.
Schwert, G.W. and Takenaka, Y. (1955) *Biochim. Biophys. Acta*, **16**, 570–575.
Wachsmuth, E.D., Fritze, I., and Pfleiderer, G. (1966) *Biochemistry*, **5**, 169–174.

3.3.3.24 Asparaginase, EC 3.5.1.1

SN: L-Asparagine amidohydrolase, ASNase,

$$\text{L-asparagine} + H_2O \rightarrow \text{L-aspartate} + NH_3$$

Bacterial enzyme (*Escherichia coli*) M_r 150 000, homotetramer, (4 × 37 000), K_m (mM): 0.0035 (L-glutamine), 0.015 (L-asparagine): pH optimum 8.6.

The released ammonia is detected with Nessler's reagent.

Assay solutions

0.05 M Tris/HCl pH 8.6

0.01 M L-asparagine solution (asparagine monohydrate, M_r =150.1; 150 mg in 100 ml 0.05 M Tris/HCl pH 8.6)

1.5 M trichloroacetic acid (TCA, M_r = 163.4; 24.5 g in 100 ml)

Nessler's reagent

Procedure

	Sample	Blank
0.05 M Tris/HCl pH 8.6	0.05 ml	–
0.01 M L-asparagine solution	0.85 ml	0.85 ml
1.5 M TCA	–	0.05 ml
Enzyme sample	0.05 ml	0.05 ml
10 min incubation at 37 °C (water bath)		
1.5 M TCA	0.05 ml	–
0.05 M Tris/HCl pH 8.6	–	0.05 ml
5 min centrifugation		
0.5 ml of the clear supernatant		
7.0 ml H_2O		
1.0 ml Nessler's Reagent		
10 min at room temperature		
Record absorption at 480 nm		

One unit is defined as the enzyme amount releasing 1 μmol NH_3/min at 37 °C, pH 8.6. Prepare an ammonium sulfate standard curve for quantification of the ammonia concentration.

Calculation

$$\text{units mg}^{-1} = \frac{\mu\text{mol NH}_3}{10\ \text{min} \times \text{mg enzyme}}$$

References

Derst, C., Henseling, J., and Röhm, K.H. (2000) *Protein Sci.*, **9**, 2009–2017.
Kushoo, A. *et al.* (2004) *Protein Expr. Purif.*, **38**, 29–36.
Sult, H.M. and Herbut, P.A. (1970) *J. Biol. Chem.*, **240**, 2234–2242.
Wriston, J.C. and Yellin, T.O. (1973) *Adv. Enzymol.*, **39**, 185–248.

3.3.3.25 Glutaminase, EC 3.5.1.2

SN: L-glutamine amidohydrolase

$$\text{L-glutamine} + H_2O \rightarrow \text{L-glutamate} + NH_3$$

Enzyme from *Escherichia coli*: M_r 90 000, K_m (mM): 0.42 (L-glutamine), 2.9 (L-glutamate), enzyme from rat mitochondria: M_r 260 000, homotetramer (4 × 63 200), K_m (mM): 2.0 (L-glutamine), pH optimum 8.6.

A. Determination of Ammonia with Nessler's Reagent

Assay solutions

0.1 M acetate buffer pH 4.9 (sodium acetate, $M_r = 82.0$, dissolve 8.2 g in 800 ml H_2O, adjust to pH 4.9 with concentrated acetic acid, fill up to 1 l)

Substrate/acetate pH 4.9 (dissolve 0.82 g sodium acetate and 1.17 g L-glutamine, $M_r = 146.2$, in 80 ml H_2O, adjust to pH 4.9 with concentrated acetic acid, bring to 100 ml with H_2O)

15% TCA (trichloroacetic acid, $M_r = 163.4$)

Nessler's reagent

1 M NH_4Cl ($M_r = 53.5$, 0.53 g in 10 ml H_2O), for standard curve

Procedure

	Concentration
0.25 ml substrate/acetate pH 4.9	40/50 mM
0.25 ml enzyme sample.	
Incubate 30 min, 37 °C	
0.5 ml 15% TCA	
Centrifuge, 5000 rpm, 5 min	
0.5 ml supernatant	
7.0 ml H_2O	
1.0 ml Nessler's reagent	

Read absorption at 480 nm against a blank with 0.1 M acetate buffer instead of the enzyme sample. Quantification with a standard curve prepared with different amounts of NH_4Cl.

B. pH-Stat Assay

Assay solutions

0.1 M L-glutamine ($M_r = 146.2$, 14.6 gl^{-1})

0.33 M KCl ($M_r = 74.6$, 24.6 g in 1 l H_2O)

0.01 M EDTA (ethylenediaminetetraacetic acid, $M_r = 292.2$; 29.2 mg in 10 ml)

1% BSA (bovine serum albumin, 100 mg in 10 ml)

5 mM HCl, as titrant

Assay mixture

Components	*Concentration*
6.1 ml 0.33 M KCl	200 mM
3.0 ml 0.1 M L-glutamine	30 mM
0.2 ml 0.1 M EDTA	0.2 mM
0.5 ml 1% BSA	0.05%

Procedure

Fill the automatic burette with 5 mM HCl

Add 0.98 ml assay mixture and

0.02 ml enzyme solution into the sample compartment of the pH stat

Keep the pH at 5.0 by the pH stat, 25 °C

Record the volume per unit time of the titrant added by the autotitrator for about 10 min

References

Arwadi, M.S.M. and Newsholme, E.A. (1984) *Biochem. J.*, **217**, 289–296.
Hartman, S.C. (1968) *J. Biol. Chem.*, **243**, 853–863.
Hartman, S.C. (1971) in *The Enzymes*, 3rd edn, vol. 4 (ed. P.D. Boyer), Academic Press, New York, London, pp. 79–100.

3.3.3.26 Urease, EC 3.5.1.5

SN: Urea amidohydrolase

$$\text{urea} + H_2O + 2H^+ \rightarrow CO_2 + 2NH_4^+$$

Nickel-containing enzyme; mammalian enzyme (bovine): M_r 130 000, K_m 0.83 mM (urea), pH optimum 8: *Klebsiella aerogenes*, K_m (mM):2.9 (urea), pH optimum 7.8. The release of ammonia can be determined either directly with the aid of a pH stat or by a photometric assay system coupled with the glutamate dehydrogenase reaction.

A. pH Stat Assay

Assay solutions

0.2 M urea ($M_r = 60.0$; 1.2 g in 100 ml H_2O, adjust to pH 6.1)

0.1 M HCl, standard solution

Urease solution (1 mg ml^{-1})

Procedure

Fill the automatic burette with 0.1 N HCl

Give 9 ml 0.2 M urea and

0.2 ml urease into the sample compartment of the pH stat

Keep the pH at 6.1 by the pH stat

Record the volume per unit time of the titrant added by the autotitrator for about 10 min

B. Photometric Assay

$$\text{urea} + H_2O + 2H^+ \rightarrow CO_2 + 2NH_4^+$$

$$2NH_4^+ + 2\alpha\text{-oxoglutarate} + 2NADH \xrightarrow{\text{GluDH}} 2\ \text{glutamate} + 2\ NAD^+ + 2H_2O$$

Assay solutions

0.1 M potassium phosphate pH 7.6

2.0 M urea ($M_r = 60.0$; 12 g in 100 ml 0.1 M potassium phosphate pH 7.6)

25 mM ADP (disodium salt, $M_r = 471.2$; 118 mg in 10 ml 0.1 M potassium phosphate pH 7.6)

10 mM NADH (disodium salt, $M_r = 709.4$; 71 mg in 10 ml 0.1 M potassium phosphate pH 7.6)

0.1 M α-oxoglutarate (monosodium salt, $M_r = 168.1$; 168 mg in 10 ml 0.1 M potassium phosphate pH 7.6)

Glutamate dehydrogenase from bovine liver (GluDH, 500 IU ml^{-1} in 0.1 M potassium phosphate pH 7.6)

Urease (1 mg ml^{-1}, dilute to 0.1 U/ml in 0.1 M potassium phosphate pH 7.6 prior to use)

Assay mixture

Components	*Concentration*
8.6 ml 0.1 M potassium phosphate pH 7.6	86 mM
0.3 ml 25 mM ADP	0.75 mM
0.2 ml 10 mM NADH	0.2 mM
0.1 ml 0.1 M α-oxoglutarate	1.0 mM
0.3 ml 2.0 M urea	60 mM
0.3 ml GluDH	15 IU ml^{-1}

Procedure

0.98 ml assay mixture

0.02 ml urease

Record the absorbance decrease at 340 nm, 25 °C, $\varepsilon_{340} = 6.3 \times 10^3$ l mol^{-1} cm^{-1}.

References

Mahadevan, S., Sauer, F.D., and Erfle, J.D. (1977) *Biochem. J.*, **163**, 495–501.
Martin, P.R. and Hausinger, R.P. (1992) *J. Biol. Chem.*, **267**, 20024–20027.
Sumner, J.B. and Hand, D.B. (1928) *J. Biol. Chem.*, **76**, 149–162.
Varner, J.E. (1960) in *The Enzymes*, 2nd edn, vol. 4 (eds P.D. Boyer, H. Lardy, and K. Myrbäck), Academic Press, New York, pp. 247–256.

3.3.3.27 **Adenosinetriphosphatase, EC 3.6.1.3**

SN: ATP phosphohydrolase, NTPase/helicase, ATPase

$$ATP + H_2O \rightarrow ADP + P_i$$

ATPase activities are connected with membrane components, such as the oxidative phosphorylation or transport systems. The nonsoluble activity must be measured in cell homogenates or membrane suspensions. Requirement for Mg^{2+}, M_r 67 000 (*Escherichia coli*), K_m 0.1 mM (ATP, hepatitis C virus), pH optimum 7 (human).

Assay solutions

0.1 M TES/Tris pH 7.5 (TES, *N*-tris(hydroxymethyl)methyl-2-aminoethane-sulfonic acid, M_r = 229.3; 3.44 g in 100 ml, adjust to pH 7.5 with 0.2 M TRIS {tris(hydroxymethyl)aminomethane, M_r 121.1; 4.84 g in 200 ml} and fill up to 200 ml)

0.1 M ATP (disodium salt trihydrate, M_r = 605.2; 605 mg in 10 ml)

0.1 M $MgCl_2$ ($MgCl_2 \cdot 6H_2O$, $M_r = 203.3$; 203 mg in 10 ml)

BSA solution, 100 mg in 100 ml

10% (w/v) sodium dodecylsulfate (SDS)

Assay mixture

Components	*Concentration*
7.5 ml 0.1 M TES/Tris pH 7.5	75 mM
1 ml 0.1 M $MgCl_2$	10 mM
0.5 ml 0.1 M ATP	5.0 mM
1 ml BSA	0.1 mg ml^{-1}

Procedure

Incubate 1 ml assay mixture with samples of 5–20 μl of the membrane suspension under gentle shaking at 37 °C

Add 1 ml SDS solution after a distinct time (1, 5, 10 min, depending on the enzyme activity) to stop the reaction

Measure inorganic phosphate formed by the phosphate determination method (cf. Section 3.4.2)

The *enzyme activity* is defined as 1 μmol inorganic phosphate formed per minute.

References

Borowski, P. *et al.* (2003) *Eur. J. Biochem.*, **270**, 1645–1653.
Gelfand, V.I., Gyoeva, F.K., Rosenblat, V.A., and Shanina, N.A. (1978) *FEBS Lett.*, **88**, 197–200.
Martin, S.S. and Senior, A.E. (1980) *Biochim. Biophys. Acta*, **602**, 401–418.
Takahashi, K. *et al.* (2006) *J. Biol. Chem.*, **281**, 10760–10768.
Vineyard, D., Patterson-Ward, J., and Lee, I. (2006) *Biochemistry*, **45**, 4604–4610.

3.3.4
Lyases, EC 4

3.3.4.1 Pyruvate Decarboxylase, EC 4.1.1.1

SN: 2-oxo-acid carboxy-lyase (aldehyde-forming). PDC

$$\text{pyruvate} \rightarrow \text{acetaldehyde} + CO_2$$

Thiamine diphosphate and Mg^{2+} as cofactors; bacterial enzyme (*Zymomonas mobilis*): M_r 240 000, homotetramer (4 × 59 000), K_m 0.3 mM (pyruvate), pH optimum 6. The carbon dioxide released can be determined manometrically or with $^{14}C1$-pyruvate as substrate. Here, an assay coupled to ADH (EC 1.1.1.1) is described:

$$\text{acetaldehyde} + \text{NADH} + \text{H}^+ \xrightarrow{\text{ADH}} \text{ethanol} + \text{NAD}^+$$

Assay solutions

0.2 M citrate buffer pH 6.0 (prepare 0.2 M citric acid, $M_r = 192.1$, 38.4 g in 1 l, and 0.2 M trisodium citrate, $M_r = 258.1$, 51.6 g in 1 l, measure pH in the salt solution, and adjust to pH 6.0 with the citric acid solution)

1 M pyruvate (sodium salt, $M_r = 110.0$; 1.1 g in 10 ml H_2O)

0.01 M thiamin diphosphate (ThDP, cocarboxylase, $M_r = 460.8$; 46.1 mg in 10 ml H_2O)

0.1 M $MgCl_2$ ($MgCl_2 \cdot 6H_2O$, $M_r = 203.3$; 203 mg in 10 ml)

0.01 M NADH (disodium salt, $M_r = 709.4$; 71 mg in 10 ml)

ADH from yeast (20 mg ml^{-1})

Assay mixture

Components	*Concentration (mM)*
8.9 ml 0.2 M citrate buffer pH 6.0	178
0.3 ml 0.01 M NADH	0.3
0.3 ml 1 M pyruvate	30
0.20 ml 0.01 M ThDP	0.2
0.10 ml 0.1 M $MgCl_2$	1.0
0.01 ml ADH	–

Procedure

0.98 ml assay mixture

0.02 ml enzyme sample

Record the absorbance decrease at 340 nm, 25 °C, $\varepsilon_{340} = 6300\,l\,mol^{-1}\,cm^{-1}$.

References

Bergmeyer, H.U. (1983) *Methods of Enzymatic Analysis*, 3rd edn, vol. 2, Verlag Chemie, Weinheim, pp. 302–303.

Neale, A.D. *et al.* (1987) *J. Bacteriol.*, **169**, 1024–1028.

Ullrich, J., Wittorf, J.H., and Gubler, C.J. (1966) *Biochim. Biophys. Acta*, **113**, 595–604.

Utter, M.F. (1961) in *The Enzymes*, 2nd edn, vol. 5 (eds P.D. Boyer, H. Lardy, and K. Myrbäck), Academic Press, New York, pp. 319–340.

3.3.4.2 Glutamate Decarboxylase, EC 4.1.1.15

SN: L-glutamate 1-carboxy-lyase (4-aminobutanoate-forming), GAD, aspartate 1-decarboxylase

$$\text{L-glutamate} \rightarrow \text{4-aminobutanoate} + CO_2$$

Mammalian enzyme (human): M_r 140 000, dimer (2 × 67 000), cofactor: pyridoxal 5-phosphate, K_m (mM): 1.28 (L-glutamate), pH optimum 7.4. The following assay measures the release of $^{14}CO_2$ from L-[1-^{14}C]glutamic acid.

Assay solutions

0.1 M potassium phosphate pH 7.0

L-glutamic acid solution, containing 50 mM L-glutamate (83.6 mg monosodium salt, M_r = 169.1) and 1 μCi L-[1-^{14}C]glutamic acid in 10 ml 0.1 M potassium phosphate pH 7.0

0.02 M AET (2-aminoethylisothiouronium bromide hydrobromide, M_r = 281.0, 54.2 mg in 10 ml)

0.02 M EDTA (ethylenediaminetetraacetic acid, M_r = 292.2; 58.4 mg in 10 ml)

5 mM PLP (pyridoxal 5-phosphate, M_r = 247.1; 12 mg in 10 ml potassium phosphate pH 6.8)

0.02 M DTT (dithiothreitol, Cleland's reagent, M_r = 154.2, 31 mg in 10 ml)

Hyamine hydroxide

Scintillation fluid

0.25 M H_2SO_4

Assay mixture

Components	*Concentration*
0.6 ml L-glutamic acid solution	30 mM, 0.06 μCi ml^{-1}
0.1 ml 0.02 M AET	2 mM
0.05 ml 0.02 M EDTA	1 mM
0.1 ml PLP	0.5 mM
0.05 ml 0.02 M DTT	1 mM

Procedure

The reaction is carried out in disposable tubes (~12 × 80 mm), tightly closed by a rubber stopper. CO_2 released is adsorbed either on a scintillation pad, containing 0.25 mmol KOH, fixed ~6 cm above the reaction mixture, or by 0.2 ml hyamine hydroxide, placed in a separate, open central well in the reaction tube.

0.09 ml reaction mixture

0.01 ml enzyme sample

Incubate 60 min at 37 °C in a shaking temperature bath

Stop the reaction by injecting 0.2 ml 0.25 M H_2SO_4 through the stopper into the reaction mixture

60 min 37 °C to establish complete CO_2 release and adsorption

Transfer the scintillation pad, respectively the central well with hyamine into 5 ml scintillation fluid, and count radioactivity in a scintillation counter

References

Alberts, R.W. and Brady, R.O. (1959) *J. Biol. Chem.*, **237**, 926–928.
Blindermann, J.M. *et al.* (1978) *Eur. J. Biochem.*, **86**, 143–152.
Hao, R. and Schmit, J.C. (1991) *J. Biol. Chem.*, **266**, 5135–5139.
Nathan, B. *et al.* (1994) *J. Biol. Chem.*, **269**, 7249–7254.
Tong, J.C. *et al.* (2002) *J. Biotechnol.*, **97**, 183–190.

3.3.4.3 Aldolase, EC 4.1.2.13

SN: D-fructose-1,6-bisphosphate D-glyceraldehyde-3-phosphate-lyase (glycerone-phosphate-forming), fructose-bisphosphate aldolase (ALDC).

$$\text{D-fructose-1,6-bisphosphate} \rightleftharpoons \text{glyceronephosphate} + \text{D-glyceraldehyde-3-phosphate}$$

Enzyme from human liver: M_r 158 000, dimer, K_m (mM): 0.0027 (D-fructose-1,6-bisphosphate). 0.88 (D-fructose-1-phosphate). D-Glyceraldehyde-3-phosphate forms a hydrazone with hydrazine, absorbing at 240 nm.

Assay solutions

3.5 mM hydrazine sulfate in 1 mM EDTA, pH 7.5 (dissolve 455 mg hydrazine sulfate, $M_r = 130.1$, and 372 mg EDTA · Na_2 · $2H_2O$, $M_r = 372.2$, in 1 l H_2O, adjust to pH 7.5 with 1 N NaOH)

12 mM fructose-1,6-bisphosphate (FDP-Na_3H, $M_r = 406.1$; 487 mg in 100 ml H_2O, adjust to pH 7.5)

Aldolase solution, about 1 IU ml^{-1} (prepare just before use)

10 mM D,L-glyceraldehyde-3-phosphate ($M_r = 170.1$, 17 mg in 10 ml)

Assay mixture

Components	*Concentration (mM)*
3.0 ml 12 mM fructose-1,6-bisphosphate	3.6
6.0 ml 3.5 mM hydrazine sulfate solution	2.1

Procedure

0.9 ml assay mixture

0.1 ml aldolase solution

Record the absorption change at 240 nm, 25 °C for about 10 min. For the blank fructose-1,6-bisphosphate is substituted by water

For quantification, prepare a standard curve with aliquots of 0.6 ml 3.5 mM hydrazine sulfate solution and varying concentrations of D,L-glyceraldehyde-3-phosphate in 0.4 ml

References

Abraham, M. *et al.* (1985) *Appl. Biochem. Biotechnol.*, **11**, 91–100.
Jagannathan, V., Sing, K., and Damodaran, M. (1956) *Biochem. J.*, **63**, 94–105.
Malay, A.D. *et al.* (2002) *Arch. Biochem. Biophys.*, **408**, 295–304.
Rutter, J.W. (1975) in *The Enzymes*, 2nd edn, vol. 5 (eds P.D. Boyer, H. Lardy, and K. Myrbäck), Academic Press, New York, pp. 341–366.

3.3.4.4 Anthranilate Synthase, EC 4.1.3.27

SN: chorismate pyruvate-lyase (amino-accepting, anthranilate-forming), chorismate lyase

$$\text{chorismate} + \text{L-glutamine} \rightarrow \text{anthranilate} + \text{pyruvate} + \text{L-glutamate}$$

Key enzyme in the biosynthesis of aromatic amino acids. Bacterial enzyme (*Serratia marcescens*): M_r 141 000, tetramer (2 × 21 000, 2 × 60 000), K_m (mM): 0.015 (chorismate), 0.5 (glutamine), inhibited by L-tryptophan, pH optimum 8.5. Chorismate is unstable and should be stored at −70 °C. A procedure for isolation of chorismate is described in Gibson (1964).

Assay solutions

10 mM chorismate (M_r = 361.5, barium salt, 18.1 mg in 5 ml)

0.2 M L-glutamine (M_r = 146.2, 292 mg in 10 ml)

0.1 M $MgCl_2$ (hexahydrate, M_r = 203.3; 203 mg in 10 ml)

0.1 M dithioerythritol (DTE, M_r = 154.2, 154 mg in 10 ml)

0.01 M anthranilic acid ($M_r = 137.1$, 13.7 mg in 10 ml)

Assay mixture and procedure

	Concentration (mM)
1.7 ml 50 mM potassium phosphate pH 7.4	42.5
0.1 ml 10 mM magnesium chloride	5.0
0.1 ml 0.2 M L-glutamine	10
0.02 ml 10 mM chorismate	0.1
0.03 ml 0.1 M DTE	1.5
0.05 ml enzyme sample	–

Measure the fluorescence emission at 400 nm, excitation is at 325 nm, 37 °C. For quantification, determine the fluorescence of various dilutions of the anthranilic acid solution within the concentration range 1–100 μM.

References

Crawford, I.P. (1987) *Meth. Enzymol.*, **142**, 300–307.
Gibson, F. (1964) *Biochem. J.*, **90**, 256–261.
Goto, Y., Zalkin, H., Keim, P.S., and Heinrikson, R.L. (1976) *J. Biol. Chem.*, **251**, 941–949.
Queener, S.W., Queener, S.F., Meeks, J.R., and Gunsalus, I.C. (1973) *J. Biol. Chem.*, **248**, 151–161.

3.3.4.5 Carbonic Anhydrase, EC 4.2.1.1

SN: carbonate hydro-lyase (carbon dioxide-forming), carbonate dehydratase, CA

$$H_2CO_3 \rightarrow CO_2 + H_2O$$

Zn-containing glycoprotein; mammalian enzyme (human, bovine stomach): M_r 44 000, K_m (mM) 2.9 (4-nitrophenylacetate), 2.3 (CO_2), pH optimum 7.6 (9.5 for hydrolysis of 4-nitrophenylacetate). An assay for a pH meter or pH stat and an assay to test the esterase activity of carbonic anhydrase with 4-nitrophenylacetate are described.

A. pH-Stat Assay

Assay solutions

0.02 M Tris–HCl, pH 8.0

CO_2-saturated H_2O (pass CO_2 gas through 200 ml of ice-cold water for 30 min)

Procedure

All solutions must be chilled to 0–4 °C

	Sample	Blank
0.02 M Tris-HCl pH 8.0	6 ml	6 ml
CO_2 water (ice cold)	4 ml	4 ml
	–	Measure pH and determine time T_0 required for the pH to drop from 8.3 to 6.3
Enzyme sample	0.1 ml	–
	Measure pH and determine time T required for the pH to drop from 8.3 to 6.3	–

Calculation

A unit is defined as $2 \times (T_0 - T)/T$:

$$\text{units mg}^{-1} = \frac{2 \times (T_0 - T)}{T \times \text{mg enzyme}}$$

B. Esterase Assay with 4-Nitrophenylacetate

4-nitrophenylacetate → 4-nitrophenol + acetate

Assay solutions

0.1 M HEPES pH 7.2

3 mM 4-nitrophenylacetate (M_r = 181.1, dissolve 27.2 mg in 1 ml acetone, dilute rapidly under vigorous shaking with H_2O to 50 ml)

Procedure

	Concentration (mM)
0.5 ml 0.1 M HEPES pH 7.2	50
0.5 ml 3 mM 4-nitrophenylacetate	1.5

Measure the absorption increase at 405 nm, $\varepsilon_{405} = 18\,500\,\mathrm{l\,mol^{-1}\,cm^{-1}}$. The reaction can also be followed at 348 nm, the isobestic point of *p*-nitrophenol and the conjugated nitrophenolate ion, difference absorption coefficient: $\varepsilon_{348} = 5000\,\mathrm{l\,mol^{-1}\,cm^{-1}}$.

References

Demir, Y. *et al.* (2005) *J. Enzyme Inhib. Med. Chem.*, **20**, 75–80.
Engstrand, C., Jonsson, B.H., and Lindskog, S. (1995) *Eur. J. Biochem.*, **229**, 696–702.
Keilin, D. and Mann, T. (1939) *Nature*, **144**, 442–443.
Murakami, H. and Sly, W.S. (1987) *J. Biol. Chem.*, **262**, 1382–1388.
Ulmasov, B. *et al.* (2000) *Proc. Natl. Acad. Sci. U.S.A.*, **97**, 14212–14217.
Verpoorte, J.A., Meiita, S., and Edsall, J.T. (1967) *J. Biol. Chem.*, **242**, 4221–4229.

3.3.4.6 Fumarase, EC 4.2.1.2

S-malate hydro-lyase (fumarate-forming), fumarate hydratase

$$\text{L-malate} \rightleftharpoons \text{fumarate} + H_2O$$

Fe enzyme, exists in *Escherichia coli* in three forms, FUMA, FUMB, FUMC dependent on the oxygen level. FUMA : M_r 120 000, dimer, 2 × 61 000, K_m (mM): 1.1 (malate), 0.15 (fumarate), FUMC, M_r 200 000, tetramer (4 × 50 000), K_m (mM): 2.9 (malate) 0.39 (fumarate), pH optimum 8.5.

Assay solutions

0.1 M potassium phosphate pH 7.5

0.05 M L-malate solution (L-malic acid, $M_r = 134.1$; suspend 134 mg in 10 ml 0.1 M potassium phosphate pH 7.5, neutralize with 3.5 ml 1 N NaOH and fill up to 20 ml with 0.1 M potassium phosphate pH 7.5)

0.1% serum albumin (w/v) in 0.1 M potassium phosphate pH 7.5

Procedure

0.98 ml 0.05 M L-malate solution

0.02 ml fumarase solution (diluted with 0.1% serum albumin)

Follow the absorption increase at 240 nm at 25 °C, $\varepsilon_{240} = 2.44 \times 10^3\,\mathrm{l\,mol^{-1}\,cm^{-1}}$.

References

Bergmeyer, H.U. (1983) *Methods of Enzymatic Analysis*, 3rd edn, vol. 2, Verlag Chemie, Weinheim, pp. 189–191.

Hill, R.L. and Bradshaw, R.A. (1969) *Meth. Enzymol.*, **13**, 91–99.
Hill, R.L. and Teipel, J.W. (1971) in *The Enzymes*, 3rd edn, vol. 5 (ed. P.D. Boyer), Academic Press, New York, London, p. 539.
Tseng, C.P. *et al.* (2001) *J. Bacteriol.*, **183**, 461–467.
Woods, S.A., Schwartzbach, S.D., and Guest, J.R. (1988) *Biochim. Biophys. Acta*, **954**, 14–26.

3.3.5 Isomerases, EC 5

3.3.5.1 Glucose/Xylose Isomerase EC 5.3.1.5

SN: D-xylose aldose–ketose isomerase, D-xylose ketol isomerase

$$\text{D-xylose} \rightleftharpoons \text{D-xylulose}$$

$$\text{D-glucose} \rightleftharpoons \text{D-fructose}$$

Glucose isomerase is an intensely technical applied enzyme, due to its capacity to convert glucose to the more valuable fructose, although its natural function is the isomerization of xylose, an important nutrient for bacteria. Xylose, produced by digestion of hemicelluloses, xylans, and arabans, is introduced into the pentose phosphate pathway after isomerization to xylulose. Because of its analogous conformation, glucose is isomerized by the same enzyme to fructose but with five times lower efficiency. Bacterial enzyme, dependent on divalent cations (Mn^{2+}, Co^{2+}, Mg^{2+}), inhibited by EDTA, *Thermus aquaticus*, M_r 196 000 homotetramer (4 × 50 000), K_m (mM): 8 (D-xylulose), 15, (D-xylose), 93 (D-glucose), K_d (mM): 0.0009 (Mn^{2+}), 0.001 (Co^{2+}), pH optimum 5.5; *Bacillus* sp. M_r 120 000, homodimer (2 × 58 000), K_m (mM): 0.076 (D-xylose), pH optimum 7; *Arthrobacter* sp.: K_m (mM): 3.3 (D-xylose), 110 (D-fructose), 230 (D-glucose).

A. D-Xylose Isomerase Assay

Assay solutions

0.2 M TES/NaOH pH 8.0 (*N*-tris[hydroxymethyl]-methyl-2-aminomethane sulfonic acid, M_r = 229.3; 9.2 g in 100 ml, adjust with 1 M NaOH to pH 8.0, fill up to 200 ml with H_2O)

0.4 M D-xylose (M_r = 150.1; 1.2 g in 20 ml)

0.03 M $MnSO_4$ (monohydrate, M_r = 169.0; 57 mg in 10 ml)

1.5% cysteine · HCl solution (monohydrate M_r = 175.6; dissolve 1.5 g in 100 ml H_2O)

0.12% carbazole solution (M_r = 167.2; 120 mg in 100 ml in ethanol)

70% sulfuric acid

10 mM D-xylulose (M_r = 150.1, 7.5 mg in 5 ml) for a standard curve

Assay mixture

Components	*Concentration (mM)*
0.24 ml 0.2 M TES/NaOH pH 8.0	48
0.25 ml 0.4 M D-xylose	100
0.01 ml 0.03 M $MnSO_4$	0.3

Procedure

20 µl assay mixture

20 µl enzyme solution (or 20 µl TES buffer for the blank)

- Incubate at the assay temperature (25 or 50 °C, dependent on the enzyme sample) for 15 min in tightly closed reaction vessels
- Stop by chilling on ice

40 µl cysteine · HCl

40 µl of carbazole solution

1.2 ml 70% sulfuric acid starts the development of the color

10 min, room temperature

Measure the absorption against the blank at 546 nm.

Remark: The color intensity is not constant, therefore the 10 min interval for color development must be observed carefully and all samples must be read exactly after the same time interval. The absorption of the blank should be about 0.2, if it exceeds an absorption of 0.3 the assay reagents should be prepared afresh.

For quantification, prepare a calibration curve with D-xylulose between 0 and 10 µM.

B. D-Xylose Isomerase Microplate Assay

Assay solutions

0.2 M TES/NaOH pH 8.0 (*N*-tris[hydroxymethyl]-methyl-2-aminomethane sulfonic acid, $M_r = 229.3$; 9.2 g in 100 ml, adjust with 1 M NaOH to pH 8.0, fill up to 200 ml with H_2O)

0.4 M D-xylose ($M_r = 150.1$; 1.2 g in 20 ml)

0.03 M $MnSO_4$ (monohydrate, $M_r = 169.0$; 57 mg in 10 ml)

Solution A: 0.05% resorcinol (M_r = 110.1; 50 mg in 100 ml ethanol)

Solution B: $FeNH_4(SO_4)_2 \cdot 12H_2O$ (M_r = 482.2; 216 mg in 1 l 37%. HCl)

Assay mixture

Components	*Concentrations (mM)*
0.24 ml 0.2 M TES/NaOH pH 8.0	48
0.25 ml 0.4 M D-xylose	100
0.01 ml 0.03 M $MnSO_4$	0.3

Procedure

Place 20 µl assay mixture and

20 µl enzyme solution (or 20 µl TES buffer for the blank) in a 96-well microplate

Incubate for 15 min, 50 °C

0.15 ml freshly prepared mixture (1 : 1, v/v) of solution A and solution B

Incubate at 80 °C for 40 min

Measure the absorption with a microplate reader at 630 nm

C. D-Glucose Isomerase Assay

Assay solutions

0.2 M TES/NaOH pH 8.0 (*N*-tris[hydroxymethyl]-methyl-2-aminomethanesulfonic acid, M_r = 229.3; 9.2 g in 100 ml, adjust with 1 M NaOH to pH 8.0, fill up to 200 ml with H_2O)

2.0 M D-glucose (M_r = 180.2; 7.2 g in 20 ml)

0.03 M $CoCl_2$ (hexahydrate, M_r = 237.9; 71 mg in 10 ml)

1.5% cysteine · HCl (solution cysteine · HCl monohydrate M_r = 175.6; dissolve 1.5 g in 100 ml H_2O)

0.12% carbazole solution (M_r = 167.2; 120 mg in 100 ml in ethanol)

70% sulfuric acid

0.01 M D-fructose (M_r = 180.2, 18.2 mg in 10 ml), for calibration curve

Assay mixture and procedure

	Concentration (mM)
0.24 ml 0.2 M TES/NaOH pH 8.0	48
0.25 ml 2.0 M D-glucose	500
0.01 ml 0.03 M $CoCl_2$	0.3

The assay procedure is the same as for xylose isomerase (assay A), except that – after the enzyme reaction and the addition of cysteine hydrochloride, carbazole, and sulfuric acid – the mixture must remain for 30 min for color development; the absorption is measured at 560 nm. The calibration curve must be prepared with D-fructose.

D. D-Glucose Isomerase Microplate Assay

Assay solutions

0.2 M TES/NaOH pH 8.0 (*N*-tris[hydroxymethyl]-methyl-2-aminomethane sulfonic acid, $M_r = 229.3$; 9.2 g in 100 ml, adjust with 1 M NaOH to pH 8.0, fill up to 200 ml with H_2O)

2.0 M D-glucose ($M_r = 180.2$; 7.2 g in 20 ml)

0.03 M $CoCl_2$ (hexahydrate, $M_r = 237.9$; 71 mg in 10 ml)

Solution A: 0.05% resorcinol ($M_r = 110.1$; 50 mg in 100 ml ethanol)

Solution B: $FeNH_4(SO_4)_2 \cdot 12H_2O$ ($M_r = 482.2$), 216 mg in 1 l 37% HCl

Assay mixture

Components	*Concentration (mM)*
0.24 ml 0.2 M TES/NaOH pH 8.0	48
0.25 ml 2.0 M D-glucose	500
0.01 ml 0.03 M $CoCl_2$	0.3

The assay procedure is the same as for xylose isomerase (assay B), with the exception that the absorption is measured at 490 nm.

References

Dische, Z. and Borenfreund, E. (1951) *J. Biol. Chem.*, **192**, 583–587.

Kwon, H.J. *et al.* (1987) *Agric. Biol. Chem.*, **51**, 1983–1989.
Lehmacher, A. and Bisswanger, H. (1990) *Biol. Chem. Hoppe-Seyler*, **371**, 527–536.
Rangarajan, D. and Hartley, B.S. (1992) *Biochem. J.*, **283**, 223–233.
Schenk, M. and Bisswanger, H. (1998) *Enzyme Microb. Technol.*, **22**, 721–723.

3.3.5.2 Phosphoglucomutase, EC 5.4.2.2

SN: α-D-glucose-1,6-phosphoglucomutase, PGM

$$\text{D-glucose-1-phosphate} \rightleftharpoons \text{D-glucose-6-phosphate}$$

Bacterial enzyme (*Bacillus subtilis*): M_r 130 000, dimer (2 × 60 000), K_m (mM): 0.012 (α-D-glucose), pH optimum 8.3. Fluorimetric assay, coupled with glucose 6-phosphate dehydrogenase (EC 1.1.1.49)

$$\text{D-glucose-6-phosphate} + \text{NADP}^+ \rightleftharpoons \text{6-phosphogluconate} + \text{NADPH} + \text{H}^+$$

Assay solutions

50 mM MOPS buffer pH 7.4, 1 mM DTT (MOPS, 3-*N*-morpholinopropanesulfonic acid, M_r = 209.3, dissolve 10.5 g in 0.8 l H_2O, add 154 mg DTT (dithiothreitol, Cleland's reagent, M_r = 154.2), adjust to pH 7.4 with 1 N NaOH, fill up to 1 l with H_2O)

0.1 M NAD (M_r = 663.4; 0.663 g in 10 ml 50 mM MOPS buffer pH 7.4)

0.1 M $MgSO_4$ (M_r = 120.4, 120 mg in 10 ml H_2O)

0.1 M D-glucose-1-phosphate (disodium salt, hydrate, M_r = 304.1, 304 mg in 10 ml H_2O)

0.5 mM D-glucose-1,6-bisphosphate (potassium salt, hydrate, M_r = 340.1, 1.7 mg in 10 ml H_2O)

0.01 M NADH (disodium salt, M_r = 709.4; 71 mg 10 ml^{-1} H_2O), for standard curve

Assay mixture

Components	*Concentration*
9.52 ml 50 mM MOPS buffer pH 7.4	48 mM
0.1 ml 0.1 M NAD	1.0 mM
0.02 ml 0.1 M D-glucose-1-phosphate	0.2 mM
0.01 ml 0.5 mM D-glucose-1,6-bisphosphate	0.5 μM
0.15 ml 0.1 M $MgSO_4$	1.5 mM

Procedure

0.98 ml assay mixture

0.02 ml enzyme sample

Follow the fluorescence increase at 430 nm (excitation 340 nM), for quantification prepare a standard curve with NADH.

References

Maino, V.C. and Young, F.E. (1974) *J. Biol. Chem.*, **249**, 5169–5175.
Naught, L.E. *et al.* (2003) *Biochemistry*, **42**, 9946–9951.
Regni, C. *et al.* (2006) *J. Biol. Chem.*, **281**, 15564–15571.

3.3.6 Ligases (Synthetases) EC 6

3.3.6.1 Tyrosine-tRNA Ligase, EC 6.1.1.1

SN: L-tyrosine:tRNA Tyr–ligase (AMP-forming)

$$\text{ATP} + \text{L-tyrosine} \rightarrow \text{AMP} + \text{PP}_i + \text{L-tyrosyl-tRNA-Tyr}$$

Assays for this enzyme are described exemplarily for the group of amino acid tRNA ligases. Mammalian enzyme (human): M_r 130 000, dimer; K_m (mM) 3 (ATP), 0.034 (tyrosine), pH optimum 7.5; *Escherichia coli*: M_r 90 000, dimer, K_m 0.5 mM (ATP), 0.012 mM (tyrosine), 0.52 μM (tRNA–Tyr), pH optimum 7.6.

A. Fluorimetric Assay This assay depends on the decrease of intrinsic tyrosine fluorescence due to the reaction and is thus, specific for tyrosine–tRNA ligase. Pyrophosphatase, cleaving PP_i serves to support the reaction for the product side.

Assay solutions

0.15 M Tris/HCl, 0.15 M KCl buffer pH 7.5 (tris(hydroxymethyl)amino-methane, Tris, $M_r = 121.1$; KCl, $M_r = 74.6$; dissolve 18.2 g Tris and 11.2 g KCl in 600 ml H_2O, adjust the pH with 1 M HCl to 7.5 and fill up with H_2O to 1 l)

0.1 M ATP (disodium salt, trihydrate, $M_r = 605.2$; 605 mg in 10 ml buffer)

0.1 M $MgCl_2$ ($MgCl_2 \cdot 6H_2O$, $M_r = 203.3$; 203 mg in 10 ml buffer)

2 mM L-tyrosine ($M_r = 181.2$; 36 mg in 100 ml buffer)

0.1 M DTE (1,4-dithioerythritol, $M_r = 154.3$; 154 mg in 10 ml buffer)

Inorganic pyrophosphatase (dilute to 200 unit ml^{-1} with buffer)

Assay mixture

Components	*Concentration*
6.6 ml 0.15 M Tris/HCl/KCl pH 7.5	0.15 M
1 ml 0.1 M ATP	10 mM
1 ml 0.1 M $MgCl_2$	10 mM
0.5 ml 2 mM L-tyrosine	0.1 mM
0.2 ml 0.1 M DTE	2 mM
0.5 ml inorganic pyrophosphatase	10 units

Procedure

1.95 ml assay mixture

0.4 ml enzyme sample

Measure the decrease of the fluorescence intensity (excitation 295, emission $\geq$ 320 nm) in quartz fluorescence cuvettes at 25 °C; for quantification, prepare a standard curve with various tyrosine concentrations.

B. ATP – ^{32}PP-Exchange

Assay solutions and substances

0.1 M Tris/HCl, pH 7.0

0.1 M ATP (disodium salt, trihydrate, M_r = 605.2; 605 mg in 10 ml 0.1 M Tris/HCl, pH 7.0)

0.1 M $MgCl_2$ ($MgCl_2 \cdot 6H_2O$, M_r = 203.3; 203 mg in 10 ml 0.1 M Tris/HCl, pH 7.0)

2 mM L-tyrosine (M_r = 181.2; 36 mg in 100 ml 0.1 M Tris/HCl, pH 7.0)

0.4 M sodium diphosphate (sodium pyrophosphate, $Na_4P_2O_7 \cdot 10H_2O$, M_r = 446; 17.84 g in 100 ml 15% perchloric acid)

0.1 M radiolabeled diphosphate ($^{32}PP_i$, 10^4–10^5 cpm/mol, adjust the concentration with cold sodium diphosphate, if necessary)

0.1 M DTE (1,4-dithioerythritol, M_r =154.3; 154 mg in 10 ml 0.1 M Tris/HCl, pH 7.0)

BSA (bovine serum albumine, 10 mg ml^{-1})

Charcoal (15% suspension)

Assay mixture

Components	*Concentration*
7.6 ml 0.1 M Tris/HCl pH 7.0	~0.1 M
0.2 ml 0.1 M ATP	2 mM
0.5 ml 0.1 M $MgCl_2$	5 mM
0.2 ml 0.1 M ^{32}PPi	2 mM
1.0 ml 2 mM L-tyrosine	0.2 mM
0.2 ml 0.1 M DTE	2 mM
0.1 ml BSA	0.1 mg ml^{-1}

Procedure

0.98 ml assay mixture

0.02 ml enzyme sample (~0.5 IU)

15 min, 37 °C

0.7 ml cold 0.4 M sodium pyrophosphate in 15% perchloric acid
0.1 ml 15% charcoal suspension

Filter through a glass microfiber filter (GF/A), wash five times with 5 ml portions of cold H_2O, and dry the filter.

Immerse in an appropriate scintillation liquid and count in a scintillation spectrometer.

An enzyme unit is defined as the incorporation of 1 μmol ^{32}PP into ATP in 1 min.

References

Austin, J. and First, E.A. (2002) *J. Biol. Chem.*, **277**, 14812–14820.
Austin, J. and First, E.A. (2002) *J. Biol. Chem.*, **277**, 28394–28399.
Buonocore, V. and Schlesinger, S. (1972) *J. Biol. Chem.*, **24**, 1343–1348.
Buonocore, V., Harris, M.H., and Schlesinger, S. (1972) *J. Biol. Chem.*, **247**, 4843–4849.

3.3.6.2 Glutamine Synthetase EC 6.3.1.2

SN: L-glutamate:ammonia ligase (ADP-forming)

$$\text{L-glutamate} + \text{ATP} + \text{NH}_3 \rightarrow \text{L-glutamine} + \text{ADP} + \text{P}_i$$

Enzyme from *Escherichia coli*: dodecamer (12 × 51 800), K_m (mM): 3.3 (L-glutamate), 0.2 (ATP), pH optimum 8.0. The assay is coupled with pyruvate kinase (PK) and lactate dehydrogenase (LDH)

$$\text{ADP} + \text{phosphoenolpyruvate} \overset{\text{PK}}{\rightleftharpoons} \text{pyruvate} + \text{ATP}$$

$$\text{pyruvate} + \text{NADH} + \text{H}^+ \overset{\text{LDH}}{\rightleftharpoons} \text{L-lactate} + \text{NAD}^+$$

Assay solutions

0.05 M buffer (HEPES or imidazol/HCl) pH 7.1

1 M L-glutamate (monosodium salt, $M_r = 169.1$, 1.69 g in 10 ml buffer)

0.1 M ATP (disodium salt trihydrate, $M_r = 605.2$; 605 mg in 10 ml buffer)

0.1 M PEP (phosphoenolpyruvate, tricyclohexylammonium salt, $M_r = 465.6$; 466 mg in 10 ml buffer)

0.01 M NADH (disodium salt, $M_r = 709.4$; 71 mg in 10 ml buffer)

1 M KCl ($M_r = 74.6$, 7.5 g in 100 ml H_2O)

1 M $MgCl_2$ ($MgCl_2 \cdot 6H_2O$, $M_r = 203.3$; 2.03 g in 10 ml buffer)

1 M NH_4Cl ($M_r = 53.5$, 0.53 g in 10 ml H_2O)

PK (pyruvate kinase, commercially available for various activities, prepare a solution with 300 IU 1 ml^{-1} in buffer)

LDH (lactate dehydrogenase, commercial products contain about 500 IU mg^{-1}, prepare a solution with 1500 IU ml^{-1} in buffer)

Assay mixture

Components	*Concentration*
6.8 ml 0.05 M buffer pH 7.1	45 mM
0.5 ml 0.1 M ATP	5.0 mM
0.1 ml 0.1 M PEP	1.0 mM
0.1 ml 1 M KCl	10 mM
0.5 ml 1 M $MgCl_2$	50 mM
0.4 ml 1 M NH_4Cl	40 mM
0.1 ml 10 mM NADH	0.1 mM
1.0 ml 1 M L-glutamate	100 mM
0.1 ml PK	3 IU ml^{-1}
0.2 ml LDH	30 IU ml^{-1}

Procedure

0.98 ml assay mixture

0.02 ml enzyme sample

Follow the absorption decrease at 340 nm at 25 °C, $\varepsilon_{340} = 6.3 \times 10^3$ l mol^{-1} cm^{-1}.

References

Alibhai, M. and Villafranca, J.J. (1994) *Biochemistry*, **33**, 682–686.
Ginsburg, A. *et al.* (1970) *Biochemistry*, **9**, 633–649.
Kingdon, H.S., Hubbard, J.S., and Stadtman, E.R. (1968) *Biochemistry*, **7**, 2136–2142.
Pearson, J.T. (2005) *Arch. Biochem. Biophys.*, **436**, 397–405.

3.3.7 Assays for Multienzyme Complexes

3.3.7.1 Pyruvate Dehydrogenase Complex (PDHC)

This large enzyme complex (M_r = ~5×10^6 from bacterial sources, ~8×10^6 from mammalian sources) combines three different enzyme activities catalyzing consecutive reactions: pyruvate dehydrogenase (E1p, EC 1.2.4.1), dihydrolipoamide acetyltransferase (E2p, EC 2.3.1.12), and dihydrolipoamide dehydrogenase (E3p, EC 1.8.1.4):

pyruvate + ThDP-E1p → hydroxyethyl ThDP-E1p + CO_2
hydroxyethyl ThDP-E1p + lipoyl E2p → acetyl dihydrolipoyl-E2p + ThDP-E1p
acetyl dihydrolipoyl-E2p + CoA → acetyl CoA + dihydrolipoyl-E2p
dihydrolipoyl E2p + FAD_{ox}-E3p → lipoyl E2p + FAD_{red}-E3p
FAD_{red}-E3p + NAD^+ → FAD_{ox}-E3p + NADH + H^+

---

pyruvate + NAD^+ + CoA → acetyl CoA + CO_2 + NADH + H^+

ThDP is the thiamin diphosphate cofactor of the E1p component, FAD_{ox} and FAD_{red} are the oxidized and the reduced cofactors, respectively, of E3p.

Assays for the three enzymes are described separately, while here assays for the overall activity are presented. Pyruvate dehydrogenase complexes from higher organisms possess two regulatory enzymes, the pyruvate dehydrogenase kinase which inactivates the overall activity by phosphorylating serine residues at the E1p component, and a pyruvate dehydrogenase phosphatase that reverses phosphorylation and reactivates the enzyme. Two assays for the overall reaction are described. The photometric assay observing the formation of NADH is convenient, but is disturbed in crude extracts by the LDH reaction which forms lactate and NAD from pyruvate and NADH and thus, counteracts the assay. The more laborious and less sensitive dismutation assay circumvents this problem. Alternatively, the overall activity of the pyruvate dehydrogenase complex can be determined by observing the production of CO_2 either by manometric methods, using a CO_2 electrode, or by measuring the radioactivity released from [1-^{14}C]pyruvate as a substrate.

A. Overall Activity of PDHC by NAD^+ Reduction

Assay solutions

0.1 M potassium phosphate pH 7.6

0.1 M NAD^+ ($M_r = 663.4$, free acid; 663 mg in 10 ml)

0.01 M thiamin diphosphate (ThDP, cocarboxylase, $M_r = 460.8$; 46.1 mg in 10 ml)

0.1 M $MgCl_2$ ($MgCl_2 \cdot 6H_2O$, $M_r = 203.3$; 203 mg in 10 ml)

0.1 M pyruvate (sodium pyruvate, $M_r = 110.0$; 110 mg in 10 ml)

0.1 M dithioerythritol (DTE, $M_r = 154.2$; 154 mg in 10 ml)

0.01 M coenzyme A (free acid, $M_r = 767.5$, $CoA \cdot Li_3$ $M_r = 785.4$; 23 mg in 3 ml)

Assay mixture

Components	*Concentration (mM)*
8.80 ml 0.1 M potassium phosphate pH 7.6	88
0.25 ml 0.1 M NAD^+	2.5
0.20 ml 0.01 M ThDP	0.2
0.10 ml 0.1 M $MgCl_2$	1.0
0.25 ml 0.1 M pyruvate	2.5
0.10 ml 0.1 M DTE	1.0
0.10 ml 0.01 M CoA	0.1

Procedure

0.98 ml assay mixture

0.02 ml enzyme sample

The absorption increase at 340 nm is measured at 37 °C. Absorption coefficient for NADH: $\varepsilon_{340} = 6.3 \times 10^3 \text{ l mol}^{-1} \text{ cm}^{-1}$.

Reference

Schwartz, E.R., Old, L.O., and Reed, L.J. (1968) *Biochem. Biophys. Res. Commun.*, **31**, 495–500.

B. Overall Activity of PDHC by Dismutation Assay Stopped assay, using LDH to regenerate NAD. Acetylphosphate is formed from acetyl CoA by the phosphotransacetylase (PTA) and is converted into hydroxamic acid, which forms a colored complex with Fe^{3+}.

$$\text{pyruvate} + NAD^+ + \text{CoA} \overset{\text{PDHC}}{\rightleftharpoons} \text{acetyl CoA} + CO_2 + \text{NADH} + H^+$$

$$\text{acetyl CoA} + P_i \overset{\text{PTA}}{\rightleftharpoons} \text{acetyl P} + \text{CoA}$$

$$\text{pyruvate} + \text{NADH} + H^+ \overset{\text{LDH}}{\rightleftharpoons} \text{lactate} + NAD^+$$

Assay solutions

0.1 M potassium phosphate pH 7.6

0.1 M $MgCl_2$ ($MgCl_2 \cdot 6H_2O$, $M_r = 203.3$; 203 mg in 10 ml)

0.1 M pyruvate (sodium pyruvate, $M_r = 110.0$; 110 mg in 10 ml H_2O)

0.01 M thiamin diphosphate (cocarboxylase, $M_r = 460.8$; 46.1 mg in 10 ml)

0.1 M dithioerythritol (DTE, $M_r = 154.2$; 154 mg in 10 ml)

0.01 M coenzyme A (free acid, $M_r = 767.5$, CoA $\cdot$ Li_3, $M_r = 785.4$; 23 mg in 3 ml)

0.1 M NAD^+ (free acid, $M_r = 663.4$; 663 mg in 10 ml)

phosphotransacetylase (4000 IU ml^{-1})

lactate dehydrogenase (10 000 U ml^{-1})

2 M hydroxylamine (mix equal volumes of 4 M hydroxylamine hydrochloride, $M_r = 69.49$, and 4 M KOH, $M_r = 56.11$)

$FeCl_3$ reagent:

- 100 ml 5% $FeCl_3$ (8.33 g $FeCl_3 \cdot 6H_2O$, $M_r = 270.3$, replenish to 100 ml with 0.1 N HCl)
- 100 ml 12% TCA (trichloroacetic acid, $M_r = 163.4$)
- 100 ml 3 M HCl (mix one part 37% HCl with three parts H_2O)

Mix the three solutions together.

Assay mixture

Components	*Concentrations*
4.0 ml 0.1 M potassium phosphate pH 7.6	80 mM
0.05 ml 0.1 M $MgCl_2$	1.0 mM
0.23 ml 0.1 M pyruvate	4.6 mM
0.1 ml 0.01 M ThDP	0.2 mM
0.1 ml 0.1 M DTE	2.0 mM
0.05 ml 0.01 M coenzyme A	0.1 mM
0.05 ml 0.1 M NAD^+	1.0 mM
0.01 ml phosphotransacetylase	4 IU ml^{-1}
0.01 ml LDH	10 IU ml^{-1}

Procedure

0.48 ml assay mixture

0.02 ml enzyme sample

Incubate at 30 °C for 15 min

Put the samples thereafter on ice

0.2 ml 2 M hydroxylamine.

10 min at room temperature

0.6 ml $FeCl_3$ reagent.

Centrifuge for 5 min, 5000 rpm

Measure the absorption at 546 nm against a blank with the enzyme sample substituted by buffer.

Calculation

The specific activity is expressed in special units, with reference to 1 h of reaction time.

$$\text{specific activity} = \frac{A_{546} \times 20 \times \text{assay volume (1.3 ml)} \times \text{dilution factor}}{\text{enzyme volume (ml)} \times \text{protein (mg ml}^{-1})}$$

References

Korkes, S.A., Campillo, A., Gunsalus, I.C., and Ochoa, S. (1951) *J. Biol. Chem.*, **193**, 721–735.
Reed, L.J., Leach, F.R., and Koike, M. (1958) *J. Biol. Chem.*, **232**, 123–142.
Reed, L.J. and Willms, C.R. (1966) *Meth. Enzymol.*, **9**, 247–269.

3.3.7.2 α-Oxoglutarate Dehydrogenase Complex (OGDHC)

This enzyme complex resembles in structure and reactions the pyruvate dehydrogenase complex (Section 3.3.7.1). It also consists of several (24) identical copies of three enzyme components, α-oxoglutarate dehydrogenase (E1o, 1.2.4.2), dihydrolipoamide succinyltransferase (E2o, EC 2.3.1.61), and dihydrolipoamide dehydrogenase (E3o, EC 1.8.1.4) catalyzing the following reaction sequence:

α-oxoglutarate + ThDP-E1o → α-hydroxy-γ-carboxypropyl-ThDP-E1o + CO_2
α-hydroxy-γ-carboxypropyl-ThDP-E1o + lipoyl-E2o
→ succinyl-dihydro-lipoyl-E2o + ThDP-E1o
succinyl-dihydrolipoyl-E2o + CoA → succinyl-CoA + dihydrolipoyl-E2o
dihydrolipoyl-E2o + FAD_{ox}-E3o → lipoyl-E2o + FAD_{red}-E3o
FAD_{red}-E3o + NAD^+ → FAD_{ox}-E3o + NADH + H^+

α-oxoglutarate + NAD^+ + CoA → succinyl-CoA + CO_2 + NADH + H^+

Assays for the partial reactions of the three enzyme components are described separately; here, a photometric assay for the overall reaction is presented. The assays described for the pyruvate dehydrogenase complex and its enzyme components can be applied, substituting pyruvate by α-oxoglutarate (i.e., the acetyl group by a succinyl residue). E3o and E3p are identical enzymes (assay in Section 3.3.1.29).

Overall Activity by NAD^+ Reduction

Assay solutions

0.1 M potassium phosphate pH 7.6

0.1 M NAD^+ (free acid M_r = 663.4; 663 mg in 10 ml)

0.01 M thiamin diphosphate (ThDP, cocarboxylase, M_r = 460.8, 46.1 mg in 10 ml)

0.1 M $MgCl_2$ ($MgCl_2 \cdot 6H_2O$, M_r = 203.3, 203 mg in 10 ml)

0.5 M α-oxoglutarate (α-ketoglutarate, 2-oxopentanedioic acid, disodium salt, M_r = 190.1, 0.95 g in 10 ml)

0.1 M dithioerythritol, DTE, M_r = 154.2; 154 mg in 10 ml)

0.01 M coenzyme A (free acid, M_r = 767.5, $CoA \cdot Li_3$, M_r = 785.4; 23 mg in 3 ml)

Assay mixture

Components	*Concentration (mM)*
7.95 ml 0.1 M potassium phosphate pH 7.6	80
0.25 ml 0.1 M NAD^+	2.5
0.20 ml 0.01 M ThDP	0.2
0.10 ml 0.1 M $MgCl_2$	1.0
0.10 ml 0.5 M α-oxoglutarate	5.0
0.10 ml 0.1 M DTE	1.0
0.10 ml 0.01 M CoA	0.1

Procedure

0.98 ml assay mixture

0.20 ml enzyme solution

The absorption increase at 340 nm is measured at 37 °C, $\varepsilon_{340} = 6.3 \times 10^3$ l mol^{-1} cm^{-1}.

Reference

Stepp, L.R., Bleile, D.M., McRorie, D.K., Pettit, F.H., and Reed, L.J. (1981) *Biochemistry*, **20**, 4555–4560.

3.3.8 Substrate Determination

The principle of substrate determination by enzymatic assays has already been discussed (cf. Section 2.5). If the enzyme reaction proceeds in an irreversible manner, substrate becomes quantitatively converted to product and the amount of product corresponding to the initial substrate concentration can be determined if the reaction is allowed to reach to its conclusion (Figure 2.31). Please refer to the respective enzyme assays described in the previous sections. In this section, the assays for substrate determination using the principle of enzymatic cycling are presented using the examples of NAD(H) and NADP(H), which allow detection of very low concentrations.

3.3.8.1 Determination of NADP(H) by Enzymatic Cycling

For the determination of NADP and NADPH, a coupled reaction of glutamate dehydrogenase (GluDH, EC 1.4.1.3) and glucose-6-phosphate dehydrogenase (G6PDH,

EC 1.1.1.49) is used:

$$\alpha\text{-oxoglutarate} + \text{NADPH} + \text{NH}_4^+ \xrightarrow{\text{GluDH}} \text{glutamate} + \text{H}_2\text{O} + \text{NADP}^+$$

$$\text{glucose-6-phosphate} + \text{NADP}^+ \xrightarrow{\text{G6PDH}} \text{6-phosphogluconate} + \text{NADPH} + \text{H}^+$$

Each NADP molecule catalyzes the formation of 5000–10 000 molecules of 6-phosphogluconate under the specified conditions. Concentrations of 10^{-9} M NADP (respectively 10^{-15} moles in 1 μl) can be detected. ADP functions as an activator of GluDH.

The amount of 6-phosphogluconate formed is determined by an independent reaction with 6-phosphogluconate dehydrogenase (PGDH, 6-phospho-D-gluconate: NADP$^+$ 2-oxidoreducatse, EC 1.1.1.44) after stopping the enzymatic cycling reaction with a fluorimetric assay:

$$\text{6-phosphogluconate} + \text{NADP}^+ \overset{\text{PGDH}}{\rightleftharpoons} \text{D-ribulose-5-phosphate} + \text{NADPH} + \text{H}^+ + \text{CO}_2$$

Assay solutions

0.1 M Tris/HCl pH 8.0

0.1 M α-oxoglutaric acid (disodium salt, $M_r = 190.1$; 190 mg in 10 ml)

0.1 M D-glucose-6-phosphate (disodium salt, hydrate, $M_r = 304.1$; 304 mg in 10 ml)

0.01 M ADP (disodium salt, $M_r = 471.2$; 47.1 mg in 10 ml)

1.0 M ammonium acetate ($M_r = 77.1$; 771 mg in 10 ml)

BSA (20 mg ml^{-1})

Glutamate dehydrogenase solution (GluDH, from bovine liver, 40 U mg^{-1}, prepare a solution of 40 IU in 0.1 ml 0.1 M Tris/HCl pH 8.0)

Glucose-6-phosphate dehydrogenase solution (G6PDH, from yeast, 300 IU mg^{-1}, prepare a solution of 60 IU in 0.1 ml 0.1 M Tris/HCl pH 8.0)

0.01 M 6-phosphogluconic acid (trisodium salt, $M_r = 342.1$; 34 mg in 10 ml)

0.01 M NADP (M_r =787.4; 79 mg in 10 ml)

0.1 M EDTA (ethylenediaminetetraacetic acid, $M_r = 292.2$; 292 mg in 10 ml)

6-phosphogluconate dehydrogenase solution (GPDH, from the yeast *Torula*, 20 IU mg^{-1}, prepare a solution of 0.1 IU in 0.1 ml 0.1 M Tris/HCl pH 8.0)

Cycling mixture

Components	*Concentration*
1.7 ml 0.1 M Tris/HCl pH 8.0	85 mM
0.10 ml 0.1 M α-oxoglutaric acid	5.0 mM
0.02 ml 0.1 M D-glucose-6-phosphate	1.0 mM
0.02 ml 0.01 M ADP	0.1 mM
0.06 ml 1 M ammonium acetate	30 mM
0.02 ml BSA	0.2 mg/ml
0.04 ml GluDH solution	8 IU ml^{-1}
0.04 ml G6PDH solution	12 U ml^{-1}

Procedure

0.1 ml cycling mixture

x µl NADP or NADPH solution to be determined, (x between 1 and 20; final concentration between 3 and 50×10^{-9} M)

$20-x$ µl 0.1 M Tris/HCl pH 8.0

Incubate for 30 min at 37 °C

Stop the reaction by transferring the sample for 2 min to 100 °C

Fluorimeteric assay of 6-phosphogluconate	*Concentration*
2 ml 0.02 M Tris/HCl pH 8.0	20 mM
0.004 ml 10 mM NADP	0.02 mM
0.002 ml 0.1 M EDTA	0.1 mM
0.05 ml sample (6-phosphogluconate)	–
0.02 ml 6-phosphogluconate dehydrogenase	0.01 IU

Measure fluorescence light excited at 260 nm and emitted at 470 nm. For quantification a standard curve with NADPH must be prepared.

3.3.8.2 Determination of NAD(H)

The procedure is similar to the determination of NADP(H) described in Section 3.3.8.1. NAD formed by the glutamate dehydrogenase reaction is converted to

NADH by the LDH reaction. In a second step, NADH is added to determine pyruvate by the LDH reaction.

$$\alpha\text{-oxoglutarate} + \text{NADH} + \text{NH}_4^+ \xrightarrow{\text{GluDH}} \text{glutamate} + \text{H}_2\text{O} + \text{NAD}^+$$
$$\text{lactate} + \text{NAD}^+ \xrightarrow{\text{LDH}} \text{pyruvate} + \text{NADH} + \text{H}^+$$

The excess of NADH is destroyed by acid treatment. After alkaline treatment of NAD the fluorescence is measured.

Assay solutions

0.2 M Tris/HCl pH 8.4

0.1 M α-oxoglutaric acid (disodium salt, $M_r = 190.1$; 190 mg in 10 ml)

1 M lactate (sodium salt, $M_r = 112.1$; 1.1 g in 10 ml)

0.01 M ADP (disodium salt, $M_r = 471.2$; 47.1 mg 10 ml^{-1})

1.0 M ammonium acetate ($M_r = 77.1$; 771 mg in 10 ml)

Glutamate dehydrogenase (GluDH, from bovine liver, 40 IU mg^{-1}, prepare a solution of 40 IU in 0.1 ml 0.1 M Tris/HCl pH 8.0)

lactate dehydrogenase (LDH, from pig heart, circa 550 IU mg^{-1}, 10 mg ml^{-1}, prepare a solution of 55 IU in 0.1 ml 0.1 M Tris/HCl pH 8.0)

9 M NaOH (36 g in 100 ml)

Sodium potassium phosphate buffer: dissolve 8.97 g NaH_2PO_4 (monohydrate, $M_r = 138.0$) and 2.61 g K_2HPO_4 ($M_r = 174.2$) in 100 ml H_2O

Cycling mixture (Solution I)

Components	*Concentration*
1.7 ml 0.2 M Tris/HCl pH 8.4	85 mM
0.10 ml 0.1 M α-oxoglutaric acid	5.0 mM
0.20 ml 1 M lactate	100 mM
0.06 ml 0.01 M ADP	0.3 mM
0.20 ml 1 M ammonium acetate	100 mM
0.04 ml GluDH	8 U/ml
0.04 ml LDH	11 IU/ml

Solution II

Add freshly to 1 ml of the sodium potassium phosphate buffer:

20 μl 10 mM NADH

LDH from stock solution according to 1.5 μg (dilute stock solution 100-fold and add 15 μl)

Procedure

0.1 ml cycling mixture

x μl NAD or NADH solution to be determined (*x* between 1 and 20; final concentration between 3 and 50×10^{-9} M)

$20 - x$ μl 0.2 M Tris/HCl pH 8.4

Incubate for 30 min at 37 °C

Transfer the sample for 2 min to 100 °C

Transfer to an ice bath

0.1 ml solution II

Incubate at 25 °C for 15 min

Put in an ice bath, add

25 μl 5 N HCl

Add 0.1 ml of each sample to

0.2 ml 9 M NaOH

Incubate at 60 °C for 10 min

1 ml H_2O

Measure the fluorescence emission at 470 nm, excitation is at 365 nm.

Reference

Lowry, O.H., Passonneau, J.V., Schultz, D.W., and Rock, M.K. (1961) *J. Biol. Chem.*, **236**, 2746–2753.

3.4 Assays for Enzyme Characterization

3.4.1 Protein Determination

In the following section, some frequently used protein assays are described together with some special examples regarding immobilized, membrane-bound, and carrier-fixed proteins. Box 3.3 summarizes the essential evidence such as the approximate detection limits, whereby the lower limit is more important than the higher, which can be attained by dilution.

3.4.1.1 Biuret Assay

The biuret reagent reacts with the peptide bond and is therefore specific for proteins and relatively insensitive against perturbations, with the exception of urea, Tris and Good buffers (cf. Section 2.2.7.4), which also show a positive reaction. The biuret assay is not very sensitive and requires larger amounts of protein (0.1–0.8 mg). It is suited for crude extracts and large-scale purifications but cannot be recommended for valuable enzyme preparations. Since for purification procedures the same protein assay should be used throughout, it must be decided from the beginning whether this or a more sensitive test should be applied.

Box 3.3: Protein Assays: Advantages and Disadvantages

Method	Sensitivity range	Advantages	Disadvantages
Biuret	100–800 μg	Specific for peptide bonds Minor disturbances	Low sensitivity
BCA	5–100 μg	Specific for peptide bonds Few disturbances Suited for immobilized and membrane enzymes Considerable sensitivity	Expensive reagent
Lowry	4–40 μg	High sensitivity	Limited linearity Many disturbances Relatively laborious
Bradford	10–100 μg (2 μg for microassay)	Fast and simple procedure Moderate sensitivity	Blue coloring of cuvettes
Absorption	20–1000 μg	Fast and easy procedure No modification of the protein Recovery of the sample	Low sensitivity Strong disturbance especially of nucleic acids
Fluorescence	0.5–50 μg	High sensitivity	Various disturbances (quenching) Expensive instrumentation

Reagents and solutions

Biuret solution: dissolve in 400 ml H_2O, one after the other

- 9.0 g sodium potassium tartrate (Rochelle salt, $C_4H_4KNaO_6 \cdot 4H_2O$; $M_r =$ 282.2)
- 3.0 g cupric sulfate ($CuSO_4 \cdot 5H_2O$, $M_r = 249.7$)
- 8.0 g sodium hydroxide (NaOH, $M_r = 40.0$)
- 5.0 g potassium iodide (KI, M_r 166.0)
- Adjust to 500 ml with H_2O

3 M TCA (trichloroacetic acid, $C_2HCl_3O_2$, $M_r = 163.4$; 49 g, adjust to 100 ml with H_2O)

BSA solution (20 mg ml^{-1} BSA in H_2O)

Biuret and TCA solutions are stable for months at room temperature, freeze the BSA solution at −20 °C.

Procedure

Different aliquots of the protein solution (0.01–0.1 ml, containing about 10–50 mg protein ml^{-1}) are filled up with H_2O to 1 ml. The blank contains only 1 ml water

Add 0.15 ml 3 M TCA

Centrifuge for 3 min at 5000 rpm

Discard the supernatant and dissolve the precipitate quantitatively in 1 ml solution A

Incubate 30 min at room temperature

Measure the absorption at 546 nm against the blank

The protein concentration can be calculated from a calibration curve prepared with the BSA solution in the respective concentration range or according to the equation:

$$\frac{\text{mg}}{\text{ml}}\text{protein} = \frac{A_{546} \times 3.01}{\text{volume protein sample (ml)}}$$

References

Beisenherz, G., Boltze, H.J., Bücher, T., Czok, R., Garbade, K.H., Meyer-Arendt, E., and Pfleiderer, G. (1953) *Z. Naturforsch.*, **8B**, 555–577.

Itzhaki, R.F. and Gill, D.M. (1964) *Anal. Biochem.*, **9**, 401–410.

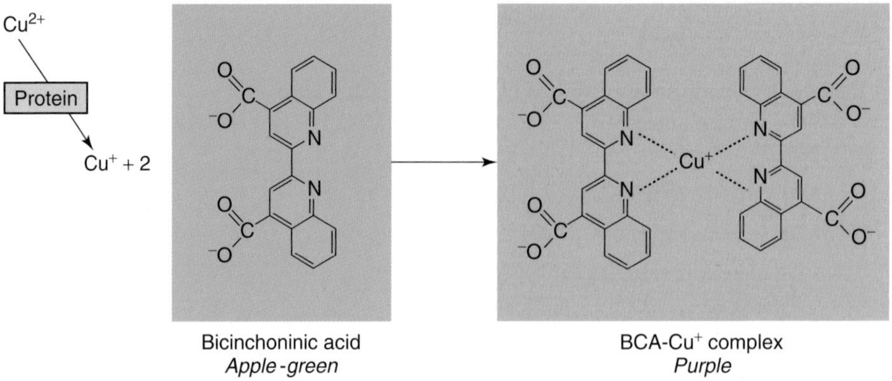

Cu^{2+} Protein Cu^{+} + 2

Bicinchoninic acid
Apple-green

BCA-Cu^{+} complex
Purple

Figure 3.1 Structure of the BCA-Cu^{+} complex.

3.4.1.2 BCA Assay

The BCA test is based on the biuret reaction. Peptide bonds and oxidizable amino acids like tyrosine, tryptophan, and cysteine reduce Cu^{2+} ions in an alkaline solution to Cu^{+}, forming a purple complex with two molecules of the bicinchoninic acid (BCA, Figure 3.1). The BCA assay is significantly more sensitive (<10 μg) than the biuret method and also not very susceptible against disturbances and high detergent concentrations. Complex forming and reducing substances, like EDTA (>100 mM), DTT (>1 mM), Tris (>0.25 M), ammonium sulfate (>20%), and glucose disturb the assay. A further advantage of the BCA assay is that the color reaction develops with Cu^{+} ions in the solution and not directly on the protein, so that precipitation of the protein (e.g., with TCA) is not necessary and the assay is also suited for determination of immobilized and membrane-bound proteins.

A. Assay for Soluble Proteins

Reagents and solutions

BCA solution: dissolve the following substances, one after the other, in 50 ml of H_2O and adjust pH to 11.25 with 1 M NaOH (4 g in 100 ml H_2O) or $NaHCO_3$ (8.4 g in 100 ml H_2O)

- 1.0 g BCA (bicinchoninic acid, 4,4′-dicarboxy-2,2′-biquinoline, disodium salt, $M_r = 388.3$)
- 2.0 g $Na_2CO_3 \cdot H_2O$ ($M_r = 124.0$)
- 0.16 g sodium tartrate ($C_4H_4Na_2O_6 \cdot 2H_2O$, $M_r = 230.1$)
- 0.4 g NaOH ($M_r = 40.0$)
- 0.95 g $NaHCO_3$ ($M_r = 84.0$)

Bring to 100 ml in a volumetric flask, store at room temperature

$CuSO_4$ solution: 1.0 g $CuSO_4 \cdot 5H_2O$, $M_r = 249.7$, fill up to 25 ml with H_2O, store at room temperature

BSA solution: 10 mg BSA in 10 ml H_2O, store frozen at −20 °C

Procedure

Prepare a BCA/$CuSO_4$ mixture: 50 parts BCA solution, 1 part $CuSO_4$ solution (v/v)

Add 50 µl protein solution (5–50 µg protein) or x µl protein solution + 50 − x µl H_2O to

950 µg of the BCA/$CuSO_4$ mixture

Incubate for 30 min at 37 °C

Measure the absorption at 562 nm against a blank without protein.

Calibration Curve

Treat in the same manner 12 samples of BSA solution from 0 to 50 µl.

B. Modification for Immobilized Proteins

Reagents and solutions as with assay A.

Procedure

Place a defined amount of the immobilized protein, such as a distinct piece[1]) of the matrix, into 1 ml of the BCA/$CuSO_4$ mixture

Incubate for 30 min at 37 °C with gentle agitation

Remove the matrix

Measure the absorption at 562 nm against a blank without protein

Use the BSA calibration curve described for assay A. The value obtained for the sample must be related to the size of the matrix. Alternatively, a calibration curve can be prepared with a known amount of immobilized protein.

References

Redinbaugh, M.C. and Turtley, R.B. (1986) *Anal. Biochem.*, **153**, 267–271.

Smith, P.K., Krohn, R.I., Hermanson, G.T., Mallia, A.K., Gartner, F.H., Provenzano, M.D., Fujimoto, E.K., Goeke, N.M., Olson, B.J., and Klenk, D.C. (1985) *Anal. Biochem.*, **150**, 76–85.

1) The protein determined by this method depends on the size of the matrix applied to the assay. This must be clearly defined, for example as mg matrix (dry weight), or mm^2 matrix surface.

3.4.1.3 Lowry Assay

This is a very sensitive method but it is strongly susceptible to various disturbances, even in lower concentrations, such as EDTA, sucrose, glycine, Tris, detergents (SDS, Triton X-100, Lubrol, Brij 35, Chaps), and inorganic salts (ammonium sulfate >28 mM, sodium phosphate >0.1 M, sodium acetate >0.2 M). A further disadvantage is the limited linearity of absorption upon increasing amounts of protein. Care must be taken that the values obtained with the unknown samples range within the linear part of the calibration curve, which is usually performed with BSA.

Reagents and solutions

Carbonate buffer: dissolve

- 0.4 g sodium tartrate ($C_4H_4Na_2O_6 \cdot 2H_2O$, $M_r = 230.1$), and
- 20 g Na_2CO_3

in 100 ml 1 N NaOH, dilute with H_2O to 200 ml, store at room temperature

Alkaline copper tartrate solution: dissolve

- 2 g sodium potassium tartrate (Rochelle salt, $C_4H_4KNaO_6 \cdot 4H_2O$; $M_r = 282.2$) and
- 1 g $CuSO_4 \cdot 5H_2O$ ($M_r = 249.7$)

in 90 ml H_2O

Add 10 ml 1 N NaOH, store at room temperature

Diluted Folin–Ciocalteau reagent: prepare fresh

- 1 part Folin–Ciocalteau reagent
- 15 parts H_2O

BSA solution: 10 mg BSA in 10 ml H_2O, store frozen at −20 °C

Procedure

0.3 ml protein solution containing 4–40 μg protein

0.3 ml carbonate buffer

- Incubate for 10 min at 50 °C
- Bring to room temperature

33 μl alkaline copper tartrate solution

- 10 min at room temperature

1 ml diluted Folin–Ciocalteau reagent, mix immediately

- 10 min at 50 °C
- Bring to room temperature

Measure the absorption at 650 nm.

Calibration curve

Treat in the same manner 12 samples of BSA solution from 0 to 40 μl.

References

Lowry, O.H., Rosenbrough, N.J., Farr, A.L., and Randall, R.J. (1951) *J. Biol. Chem.*, **193**, 265–275.
Peterson, G.L. (1979) *Anal. Biochem.*, **100**, 201–220.

3.4.1.4 Coomassie Binding Assay (Bradford Assay)

This rapid and simple assay was adapted from protein staining in electrophoresis gels. Upon binding to proteins in acidic solution, the Coomassie® Brilliant G-250 dye shifts its absorption maximum from 465 to 595 nm. The assay has relatively few perturbations (detergents like Triton X-100 and sodium dodecyl sulfate). Strength of interaction and thus staining intensity depend on the special protein. Therefore, the BSA calibration curve yields only relative values, variations of more than 50% are possible. For higher accuracy the calibration curve must be prepared with a known solution of the protein under study. The sensitivity is within 10–100 μg, with micro assays even 2 μg protein may be detected. A disadvantage is the blue coloring of the cuvettes; instead, disposable plastic cuvettes may be used. Glass cuvettes can be cleaned with acetone or by incubating in 0.1 M HCl for a few hours.

Reagents and solutions

Coomassie reagent: dissolve

- 100 mg Coomassie® Brilliant G-250 in 50 ml ethanol
- Add 100 ml 85% phosphoric acid (88%)
- Bring to 1 l with H_2O

Filter and store for several weeks at room temperature.

BSA solution: 10 mg BSA in 10 ml H_2O, store frozen at −20 °C

Procedure

Samples from 0 (for the blank) to 50 μl (2–40 μg protein), adjust with H_2O to 50 μl

0.95 ml of the Coomassie reagent, mix immediately

Measure after at least 2 min (but not more than 1 h) the absorption at 595 nm against a blank without protein. The protein content is determined from a standard curve prepared from the BSA solution by the same procedure.

Modification for immobilized proteins

The Coomassie binding assay can principally be applied for immobilized proteins, but essential modifications are necessary. The absorption must be measured directly at the matrix surface, after washing out the soluble dye. This requires a special photometric device. For calibration, immobilized samples of known protein content must be determined. Alternatively, an approximate calibration curve can be determined from protein immobilized to filter paper pieces of a defined size.

References

Bradford, M.M. (1976) *Anal. Biochem.*, **72**, 248–254.
Friedenauer, S. and Berlet, H.H. (1989) *Anal. Biochem.*, **178**, 263–268.

3.4.1.5 Absorption Method

Proteins, although differing considerably in their primary sequence, possess nearly identical absorption spectra (cf. Section 2.3.1.1), consisting of two characteristic peaks in the UV region; a more intense one at about 205–210 nm and a less intense one at 280 nm (Figure 3.2c), which are described in detail in Section 4.3.1. The far UV peak is composed of various contributions, while the near UV peak contains contributions of the aromatic amino acids alone, tryptophan clearly dominating the graph (Figure 3.2a). The absorption maximum at 280 nm is a clear indication of the presence of this amino acid. In its absence, the maximum shifts to shorter wavelengths. Since the relative tryptophan content is similar for most large proteins, the absorption intensity at 280 nm serves as a measure of protein concentration.

The absorption method is simple. It requires only the determination of the absorption of the sample solution at 280 nm with quartz cuvettes. There is no loss of the sample. As a crude estimate, protein solutions of 1 mg ml^{-1} show absorptions of about 1 at 280 nm. Solutions with higher absorptions must be diluted, concentrations lower than 0.02 mg ml^{-1} can hardly be measured with sufficient accuracy; thus the sensitivity of the method is not very high. Substances absorbing within this UV range disturb this method. Low molecular components can be removed by dialysis before the assay, but there is a strong interference with the absorption of nucleic acids. Although their maximum absorption is at 260 nm, due to high intensity, this overlaps considerably with the protein absorption at 280 nm (Figure 3.2b).

The method of Warburg and Christian (1941) is based on the absorption ratio between 280 and 260 nm and enables the simultaneous determination of proteins

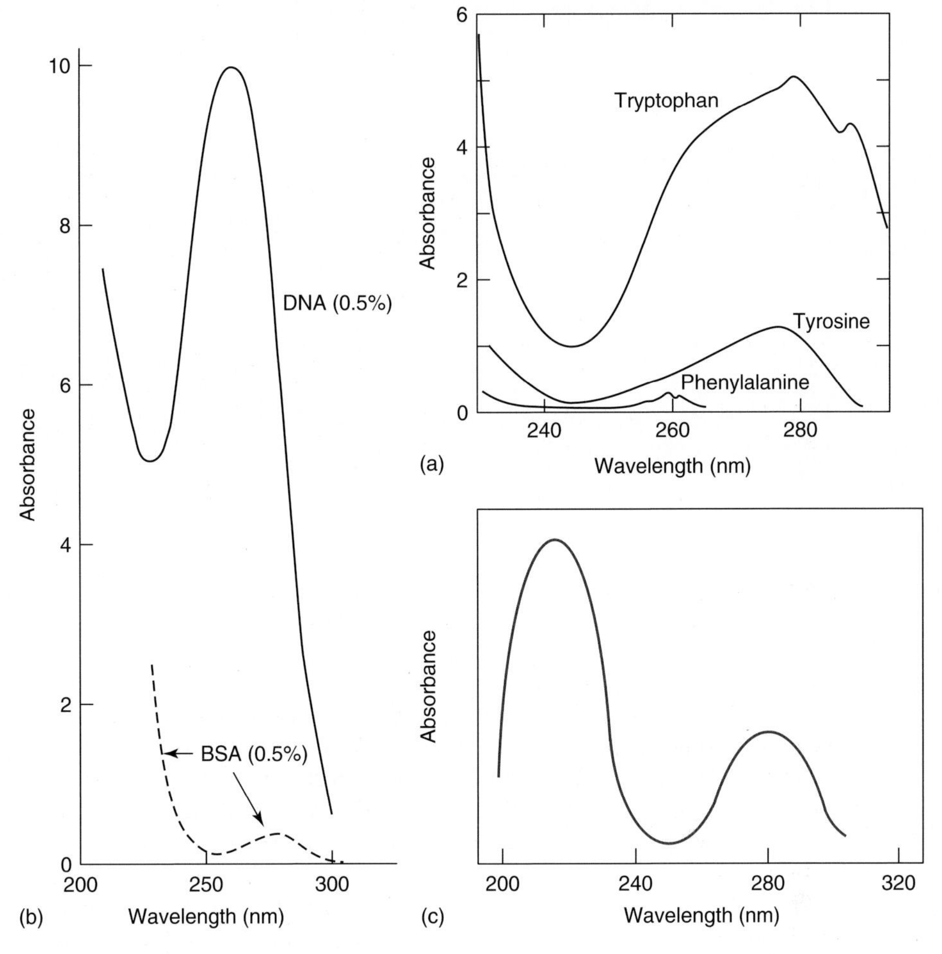

Figure 3.2 UV absorption spectra. (a) Absorption of the three aromatic amino acids in the near-UV range. (b) Comparison of absorption intensities of proteins (BSA), and of DNA in equivalent concentration ranges (0.5%) (K. L. Manchester (1997) *Biochemical Education* **25**, 214–215; with permission from *The American Society for Biochemistry & Molecular Biology*). (c) Typical protein spectrum.

and nucleic acids in the same solution. The absorption of the sample solution is measured (in quartz cuvettes) at both wavelengths against a blank. From the ratio a factor is obtained (Table 3.2), which allows the estimation of the protein concentration. Alternatively the formula

$$\text{protein (mg ml}^{-1}) = 1.55\,A_{280} - 0.76\,A_{260}$$

can be applied. In the presence of large nucleic acid concentrations, protein determination becomes difficult.

Table 3.2 Determination of the amount of protein and nucleic acids by the absorption method of Warburg and Christian (1941). The factor corresponding to the absorption ratio 280/260 nm multiplied with the absorption at 280 nm yields the protein concentration: $A_{280} \times \text{fp} = \text{mg protein/ml}$.

Absorption ratio 280/260 nm	Nucleic acids (%)	Factor (fp)
1.75	0	1.10
1.60	0.25	1.07
1.50	0.50	1.05
1.40	0.75	1.02
1.30	1.00	0.99
1.25	1.25	0.97
1.20	1.50	0.95
1.15	2.00	0.91
1.10	2.50	0.87
1.05	3.00	0.83
1.00	3.50	0.80
0.96	3.75	0.78
0.92	4.25	0.75
0.88	5.00	0.71
0.86	5.25	0.70
0.84	5.50	0.69
0.82	6.00	0.67
0.80	6.50	0.64
0.78	7.25	0.62
0.76	8.00	0.59
0.74	8.75	0.56
0.72	9.50	0.54
0.70	10.75	0.51
0.68	12.00	0.48
0.66	13.50	0.45
0.65	14.50	0.43
0.64	15.25	0.41
0.62	17.50	0.38
0.60	20.00	0.35
0.49	100.00	–

References

Layne, E. (1957) *Meth. Enzymol.*, **3**, 447–454.
Warburg, O. and Christian, W. (1941) *Biochem. Z.*, **310**, 384–421.

3.4.1.6 Fluorimetric Assay

For protein determination, the intrinsic fluorescence of tryptophan can be used (excitation 280 nm, emission 340 nm). Quantification, however, is difficult because the tryptophan fluorescence changes considerably upon integration of the amino acid into the protein structure, both with respect to intensity and wavelength (blue shift) of the fluorescence maximum. Here, a very sensitive assay with fluorescamine, applicable down to 0.5 μg protein, is described. A fluorimeter is required (filter fluorimeter is sufficient). The sensitivity can be further increased by removal of unbound dye, applying dialysis or gel filtration, for example with Sephadex G-25. This, however, complicates routine assays and should be integrated into an automatic procedure.

Reagents and solutions

0.05 M sodium phosphate pH 8.0

Fluorescamine solution (30 mg in 100 ml dioxane)

BSA solution (1 mg ml^{-1})

Procedure

Bring the protein solution (10–250 μl) to 1.5 ml with 0.05 M sodium phosphate pH 8.0

Add 0.5 ml fluorescamine solution rapidly under vigorous shaking

Measure the fluorescence at 475 nm at an excitation of 390 nm.

Prepare a standard calibration curve for quantitative determination with the BSA solution.

Reference

Böhlen, P., Stein, S., Dairman, W., and Udenfriend, S. (1973) *Arch. Biochem. Biophys.*, **155**, 213–220.

3.4.1.7 Ninhydrin Assay

The ninhydrin reaction detects free amino acids with high sensitivity. This method is especially suited for immobilized proteins by determining the released amino acids after hydrolyzing the proteins from the matrix.

Reagents and solutions

6 N HCl (49.7 ml 37% hydrochloric acid bring to 100 ml with H_2O)

6 N NaOH ($M_r = 40.0$; 24 g in 100 ml H_2O, store in a PE flask)

Acetate buffer pH 5.4 (dissolve 270 g sodium acetate trihydrate, $C_2H_3O_2Na \cdot 3H_2O$, $M_r = 136.1$, in 200 ml H_2O, add 50 ml glacial acetic acid, fill up to 750 ml)

10 mM KCN ($M_r = 65.1$; 65 mg in 100 ml H_2O)

0.2 mM KCN in acetate buffer (0.2 ml 10 mM KCN + 9.8 ml acetate buffer, prepare fresh)

Ninhydrin reagent (2,2-dihydroxy-1,3-indanedione, $M_r = 178.1$, commercially available)

50% ethanol

Standard BSA solution, 40 μg ml^{-1}, prepared by 100-fold dilution of a stock solution of 40 mg in 10 ml H_2O

10 mM leucine (D,L-leucine, $M_r = 131.2$; 131 mg in 100 ml H_2O)

Hydrolysis

Incubate defined amounts of the protein sample and an exactly determined piece of the matrix for immobilized enzymes, overnight (16 h) at 90 °C in 1 ml 6 N HCl in closely sealed cups (screw caps are preferable to avoid dehiscence of the cap and splashing of concentrated acid)

Open the cups carefully (use goggles!), remove 0.5 ml and neutralize with 0.5 ml 6 M NaOH (control the pH)

Procedure

0.2 ml hydrolysate (sample aliquots filled up with H_2O to 200 μl)

0.1 ml 0.2 mM KCN in acetate buffer

0.1 ml ninhydrin reagent

10 min incubation at 100 °C (boiling water bath or heating block)

Chill in ice for 1 min and add 0.5 ml 50% ethanol

Measure the absorbance at 570 nm against a blank where the hydrolysate is displaced by water.

Calibration curve

Generate a calibration curve with 12 aliquots (1 and 50 μl, filled up to 200 μl) of the 10 mM leucine solution.

Reference

Rosen, H. (1957) *Arch. Biochem. Biophys.*, **67**, 10–15.

3.4.1.8 Modified Ninhydrin Assay without Hydrolysis

This method is especially recommended for proteins immobilized to acid-labile matrices. The samples must be free of salt and solvents. For preparation, wash with

H_2O and dry. Matrices in organic solvents should be washed with dichloromethane and dried before use.

Reagents and solutions

Amberlite MB-3

Potassium cyanide (KCN) (*Caution!* extremely toxic)

Ninhydrin (2,3-dihydroxy-1,3-indanedione, $M_r = 178.1$)

Phenol, p.A.

Pyridine

Absolute ethanol, p.A.

0.5 M TEA chloride (tetraethylammonium chloride, $M_r = 165.7$, 0.83 g in 10 ml dichloromethane)

BSA solution (1 mg ml^{-1} in H_2O)

Solution A: dissolve

- 40 g phenol in 10 ml by gentle warming and add
- 4 g Amberlite MB-3
- Stir for 45 min and filtrate thereafter

Solution B: dissolve

- 65 mg KCN ($M_r = 65.1$) in 100 ml H_2O
- give 2 ml of this solution to 98 ml pyridine and add
- 4 g Amberlite MB-3
- Stir the mixture and filtrate

Solution C:

- Mix equal parts of solutions A and B

Solution D:

- 2.5 g ninhydrin, dissolve in 50 ml ethanol. Store in a nitrogen-filled atmosphere

Procedure

Add 200 µl of solution C and 50 µl of solution D to an aliquot of the matrix with the immobilized enzyme, corresponding to about 2–5 mg and seal the samples closely. For the blank matrix without immobilized enzyme no aliquot was used

10 min at 100 °C in a heating block

Chill on ice, add

1 ml 60% ethanol and mix carefully

Filter or centrifuge to remove the matrix

Separate the supernatant

Wash the matrix two times with 0.2 ml 0.5 M TEA chloride and give the washing solution to the supernatant

Bring the samples to 2 ml with 60% ethanol

Measure against the blank at 570 nm

Calculation

Prepare a standard curve with BSA solution. Use an absorption coefficient of $\varepsilon_{570} = 1.5 \times 10^4\ \text{l mol}^{-1}\text{cm}^{-1}$ for free amino groups.

Reference

Sarin, V.K., Kent, S.B.H., Tam, J.P., and Merrifield, R.B. (1981) *Anal. Biochem.*, **117**, 147–157.

3.4.1.9 Protein Assay with 2-Hydroxy-1-naphthaldehyde

This time-consuming assay is suitable for immobilized proteins. The aldehyde forms a Schiff's base with amino residues of the protein, which are substituted by benzylamine. The absorption of the soluble Schiff's base is determined in the supernatant.

Reagents and solutions

1.2 M 2-hydroxy-1-naphthaldehyde (2-naphthol-1-carboxyaldehyde, M_r = 172.18; 2.07 g in 10 ml)

Dimethylformamide

0.4 M benzylamine (hydrochloride, $M_r = 143.6$; 0.57 g in 10 ml ethanol)

BSA solution (1 mg ml^{-1} in H_2O)

Ethanol p.A.

Procedure

Wash the sample with immobilized enzyme thoroughly with dimethylformamide

Overlay with 0.5 ml 1.2 M 2-hydroxy-1-naphthaldehyde

Shake overnight (~14 h) at room temperature

Remove the immobilized sample and wash it thoroughly, 5–10 times with dimethylformamide and 5 times with ethanol, until no more absorption can be detected at 280 nm

Add 1 ml 0.4 M benzylamine and shake for 15 h at room temperature

Centrifuge, 5 min, 5000 rpm

Measure the absorption in the supernatant at 420 nm ($\varepsilon_{420} = 1.09 \cdot 10^4$ l mol^{-1} cm^{-1})

Calculate the protein content using a calibration curve prepared with BSA solution.

Reference

Bisswanger, H., Figura, R., Möschel, K., and Nouaimi, M. (2001) *Enzymkinetik, Ligandenbindung und Enzymtechnologie*, 2nd edn, Shaker Verlag, Aachen, p. 71.

General References for Protein Assays

Kresse, G.B. (1983) in *Methods of Enzymatic Analysis* (ed. H.U. Bergmeyer), Verlag Chemie, Weinheim, New York, pp. 86–99.

Scopes, R.K. (1987) *Protein Purification*, Springer, New York, p. 280.

Stoscheck, C.A. (1990) *Meth. Enzymol.*, **182**, 50–68.

Thorne, C.J.R. (1978) in *Techniques in Protein and Enzyme Biochemistry*, vol. B104 (ed. H.L. Kornberg), Elsevier, Amsterdam, pp. 1–18.

3.4.2
Phosphate Determination

Various enzyme assays (e.g., ATPase) need the determination of inorganic phosphate. A micromethod with a sensitivity of 2–40 μg inorganic phosphate is described.

Reagents and solutions

Sodium phosphate standard solution (dissolve 0.5853 g KH_2PO_4 in 1 l water, the solution contains 133.3 μg phosphorus per ml)

12% (w/v) TCA (trichloroacetic acid, $M_r = 163.4$; 120 g/l)

10 N H_2SO_4 (add 278 ml concentrated sulfuric acid into 0.7 l of water, slowly and carefully, *use goggles*! Bring to 1 l after cooling)

10% ammonium molybdate solution (take 50 g $(NH_4)_6Mo_7O_{24} \cdot 4H_2O$ in a beaker and add 400 ml 10 N H_2SO_4 with constant stirring. Transfer the dissolved solution quantitatively into a volumetric flask and bring to 500 ml with 10 N H_2SO_4, store in the dark)

Ferrous sulfate–ammonium molybdate reagent (prepare fresh before use)

- Dilute 10 ml 10% ammonium molybdate solution to 70 ml with H_2O
- Dissolve 5 g of $FeSO_4 \cdot 7H_2O$
- Bring to 100 ml

Procedure

1.8 ml 12% TCA

0.1 ml sample, mix vigorously

- 10 min at room temperature
- Centrifuge at 1500 rpm for 10 min

Give 1.5 ml of the supernatant and 1 ml ferrous sulfate–ammonium molybdate reagent into a cuvette

Measure the absorption at 820 nm within 2 h.
Prepare a calibration curve with the sodium phosphate standard solution, each sample containing 4–20 μg P_i.

Modification

A modification of the above method uses an ascorbic–molybdate mixture:

One part 10% ascorbic acid

Six parts 0.42% ammonium molybdate × $4H_2O$ in 1 N H_2SO_3

Procedure

0.3 ml sample (supernatant, see above)

0.7 ml ascorbic–molybdate mixture

Incubate for 20 min at 45 °C

Measure at 820 nm; 0.01 μmol phosphate yields an absorption of 0.24

References

Ames, B.N. and Dubin, D.J. (1960) *J. Biol. Chem.*, **235**, 769–775.
Chen, P.S., Toribara, T.Y., and Warner, H. (1956) *Anal. Chem.*, **28**, 1756–1758.
Roufogalis, B.D. (1971) *Anal. Biochem.*, **44**, 325–328.
Taussky, H.H. and Shorr, E. (1953) *J. Biol. Chem.*, **202**, 675–685.

3.4.3 Glycoprotein Assays

From the various assays for determination of glycoproteins reported so far, a qualitative detection method in electrophoresis gels with Schiff's reagent and a quantitative method to determine protein-bound hexoses are described.

A. Detection in Electrophoresis Gels

Assay solutions

7% acetic acid (v/v) in H_2O

40% methanol

1% periodo acid in 7% acetic acid

1% sodium disulfite (sodium metabisulfite, $Na_2S_2O_5$, $M_r = 190.1$) in 0.1 N HCl

Schiff's reagent: though commercially available, steps for preparation are given below

- Dissolve 1 g pararosaniline (basic parafuchsin) in 500 ml H_2O
- Add 20 ml saturated sodium disulfite
- 10 ml concentrated HCl
- Make up to 1 l with H_2O

Procedure

Wash the electrophoresis gel overnight, with a large amount (>1 l) of 40% methanol and 7% acetic acid

Oxidize the gels for 1 h at 4 °C with 1% periodo acid in 7% acetic acid in the dark

Wash the gels with 7% acetic acid for 24 h, change the solution two times

Incubate the gel with Schiff's reagent for 1 h at 4 °C in the dark

Wash the gel four times with 1% sodium disulfite in 0.1 N HCl. The positive bands show a violet–purple coloring

B. Determination of Protein-Bound Hexoses

Assay solutions

Solution A:

- 60 ml concentrated H_2SO_4
- 40 ml H_2O

Solution B: 1.6 g orcinol (3,5-dihydroxytoluene, monohydrate, $M_r = 142.2$, in 100 ml H_2O)

Orcinol/H_2SO_4 reagent (freshly prepared):

- 7.5 parts solution A
- 1 part solution B

Galactose/mannose standard (20 mg D-galactose, $M_r = 180.2$, and 20 mg D-mannose, $M_r = 180.2$, in 100 ml H_2O)

95% ethanol

Procedure

0.1 ml protein sample

5 ml 95% ethanol

Mix and centrifuge for 15 min at 5000 rpm

Decant and suspend the precipitate in 5 ml 95% ethanol

Centrifuge 15 min at 5000 rpm

Dissolve the precipitate in 1 ml 0.1 N NaOH

Prepare for a calibration curve 10 aliquots with increasing volumes, from 0 to 1 ml, of the galactose/mannose standard; bring all samples to 1 ml with H_2O

Add 8.5 ml orcinol/H_2SO_4 reagent to each protein and standard sample and close the tubes with a sealing film (aluminum foil or parafilm)

Incubate at 80 °C for 15 min

Cool the tubes in tap water

Measure the absorption at 540 nm; for quantification, use the calibration curve

References

Glossmann, H. and Neville, D.M. (1971) *J. Biol. Chem.*, **246**, 6339–6346.
Winzler, R.J. (1955) *Meth. Biochem. Anal.*, **2**, 279–311.

3.4.4
Cross-Linking of Proteins with Dimethylsuberimidate

The following method is one of the most intensely studied cross-linking methods and is described representatively for a large variety of cross-linking reactions. Intramolecular bridges within a protein and also between different proteins are formed by this method. The extent of cross-linking depends on the distance of the reactive groups and their reactivity. Variation of the incubation time of the cross-linking reaction, of the concentration, and of the chain length of the cross-linking reagent gives insight into the spatial arrangement of complex protein structures, such as subunit composition and distances of distinct domains. A similar method is described in Section 3.5.4.7 with glutaraldehyde. The cross-linked proteins can be analyzed by SDS electrophoresis, where combined subunits migrate, corresponding to their aggregation state.

Reagents and solutions

0.2 M triethanolamine/HCl, pH 8.5

Dimethylsuberimidate (dihydrochloride, $M_r = 273.2$, prepare a solution in 0.2 M triethanolamine/HCl, pH 8.5, e.g. 3 mg ml^{-1}, just before use according to the procedure below)

Protein solution (required concentration see procedure below)

0.1 M potassium phosphate pH 7.0

Procedure

Mix dimethyl suberimidate solution (to give a final concentration 1–12 mg ml^{-1}) with protein (final concentration 0.4–5 mg ml^{-1}) in 0.2 M triethanolamine/HCl pH 8.5

Incubate 3 h at room temperature

Dialyze against 0.1 M potassium phosphate pH 7.0

References

Carpenter, F.H. and Harrington, K.T. (1972) *J. Biol. Chem.*, **247**, 5580–5586.
Davies, G.E. and Stark, G.R. (1970) *Proc. Natl. Acad. Sci.*, **66**, 651–656.
Hunter, M.J. and Ludwig, M.L. (1970) *Meth. Enzymol.*, **25**, 585–596.

3.4.5
Concentration of Enzyme Solutions

Concentration of protein solutions and especially of sensitive enzyme solutions is sometimes a difficult task and often an appropriate method, which concentrates without serious loss of activity can hardly be found. The numerous methods described so far reflect these difficulties and a generally applicable method is still missing. Besides some miscellaneous techniques, the essential methods are based on the principles of precipitation, ultrafiltration, ultracentrifugation, and lyophilization. Here, advantages and disadvantages are discussed to facilitate the choice but general rules for a special system cannot be given (Box 3.4).

Box 3.4: Concentration Methods: Advantages, and Disadvantages

Method	Advantages	Disadvantages
Precipitation with		
• $(NH_4)_2SO_4$, PEG	Reversible procedure Partial purification	Partial activity loss (facultative)
• Urea, guanidinium chloride	Reversible procedure Dissociation of subunits	Partial activity loss (facultative)
• TCA, perchloric acid	Total removal of protein, collection for protein determination	Irreversible procedure Complete inactivation
Ultrafiltration	Protein remains in solution Purification effected by an appropriate pore size	Losses due to adherence to the filter membrane
Dialysis	No special device necessary	Large losses due to adherence of protein to the large inner membrane surface
Ultracentrifugation	Purification effect, removal of small proteins and components	Only for large proteins Losses due to incomplete resolution of the pelleted protein
Lyophilization	Dry protein samples, suited for long-term storage and shipping	Considerable activity losses with distinct enzymes
Crystallization	For long-term storage	Only for proteins easy to crystallize Pure enzyme preparations

3.4.5.1 Precipitation

Precipitation, still widely applied, is the oldest concentration principle for enzymes. It combines two advantages: concentration and purification. However, precipitation can distort the native state of the enzyme even after removing the precipitating agent. On the other hand, even such fragile enzymes such as multienzyme complexes are treated by this method without discernible alterations. The method may be applied as long as no impairment is detectable. In the early periods of enzyme research, especially for purification, various principles were used: precipitation by heat, acid pH, ethanol, or acetone. However, such methods are rather denaturating and have largely been displaced by modern methods. An exception with respect to pH is precipitation at the *isoelectric point*, where the protein has its lowest solubility (cf. Section 2.2.7.3). Precipitation with perchloric acid and trichloroacetic acid (TCA) causes irreversible denaturation and is applied only to remove proteins quantitatively from solutions or to collect them for protein determination.

For concentration conserving the native structure, *chaotropic* substances are used. Urea (up to 6–8 M) and guanidinium chloride (4–6 M) are used especially for dissociating protein aggregates and separating subunits under more or less native conditions (in contrast, for e.g., to sodium dodecyl sulfate). $MgCl_2$ in higher concentrations is also sometimes used for enzyme concentration, but mostly ammonium sulfate and polyethylene glycol (PEG) are applied. These reagents destabilize hydrogen bonds and destroy the hydrate shell around the protein surface, causing protein aggregation and subsequently, precipitation. It is assumed that removal of the reagent will reverse completely these effects and restore the native structure.

Each protein precipitates at a characteristic concentration of ammonium sulfate or polyethylene glycol. This feature is used to separate the desired protein from foreign proteins. For each special protein, the concentration limits of the reagent, at which the precipitation occurs, must be tested out. Generally with ammonium sulfate, precipitation happens between 20 and 100% saturation, with polyethylene glycol between 0 and 15% saturation (w/v).

Two different modes for addition of ammonium sulfate are used: solid form or saturated solution. The first procedure is simpler and has the further advantage of not enlarging the final volume severely in contrast to the second method, where especially for higher saturation degrees, exceptional volume increase must be accepted and 100% saturation cannot be reached. On the other hand, this is the more gentle procedure. Upon dissolving solid ammonium sulfate, air bubbles enclosed in the salt crystals become released. Their strong surface tension is harmful for proteins. Beyond that, zones of high concentrations are generated around the dissolving salt crystals and proteins precipitate in such a microenvironment even if the final concentration of the solution remains below the precipitation limit. The protein suffers repeated precipitation and dissolving cycles. To reduce such unfavorable effects, the calculated amount of salt crystals should be added very slowly within about 1 h. Special devices are described for slow additions of salts (Beisenherz *et al.*, 1953). Saturated ammonium sulfate solutions and fast stirring minimize such effects. Temperature is important for solubility of both the salt and the protein, and it must be decided beforehand whether the procedure should be performed at room temperature or (preferably) in the cold (4 °C). For concentration of the protein, a surplus of the salt may be added and the precipitate collected by centrifugation. If the method should be used for purification as well, the respective precipitation limits must be regarded. In a first step, salt is added to the lower limit, where the respective protein or enzyme remains just in solution, while some foreign proteins precipitate and can be removed by centrifugation. In a second step, ammonium sulfate is added to reach the upper limit so that the protein of interest precipitates completely, leaving foreign proteins in solution. The equation

$$S_2 = \frac{(S_2 - S_1) \times 515}{100 - 0.27 \times S_2} \times V \tag{3.1}$$

allows the calculation of the amount of ammonium sulfate (in grams) to be added to a protein solution to obtain the desired saturation degree (S_2) at 0 °C. S_1 is the initial saturation degree and V the volume (in milliliters) of the protein solution. Tables both for addition of solid ammonium sulfate and of saturated solutions are available (Table 3.3 Green and Hughes, 1955; Brewer, Pesce, and Ashworth,

Table 3.3 Precipitation of protein with ammonium sulfate. The values indicate the amount of solid ammonium sulfate to be added to 1 l of the protein solution to give the desired saturation degree S2 (upper line) at 0 °C. The initial saturation degree of the protein solution (S1) is indicated at the left column (after Green & Hughes, 1955, Brewer et al., 1977).

S1/S2	10	15	20	25	30	33	35	40	45	50	55	60	65	67	70	75	80	85	90	95	100
0	53	80	106	134	164	187	194	226	258	291	326	361	398	421	436	476	516	559	603	650	697
5	27	56	79	108	137	162	166	197	229	262	296	331	368	390	405	444	484	526	570	615	662
10		28	53	81	109	133	139	169	200	233	266	301	337	358	374	412	452	493	536	581	627
15			26	54	82	87	111	141	172	204	237	271	316	327	343	381	420	460	503	547	592
20				27	55	75	83	113	143	175	207	241	276	296	312	349	387	427	469	512	557
25					27	46	56	84	115	146	179	211	245	264	280	317	355	395	436	488	522
30						17	28	56	86	117	148	181	214	233	249	285	323	362	402	445	488
33							11	40	70	101	133	166	200	214	235	271	309	347	387	429	472
35								28	57	87	118	151	184	201	218	254	291	329	369	410	453
40									29	58	89	120	153	170	182	212	258	296	335	376	418
45										29	59	90	123	138	156	190	226	263	302	342	383
50											30	60	92	107	125	159	194	230	268	308	348
55												30	61	76	93	127	161	197	235	273	313
60													31	44	62	95	129	164	201	239	279
65														13	31	63	97	132	168	205	244
67															19	52	85	120	156	194	233
70																32	65	99	134	171	209
75																	32	66	101	137	174
80																		33	67	103	139
85																			34	68	105
90																				34	70
95																					35

1974). After precipitation, the pellet is collected by centrifugation and resolved in an appropriate buffer. The smallest possible volume should be used by adding low amounts of the respective buffer, until the precipitate is just dissolved. To avoid addition of larger amounts of buffer than necessary, a special device has been constructed: a glass or plastic cylinder with a fitting pestle, where the precipitate can be treated thoroughly with small amounts of buffer (Beisenherz *et al.*, 1953). Since a considerable amount of ammonium sulfate remains in the protein solution after resolution, extensive dialysis must follow the precipitation.

3.4.5.2 Ultrafiltration and Dialysis

Ultrafiltration is the most frequently used concentration method. It is easy in manipulation, needs only simple devices, and causes no undesirable effects like structural changes of the protein or increase in salt concentration the way precipitation does. Low molecular substances such as buffer ions pass the ultrafiltration membrane and are removed. This effect is used for changing buffer systems, for example during purification procedures. A severe disadvantage, however, is accumulation of the concentrated protein on the membrane surface, forming a closed layer and blocking the flow through the membrane. Accumulation of the protein on the membrane favors aggregation and often, the protein cannot be resolved, causing considerable losses in enzyme activity. In such cases, some of the protein may be recovered by treatment of the membrane with detergents, like Triton X-100.

Ultrafiltration membranes prepared from different materials such as cellulose, nitrocellulose, and synthetic polymers are available. For protein concentration, the membrane should possess a hydrophilic surface. According to pore size, membranes with distinct retention limits can be obtained; for example, for 3, 10, 30, and 100 KDa. This feature can be used to separate components of different sizes. However, pores of ultrafiltration membranes of any type are not homogeneous in their diameter like the meshes of a sieve and the retention limit must be considered more as a preference than as a strict cutoff.

To avoid membrane blocking, the solution near the ultrafiltration membrane should be continuously stirred. However, if the stirrer touches the fragile membrane, it may damage it and destroy the small pores. Therefore, a perpendicular axis fixed to the top of the ultrafiltration cell keeps the magnetic stirrer above the membrane. Ultrafiltration cells are available in distinct sizes for volumes ranging from a few milliliters up to several liters. The solution is forced through the membrane by compressed air or better, by nitrogen gas, and the ultrafiltrate is collected below the membrane. The blocking of the membrane by the protein progressively reduces the flow rate. Increase in the air pressure causes more blocking than the flow rate does. To continue, the filter must be replaced. Tangential flow devices counteract this blocking with a permanent flow of the solution over the membrane.

Various ultrafiltration systems have been developed. Instead of air pressure, some use the centrifugal force, while others suck the solution through the ultrafiltration membrane with the aid of a vacuum. With filter devices mounted on the tip of a syringe, the solution can be manually concentrated.

Dialysis can be regarded as a special type of ultrafiltration. A dialysis bag containing the protein solution is placed in an outer, usually buffered, solution. Only small compounds, but not the protein, penetrate the dialysis membrane and are diluted in the large outer volume. Dialysis is used to remove low molecular weight components, for example from crude homogenates, or reduce high salt after gradients or ammonium sulfate precipitation. For complete removal of the low molecular weight components, the outer solution must be repeatedly exchanged.

To concentrate proteins, the dialysis bag containing the protein solution is embedded in a hygroscopic powder, polyethylene glycol or Sephadex G-25, which withdraws water together with low molecular weight components from the dialysis bag and, thus concentrates the protein solution inside. No special device is required, but relatively high losses in protein and enzyme activity can occur due to interaction of the protein with the large inner surface of the bag. The hygroscopic medium adhering to the outside of the dialysis membrane renders the recovery of proteins attached to the inside difficult.

3.4.5.3 Ultracentrifugation

Ultracentrifugation is especially suited for very large proteins, such as multienzyme complexes, which sediment at a high speed within a reasonable time, while most proteins require very long centrifugation times. Given the presence of a preparative ultracentrifuge, this method is easy but when compared to the membrane effect of ultrafiltration, the concentrated protein forms a dense pellet at the bottom of the centrifugation tube, which may agglutinate to insoluble aggregates. A cushion of 30% sucrose solution layered at the tube bottom can reduce this effect. The time required for concentration depends, besides the protein size, on the rotor dimension and the centrifugation speed. Several hours are needed with a preparative centrifuge, but the time can be considerably reduced by applying a bench-top ultracentrifuge or an air-driven ultracentrifuge (Airfuge®) with a small rotor. The ultracentrifugation method is also useful to segregate smaller proteins and components that remain in the supernatant.

3.4.5.4 Lyophilization

In contrast to the methods described hitherto, which yield concentrated protein solutions, lyophilization produces dry protein, a convenient storage form. However, this technique often causes considerable loss of enzyme activity. Therefore, some precautions must be regarded for successful performance.

Unlike other methods like ultrafiltration or ultracentrifugation, lyophilization removes only the solvent, while all other components are concentrated together with the protein. This holds for the buffer as well, resulting in high salt concentrations. Volatile buffers containing acetate, carbonate, triethylamine, or triethanolamine ions should be used, but both buffer components, the acidic and the basic one, must be equally volatile; otherwise the pH changes severely. Additives like glycerol or BSA can protect the protein from denaturation. After lyophilization, the dry protein sample possesses a fluffy consistency and even faint drafts can blow away the valuable material. Vacuum centrifuges are specially suited for the-concentration of small samples, but cooling is often a problem.

In some cases, vacuum concentration of enzyme solutions in a rotatory evaporator has been reported, although this method cannot be recommended. Elevated temperature is necessary for evaporation and foaming may occur; this is harmful for proteins due to the high surface tension.

3.4.5.5 Other Concentration Methods

Crystallization can be regarded as one of the best methods for concentration and is also an excellent storage form. However, only few enzymes, like catalase or urease, show a particular tendency to form crystals. The method requires relatively pure preparations but can also be used to improve the purity of the enzyme. A variety of elaborate methods exist to obtain crystals suited for X-ray crystallography, but for normal crystallization, such efforts are not necessary. Unlike chemical substances, proteins do not crystallize from saturated solutions; rather, an additive is required to induce crystallization, a chaotropic agent such as ammonium sulfate. Its concentration should be slightly below the precipitation limit of the respective protein, which must be present in high concentrations ($\geq$10 mg ml^{-1}). Crystals will be formed within some days in the cold. Slow vaporizing during this time and thus, a continuous concentration increase, promotes the crystallization process. The crystals should be kept in the crystallization solution. Upon dilution or dialysis, the crystals dissolve again.

Column chromatography usually causes dilution of the applied protein solution, but a certain concentration can be achieved by ion exchange and hydroxylapatite chromatography. This effect is useful for collection of proteins from large volumes of highly diluted solutions, where other concentration methods are too laborious. A considerable concentration effect ($\sim$5 mg ml^{-1}) resulting in high yield from very diluted solutions has been attained with aminohexyl agarose which, in this case, acts as a gentle ion exchanger (Schmincke-Ott and Bisswanger, 1980). The gel bed must be so small that it becomes nearly overloaded with protein. Most proteins bind to the material at low ionic strength (10 mM potassium phosphate pH 7.6). Elution is achieved in one step with high ionic strength (0.1 M or higher potassium phosphate pH 7.6). The concentrated protein solution appears immediately thereafter and must be collected in one single fraction. Preliminary studies should be undertaken to find out the optimum binding and elution conditions.

References

Beisenherz, G., Boltze, H.J., Bücher, T., Czok, R., Garbade, K.H., Meyer-Arendt, E., and Pfleiderer, G. (1953) *Z. Naturforsch.*, **8B**, 555–577.

Brewer, J.M., Pesce, A.J., and Ashworth, R.B. (1974) *Experimental Techniques in Biochemistry*, Prentice-Hall, Englewood Cliffs.

Cooper, T.C. (1977) *The Tools of Biochemistry*, John Wiley & Sons, Inc., New York.

Green, A.A. and Hughes, W.L. (1955) *Meth. Enzymol.*, **1**, 67–90.

Harris, E.L.V. and Angal, S. (1989) *Protein Purification Methods*, IRL Press, Oxford.

Schmincke-Ott, E. and Bisswanger, H. (1980) *Prep. Biochem.*, **10**, 69–75.

Scopes, R.K. (1993) *Protein Purification, Principles and Practice*, 3rd edn, Springer-Verlag, New York.

3.5 Enzyme Immunoassays

3.5.1 Radioimmunoassays

Radioimmunoassays (RIA) are highly sensitive methods used to determine an ^{131}I-radiolabeled antigen with a specific, often monoclonal, antibody (or conversely, a radiolabeled antibody with an antigen). Small, not immunogenic compounds, such as drugs or metabolites, can also be measured by coupling as *hapten* to an immunogenic protein. The assay is often performed in a competitive manner, where a known amount of radiolabeled antigen is given to the unknown sample of the same, but not labeled, antigen (Figure 3.3a). A limiting amount of a first antibody specific for the antigen traps part of the antigen and is, together with its bound antigen, precipitated by a second antibody, directed against the first one. In this way, the unbound antigen remains in solution and becomes separated from the bound one. Radioactivity is determined in the precipitate and from this value, the amount of unlabeled antibody can be derived: the lower the radioactivity found, the higher the amount of unlabeled antigen.

Notwithstanding its high sensitivity, working with radioactivity is troublesome and needs experience, specialized laboratory facilities, and expensive instruments. Alternative targets with high sensitivity are chemiluminescence and fluorescence labels. Comparably as sensitive as RIA and easier to perform, are **enzyme immunoassays (EIA)**, namely **enzyme-linked immunoadsorbent assays (ELISA)**. They need only the common equipment found in biochemical laboratories, such as UV/Vis photometers and centrifuges. These assays are based on the reaction of the antibody with its antigen and make use of the high catalytic efficiency of enzymes, which are labeled as indicators to the reactants. In the **solid-phase EIA**, one of the immunoreactants, the antigen or the antibody, is immobilized on a solid support. The immobilized reactant captures in the following *immunoextraction* step the complementary reactant from the sample. With EIA two alternative strategies can be pursued, *activity amplification* (AA), or *activity modulation* (AM). In the AA procedure, the antibody is present in large excess to obtain an intense maximum signal for the assayed compound. Thus, the antigen is complexed proportional to its concentration. In the AM method, the modulation of the enzyme signal by competition of the test molecule for the same immunoreactant is observed. Here, the sensitivity increases with lower immunoreactant concentrations. AM-type EIA are more specific than AA assays.

Sensitivity is determined by the **dose-response curve** $\mathrm{d}R/\mathrm{d}C$, corresponding to the change in response ($\mathrm{d}R$) per unit amount of reactant ($\mathrm{d}C$). The response plotted against the reactant concentration should result in a straight line. Its steepness reflects the sensitivity: the steeper the curves, the more sensitively the

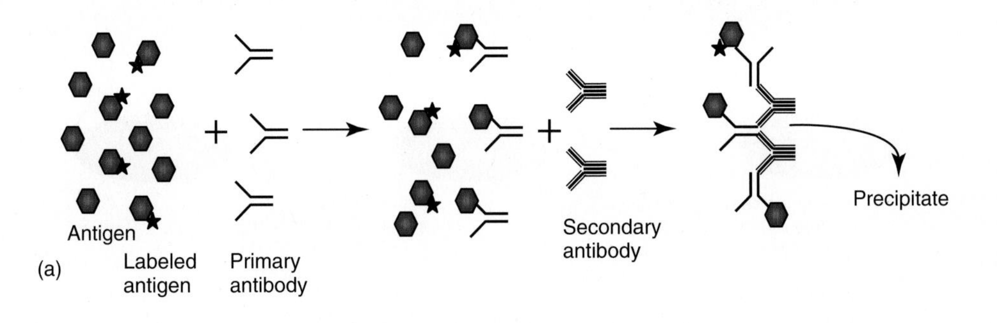

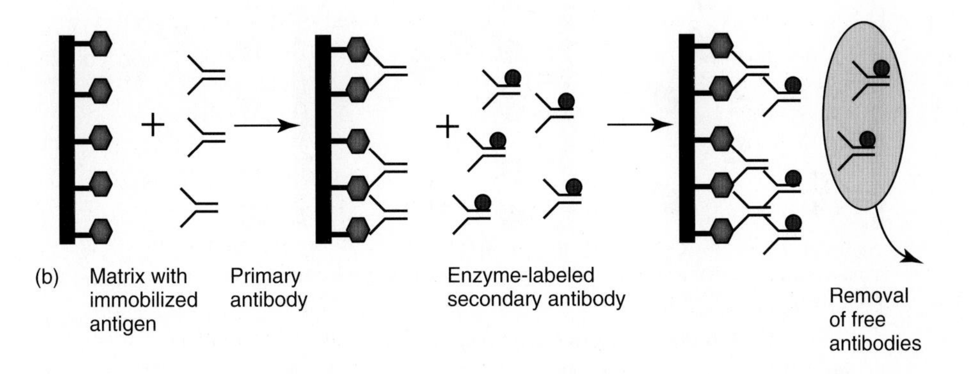

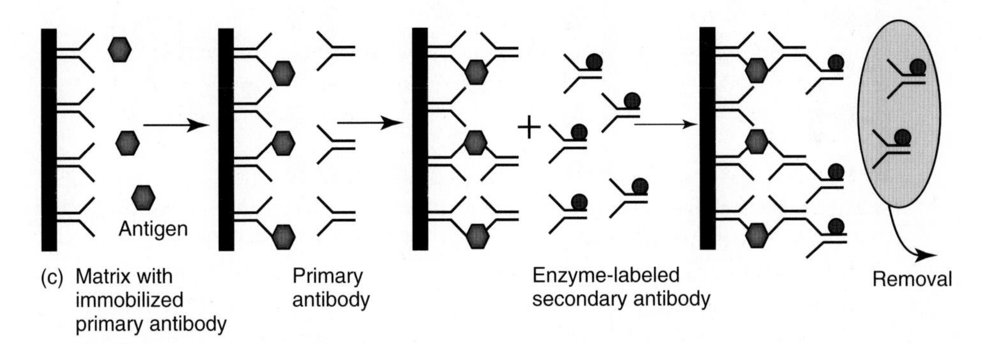

Figure 3.3 Schemes of different immunoassays: (a) competitive radioimmonoassay; (b) noncompetitive solid-phase immunoassay; (c) sandwich method; (d) noncompetitive homogeneous enzyme immunoassay (S_1, substrate of enzyme 1; P_1, product of enzyme 1 = substrate of enzyme 2; P_2, product of enzyme 2); and (e) competitive solid-phase enzyme immunoassay.

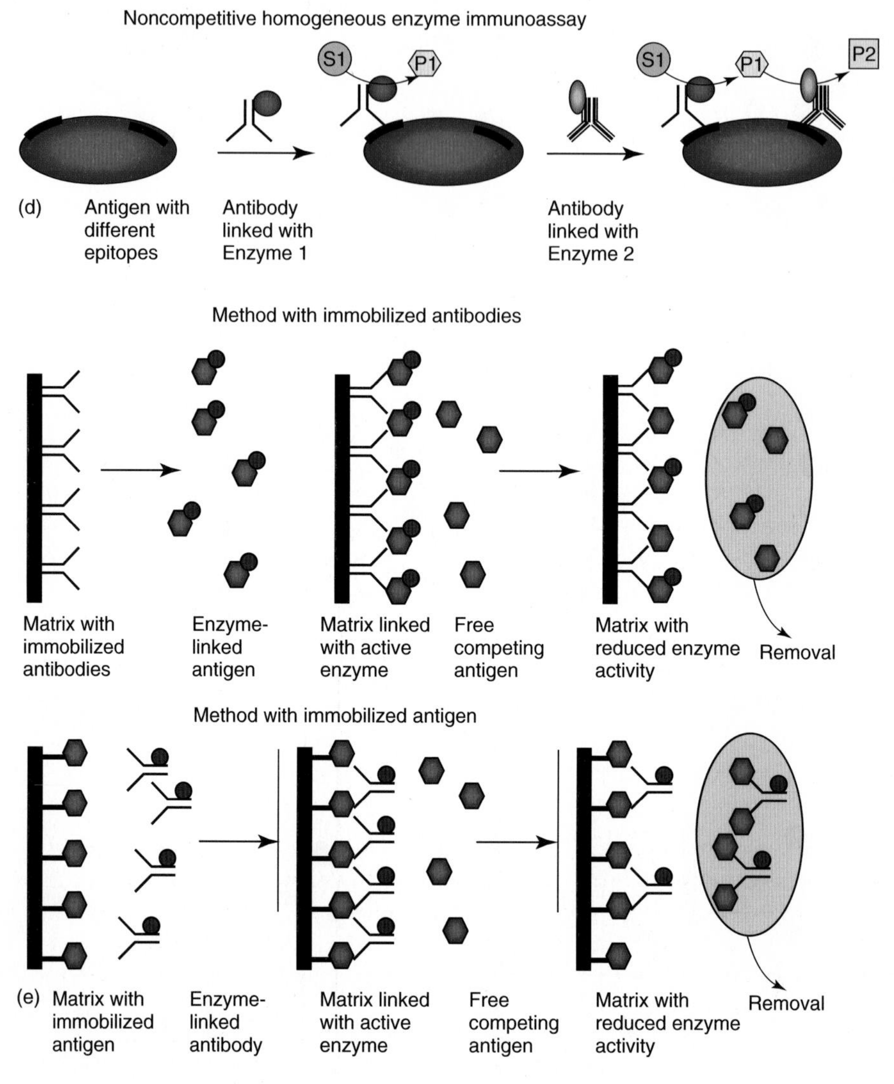

Figure 3.3 (Continued)

system responds to small changes in reactant concentration. Small amounts can be detected more easily by a less steep dependency (detection limit).

3.5.2 Noncompetitive Solid-Phase Enzyme Immunoassay

An immunoreactant, either antigen or antibody, is immobilized to a solid surface (Figure 3.3b). This captures the opposed (antibody or antigen) free immunoreagent from the sample, forming a second layer on the surface. Different methods can be applied to detect quantitatively the opposed immunoreagent. If it is an antibody, an enzyme-labeled secondary antibody directed against it can be used. For instance, if the primary antibody is a mouse-IgG, anti-mouse-IgG from rabbit or goat may be used. This secondary antibody should exclusively react with the mouse-IgG fraction, but not with the antigen fixed to the support. The enzyme activity fixed to the secondary antibody is measured after removing the fraction of the unbound antibody. For the *bridge method*, nonimmunological recognition systems are used, for example biotin–avidin or protein A, a polypeptide from *Staphylococcus aureus*, binding to the Fc region of IgG molecules.

Instead of the antigen, the primary antibody may be directly fixed to the support, capturing antigens as the opposed free immunoreactants from the sample. In a second step, the unbound primary antibody attaches to the captured antigen (*sandwich method* Figure 3.3c) and finally, in a third step, the secondary enzyme-labeled antibody is bound.

A **noncompetitive homogeneous enzyme immunoassay** without need of a solid phase is achieved by the reactions of two types of monoclonal antibodies directed against different epitopes of the same antigen (Figure 3.3d). The antibodies are labeled with two different enzymes, selected in a manner such that the product of the first enzyme (e.g., hydrogen peroxide from glucose oxidase) serves as a substrate for the second (peroxidase). Its product will be formed in considerable amounts only if both enzymes are brought together in direct contact by binding to the same antigen.

3.5.3 Competitive Solid-Phase Enzyme Immunoassay

In this assay, the antibody is immobilized to a solid support and the free, enzyme-labeled antigen is bound to it. In a second step, the nonlabeled antigen from the sample competes with the bound enzyme-labeled antigen and displaces it. Thus, support-fixed enzyme activity will be reduced and after washing out the displaced enzyme-labeled antigen, the amount of product formed is inversely proportional to the amount of free antigen (Figure 3.3e).

In the opposite procedure, the antigen is immobilized to the support, to which the free, enzyme-labeled antibody will be bound. Free antigen in the sample competes with the enzyme-labeled antibody, thus reducing the amount of enzyme activity to be fixed to the support.

3.5.4 Methods for Enzyme Immunoassays and Immobilization Techniques

3.5.4.1 Protein Coupling to Cyanogen Bromide–Activated Agarose

This is a very useful method for covalent immobilization of enzymes or proteins such as peroxidase and antibodies to agarose or dextrans, for example Sepharose® or Sephadex®. Due to the high toxicity of cyanogen bromide, commercially prepared material (e.g., CNBr-activated Sepharose®) is emphatically recommended. The procedure must be performed quickly because the strong alkaline condition needed for activation destabilizes the structure of the gel.

Reagents and solution

Sepharose 4B, suspension, 6.6 g in 20 ml H_2O

3 M NaOH ($M_r = 40.0$; 12 g in 100 ml)

Cyanogen bromide (CNBr, $M_r = 105.9$)

$NaHCO_3$/NaCl solution: 0.1 M $NaHCO_3$ ($M_r = 84.0$; 0.84 g in 100 ml) pH 8.2, containing 0.5 M NaCl ($M_r = 58.4$; 2.92 g in 100 ml)

1 M glycine ($M_r = 75.1$; 1.5 g in 20 ml H_2O)

0.1 M sodium acetate/acetic acid pH 4.0

0.1 M potassium phosphate pH 7.6

Preparation of CNBr-activated Sepharose 4B

The following procedure must be done in a fume hood with the utmost care.

Adjust the Sepharose 4B suspension to pH 11.2 with 3 M NaOH, using a pH-meter equipped with a pH-electrode

A dropping funnel is mounted on a balance and tared; add 0.67 g solid CNBr and dissolve in a small volume of water; close the funnel tightly with a stopper

Drop the CNBr solution slowly to the gel suspension under stirring (use a KPG or propeller stirrer, not a magnetic stirrer); keep the pH constant at pH 11.2 by adjusting with 3 M NaOH at the pH meter. If the pH remains constant for about 6 min, wash the gel thoroughly on a Buchner funnel or a frit with about 1 l of water

Coupling of protein

Transfer the CNBr-activated Sepharose into 20 ml $NaHCO_3$/NaCl solution and add 0.2 g of the protein

Shake the suspension gently for 3 h at room temperature or overnight at 4 °C

Determine the remaining protein content in the supernatant with a protein assay; the difference to the total protein is supposed to be immobilized

Add 5 ml 1 M glycine to block free reactive groups

Wash the gel with 0.1 M sodium acetate/acetic acid pH 4.0 and thereafter, with 0.1 M potassium phosphate pH 7.6

For storage add sodium azide (NaN_3) to a final concentration of 0.1 M to the gel

3.5.4.2 Coupling of Diaminohexyl Spacer

Instead of direct coupling of the protein to the matrix, a spacer can be used as a connecting bridge between matrix and protein.

Reagents and solutions

0.1 M potassium phosphate pH 7.6

1.6-hexane diamine (1.6 diaminohexane, M_r 116.2)

0.5 M NaCl ($M_r = 58.4$; 2.92 g in 100 ml)

EDC · HCl (*N*-ethyl-*N'*-(3-dimethylaminopropyl)carbodiimide hydrochloride, $M_r = 191.7$)

1 N HCl

Procedure

The procedure for coupling the spacer is the same as for the protein (see above), instead of protein take the following:

1) 1 mg ml^{-1} (20 mg in 20 ml for the described procedure) of 1.6 hexane diamine
2) Wash the resulting aminohexyl derivative (AH-Sepharose) with 0.5 M NaCl
3) For coupling of the protein to the spacer, adjust the pH to 4.5 with 1 N HCl
4) Add 10 mg ml^{-1} gel EDC · HCl and
5) 10 mg ml^{-1} gel of the protein, keep the final pH at 4.5
6) Control the pH after 1 h and shake the mixture overnight at 4 °C
7) Wash the gel with 0.1 M potassium phosphate pH 7.6

3.5.4.3 Periodate Activation of Cellulose

This procedure is applicable for materials with vicinal glycols, like cellulose, agarose, dextrans (Separose, Sephadex, Sephacryl). Aldehyde groups are formed by periodate oxidation. They react with amino groups to Schiff's bases, which can be stabilized by borohydride reduction.

Reagents and solutions

Cellulose powder

0.1 M $NaHCO_3$/0.06 M $NaIO_4$ ($NaHCO_3$, $M_r = 84.0$; $NaIO_4$, $M_r = 213.9$; 8.4 g $NaHCO_3$, 12.8 g $NaIO_4$ in 100 ml H_2O)

0.1 M sodium carbonate buffer ($NaHCO_3$/$NaCO_3$) pH 9.0

$NaBH_4$ solution ($NaBH_4$, M_r = 37.8; 0.1 g in 10 ml 0.1 M NaOH, freshly prepared)

0.1 M sodium acetate/acetic acid pH 4.0

PBS (phosphate buffered saline: 10 mM sodium/potassium phosphate pH 7.2, 0.8% NaCl, 0.02% KCl)

0.05% Tween 20 in PBS

Procedure

Suspend 2 g cellulose in 40 ml 0.1 M $NaHCO_3$/0.06 M $NaIO_4$ and keep for 2 h in the dark

Wash the cellulose on a frit with 0.1 M sodium carbonate buffer pH 9.0 and suspend in 40 ml of this buffer

Add 40 mg of the protein (final concentration: 1 mg ml^{-1}) and shake overnight at room temperature

Add 2 ml $NaBH_4$ solution to reduce the Schiff's base and shake for 30 min at room temperature

Add another 2 ml $NaBH_4$ solution and shake for a further 30 min

Wash the solid material with 0.1 M sodium acetate/acetic acid pH 4.0, thereafter with PBS, and finally with 0.05% Tween 20 in PBS

3.5.4.4 Introduction of Thiol Groups into Proteins (Antibodies)

To link enzyme activities to proteins such as antibodies, free thiol groups should be available. They can be introduced with specific reagents, for example AMSA, which is described here.

Reagents and solutions

AMSA-solution (*S*-acetylmercaptosuccinic anhydride; M_r = 174.2, 60 mg in 1 ml) *N,N'*-dimethylformamide

0.1 M potassium phosphate pH 6.0

0.1 M Tris/HCl pH 7.0

Protein (antibody) solution (10 mg ml^{-1} in 0.1 M potassium phosphate pH 7.0)

Nitrogen gas

0.5 M hydroxylamine/0.01 M EDTA (3.46 g $NH_2OH \cdot HCl$, M_r = 69.5; 0.29 g EDTA, M_r = 292.3, in 100 ml 0.1 M Tris/HCl pH 7.0)

Sephadex G-25-column, 1.5 × 30 cm, equilibrated with 0.1 M potassium phosphate pH 6.0, 5 mM EDTA

Procedure

Add 0.1 ml AMSA solution to 1 ml protein solution under nitrogen gas, seal the tube, and stir gently for 0.5 h at room temperature

Add 0.2 ml 0.5 M hydroxylamine/0.01 M EDTA and adjust to pH 7.0

Incubate at 30 °C for 4 min

Apply the mixture to the Sephadex column and elute with 0.1 M potassium phosphate pH 6.0, 5 mM EDTA

Measure absorption at 280 nm to detect the protein fractions, collect the strongly absorbing fractions, and concentrate to 1 ml by ultrafiltration

Reference

Klotz, I.M. and Heiney, R.E. (1962) *Arch. Biochem. Biophys.*, **96**, 605–612.

3.5.4.5 Conjugation of a Protein (Antibody) with an Enzyme (Peroxidase)

Reagents and solutions

Peroxidase (horse radish)

SMCC solution (succinimidyl 4-(*N*-maleimidomethyl)cyclohexane-1-carboxylate; 4-(*N*-maleimidomethyl)cyclohexane-1-carboxylic acid *N*-hydroxysuccinimide ester; $M_r = 334.3$; dissolve 5 mg in 60 µl *N,N'*-dimethylformamide, at 30 °C)

0.1 M potassium phosphate pH 7.0

0.1 M potassium phosphate pH 6.0

Sephadex G-25-column, about 1.5 × 30 cm, equilibrated with 0.1 M potassium phosphate pH 6.0

0.1 M cysteamine ($M_r = 77.2$; 77.2 mg in 10 ml)

Procedure

Dissolve 10 mg peroxidase in 1 ml 0.1 M potassium phosphate pH 7.0

Warm up to 30°C and add 60 µl SMCC solution (final concentration 15 mM)

Stir gently for 1 h

Centrifuge to separate the precipitate

Apply the supernatant onto the Sephadex-G25 column and elute with 0.1 M potassium phosphate pH 6.0

Peroxidase elutes first, determine the enzyme activity (cf. assay in Section 3.3.1.32), or absorption at 280 nm, concentrate the active fractions to about 1 ml by ultrafiltration

Add 10 mg of thiolated protein (antibody) in 1 ml and incubate 1 h at 30 °C

Add 0.15 ml 0.1 M cysteamine to block the unreacted maleimide groups

The reaction products can be further purified by gel filtration on the Sephadex-G25 column using the conditions described above.

Reference

Yoshitake, S., Imagawa, M., Ishikawa, E., Niitsu, Y., Urushizaki, I., Nishiura, M., Kanazawa, R., Kurosaki, H., Tachibana, S., Nakazawa, N., and Ogawa, H. (1982) *J. Biochem.*, **92**, 1413–1424.

3.5.4.6 Conjugation of β-Galactosidase to Proteins (Antibodies) by MBS

Reagents and solutions

MBS solution (3-maleimidobenzoyl-*N*-hydroxysuccinimide ester, $M_r = 314.3$; 20 mg/ml dioxane)

0.1 M potassium phosphate pH 7.0, 10 mM $MgCl_2$, 50 mM NaCl (2 g $MgCl_2 \cdot 6H_2O$, 2.9 g NaCl per liter of buffer)

Protein (antibody) solution (1 mg ml^{-1} in 0.1 M potassium phosphate pH 7.0, 50 mM NaCl)

β-Galactosidase (from *Escherichia coli*)

Sephadex G-25-column, 1 × 30 cm, equilibrated with 0.1 M potassium phosphate pH 7.0, 10 mM $MgCl_2$, 50 mM NaCl

1 M 2-mercaptoethanol ($M_r = 78.1$; 0.78 g in 10 ml)

0.1 M dithioerythritol, (DTE, $M_r = 154.2$, 154 mg in 10 ml)

Procedure

Add 1 µl MBS solution to 1 ml protein solution

Incubate 25 °C for 1 h

Apply to the Sephadex G-25 column, elute with 0.1 M potassium phosphate pH 7.0, 10 mM $MgCl_2$, 50 mM NaCl

Collect protein fractions (absorption at 280 nm) and add immediately β-galactosidase (sample containing 2 mg)

Incubated at 30 °C for 1 h

Stop the reaction by addition of 50 μl ml^{-1} 0.1 M dithioerythritol

Reference

Kitagawa, T. and Aikawa, T. (1976) *J. Biochem.*, **79**, 233–236.

3.5.4.7 Conjugation of Alkaline Phosphatase to Antibodies by Glutaraldehyde

This method is generally applicable for cross-linking of proteins. A similar technique with dimethylsuberimidate, also suitable for enzyme immunoassays is described in Section 3.4.4.

Reagents and solvents

Glutaraldehyde (glutardialdehyde, pentane-1,5-dial, $M_r = 100.1$; 50% aqueous solution, commercially available)

Alkaline phosphatase (from bovine intestinal mucosa, 10 mg ml^{-1})

0.5 M Tris/HCl pH 8.0

Antibody solution (10 mg ml^{-1} in buffer, PBS, or 0.5 M Tris/HCl pH 8.0)

PBS (phosphate buffered saline: 10 mM sodium/potassium phosphate pH 7.2, 0.8% NaCl, 0.02% KCl)

BSA, bovine serum albumine

Dialysis tubes (Visking)

Procedure

Mix appropriate volumes of alkaline phosphatase solution (containing 25 mg) and antibody solution (10 mg) and dialyze overnight at 4 °C against PBS, changing the PBS solution several times

Add glutaraldehyde to the dialyzed solution, to a final concentration of 2% (v/v)

Stir gently for 2 h at 4 °C

Dialyze against PBS for 6 h and finally against 0.5 M Tris/HCl, pH 8.0, overnight at 4 °C. For storage of the solution, add 1% BSA

General References

Tijssen, P. (1985) *Practice and Theory of Enzyme Immunoassays*, Elsevier, Amsterdam.
Kemeny, D.M. and Challacombe, S.J. (1988) *ELISA and Other Solid Phase Immunoassays*, John Wiley & Sons, Ltd, Chichester.
Nowotny, A. (1979) *Basic Exercises in Immunochemistry*, Springer-Verlag, Berlin.

4 Binding Measurements

4.1 Different Types of Binding

4.1.1 General Considerations

Any specific interaction within the cell among proteins, macromolecules, aggregates, membranes, and metabolites is initiated by a binding step, such as binding of substrate or inhibitors to enzymes, antibody binding, or hormone action. Thus, binding must be regarded as one of the most important biological processes and its perception is absolutely necessary for understanding cell functions. However, binding processes are more difficult to observe than enzyme reactions, which are accompanied by the chemical conversion of substances, whereas binding components remain essentially unchanged. In the following sections, different types of binding are discussed together with some theoretical derivations, followed by the description of experiments to observe binding processes.

Binding can simply be characterized by the following relationship:

$$A + E \rightleftharpoons EA \tag{4.1}$$

E is the target of the binding process, an enzyme or, more generally, a **macromolecule**, both terms are used synonymously in this chapter. A is the (mostly small) binding component, the **ligand**, a substrate, cofactor, activator, or inhibitor binding to an enzyme, an antibody, a hormone, neurotransmitter, or a drug binding to a macromolecule. The essential difference between binding processes and enzyme reactions is the fact that, even with substrates, only the direct contact, but no subsequent reaction, is considered. This makes theoretical considerations easier, but in practice care must be taken that only the binding step and no following step is observed, that is, enzyme catalysis must be completely excluded.

Binding can be defined as the contact between two originally separated compounds. This contact can persist either for a distinct, often short, time (**reversible binding**) or permanently (**irreversible binding**). It is assumed that even irreversible binding is initiated by a reversible binding step, followed by irreversible fixation. Further, it must be discerned between **specific** and **unspecific binding**. The latter case is present if binding is solely caused by electrostatic or hydrophobic

Practical Enzymology, Second Edition. Hans Bisswanger.

interactions; any compound possessing such features can bind. Examples are binding of detergents or dyes (SDS, Coomassie blue) to proteins. In the case of specific binding, only a certain compound will be accepted by the target molecule (enzyme or macromolecule) and no other compound can bind (although structurally related substances are often accepted). This exclusive binding is responsible for the high specificity, for example, of enzyme reactions and antibody actions. Specific reversible binding is decisive for all biological processes and, therefore, only this type of binding is treated in detail.

Irreversible binding can also be either specific or unspecific. It is mostly a covalent interaction, especially in the case of *unspecific binding*, when reactive substances, for example, with halogen or thiol residues, attack functional groups of proteins such as sulfhydryl or amino groups. **Suicide substrates** are characteristic for *specific irreversible binding*, which are reactive substances structurally resembling the substrate. They imitate the catalytic mechanism, but during the course of the reaction a covalent bound to the enzyme is formed and the active site is blocked. **Transition state analogs** belong to another group of irreversible inhibitors, which mimic the transition state between substrate and product. During the transition state the substrate binds, even noncovalent, extremely strong to the active site. Thiamin thiazolon diphosphate, a transition state analog of thiamin diphosphate, the cofactor in catalysis of oxidative decarboxylation of α-oxo acids, cannot be removed from the enzyme by dialysis for several days. Antigen–antibody and biotin–avidin interactions are further examples of this type of noncovalent irreversible binding; it is characterized by extremely low binding constants ($K_d \leq 10^{-10}$ M).

4.1.2
How to Recognize Specific Reversible Binding?

For binding studies, it must be first established that the interaction of the ligand with the macromolecule is both specific and reversible. **Competition** experiments are most efficient to discern specific binding from unspecific binding. A physiological ligand can be displaced from its binding site on the enzyme by high concentrations of a structurally related compound, a competitor which is often a synthetic compound. Competition can be analyzed by binding assays (see following text). One of the two compounds, a ligand or competitor, should be labeled either with a radioactive isotope or with a fluorescent group. The labeled compound is incubated with the enzyme or macromolecule, and the fraction of unbound compound is separated. The first result of this experiment is the observation of a detectable signal of the label in the enzyme fraction, reflecting the interaction between both components. The absence of such a signal indicates the lack of any binding, while the intensity of the signal is a direct measure for the binding of the labeled compound. A decrease of this signal on adding the unlabeled compound is a clear indication of a specific binding site at the macromolecule, which can be occupied by both compounds. If, however, the signal remains essentially unchanged, the unlabeled compound cannot displace the labeled one and it can be assumed that there is no specific binding site for both compounds.

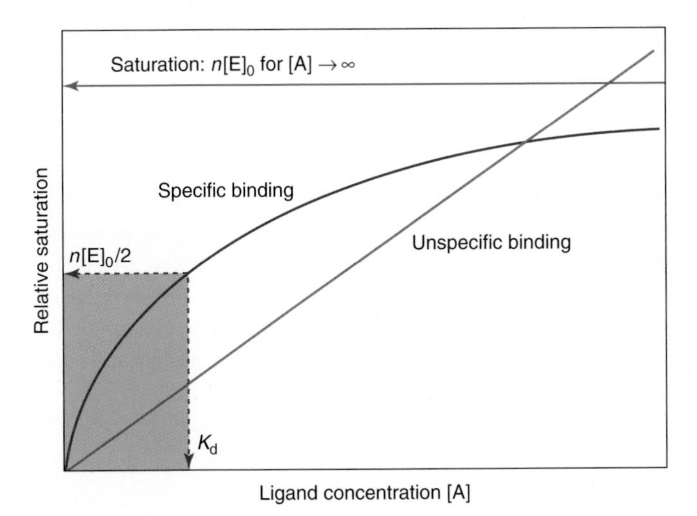

Figure 4.1 Curves for specific and unspecific binding. The determination of the dissociation constant K_d is shown.

Such experiments are laborious and require labeling of one compound. It must also be considered that labeling of a compound changes its structure and, therefore, its binding capacity, lack of binding can be caused by this modification. This problem does not exist when radioactive isotopes are used. In some cases, an intrinsic feature of the compound can be used, for example, its fluorescence. Specific binding can also be recognized from the saturation behavior of the ligand. Specific binding sites must be saturable, while unspecific ones must not. Upon enhancing the ligand concentration, the extent of binding increases only to a finite saturation value, where all bindings sites are occupied (Figure 4.1). In contrast, when there are no specific binding sites, no defined saturation value can be reached and binding increases in a linear manner.

Irreversible binding can be discerned from reversible binding by the removal of the ligand after incubation with the macromolecule. This can be achieved by dialysis, ultrafiltration, gel filtration, or ultracentrifugation. A prerequisite is a detectable signal for the ligand to measure its binding to the macromolecule, which can be either an intrinsic feature (absorption, fluorescence) or a label as described above. If there is no such signal, the amount of the free ligand after separation can be determined, which, in the case of reversible binding, should be the same as the initial amount given to the macromolecule, whereas in the case of irreversible binding no ligand should be regained (provided that the macromolecule was present in a higher concentration than the ligand). This procedure is simple, but not very accurate, since the amount of regained ligand can be reduced also by unspecific losses, such as adsorptions to the material of the respective device.

A crude estimation for specific binding is the value of the binding constant. Values higher than 10^{-3} M are hints for unspecific binding, but there are numerous exceptions, for example, binding constants for glucose or for H_2O_2 with catalase.

If irreversible binding of a ligand to an enzyme causes inactivation, this can be detected by its time dependency. The progression of inactivation usually occurs in an exponential manner (first-order kinetics). In contrast, binding of a reversible inhibitor reduces the enzyme activity instantly and the reduced activity remains constant and no time dependency can be observed. For this experiment, the concentration must be considered; at high ligand concentrations, even irreversible binding may occur instantly. If the first experiment shows no time dependency, the ligand concentration should be reduced. For irreversible inhibition, as binding of one molecule of the inhibitor already results in inactivating the enzyme, an equimolar amount of both will be sufficient to inactivate the enzyme completely, and now the time dependency should become visible; otherwise, inhibition should be reversible.

4.1.3 Experimental Aspects

From the relationship shown for a simple binding process (Eq. (4.1)), a binding equation can be derived, which is mathematically equivalent to the Michaelis–Menten equation (Section 2.2.4) and, accordingly, all derivations discussed there, such as the diagrams, are principally applicable. The similarities between both equations are summarized in Box 4.1. Direct plotting of the ligand concentration against the relative saturation degree yields a hyperbolic function, from which the dissociation constant can be obtained, as shown in Figure 4.1. Often, it is sufficient to learn from a binding experiment that the ligand interacts specifically with the enzyme. For a more detailed analysis, the dissociation constant as a measure of the binding affinity should also be determined. The accordance of the measured values with the hyperbolic binding function is an indication for a normal binding behavior; deviations point toward more complex binding processes.

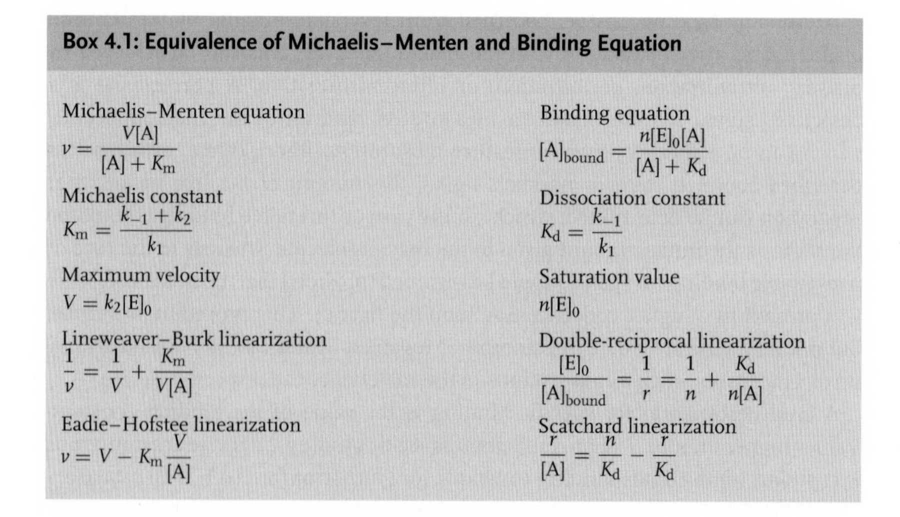

Box 4.1: Equivalence of Michaelis–Menten and Binding Equation

Michaelis–Menten equation	Binding equation
$v = \frac{V[\mathrm{A}]}{[\mathrm{A}] + K_\mathrm{m}}$	$[\mathrm{A}]_\mathrm{bound} = \frac{n[\mathrm{E}]_0[\mathrm{A}]}{[\mathrm{A}] + K_\mathrm{d}}$
Michaelis constant	Dissociation constant
$K_\mathrm{m} = \frac{k_{-1} + k_2}{k_1}$	$K_\mathrm{d} = \frac{k_{-1}}{k_1}$
Maximum velocity	Saturation value
$V = k_2[\mathrm{E}]_0$	$n[\mathrm{E}]_0$
Lineweaver–Burk linearization	Double-reciprocal linearization
$\frac{1}{v} = \frac{1}{V} + \frac{K_\mathrm{m}}{V[\mathrm{A}]}$	$\frac{[\mathrm{E}]_0}{[\mathrm{A}]_\mathrm{bound}} = \frac{1}{r} = \frac{1}{n} + \frac{K_\mathrm{d}}{n[\mathrm{A}]}$
Eadie–Hofstee linearization	Scatchard linearization
$v = V - K_\mathrm{m}\frac{V}{[\mathrm{A}]}$	$\frac{r}{[\mathrm{A}]} = \frac{n}{K_\mathrm{d}} - \frac{r}{K_\mathrm{d}}$

Another important information to be gained from binding experiments is the number of binding sites for the ligand at the enzyme. Many enzymes possess oligomeric structures composed of several, mostly identical, subunits n. This value can be obtained from saturation, which has the meaning $n[E]_0$, provided that the molarity of the enzyme is known.

Various methods for the determination of binding equilibria exist, but to find out the appropriate method for a special binding problem is not always easy. Compared with enzymes assays, binding measurement are much more difficult; the main differences between them are listed in Box 4.2. Unlike enzyme reactions proceeding with the chemical conversion of the substrate, the binding components remain unchanged and the binding complex exists in a rapid equilibrium with the free components. Any attempt to isolate the binding complex for the analysis influences the equilibrium. Therefore, the experiments must be performed in a manner such that the equilibrium is not disturbed. Two principles allow the performance of binding experiments observing this prerequisite: (i) difference in the size between the free, mostly low-molecular ligand and the bound one, which admits the size of the macromolecule and (ii) binding which can influence the spectral features of either the ligand or the enzyme, although such effects are rather small and require accurate measurements and exact instruments. Box 4.3 summarizes the most important methods. A new and sophistic technique is the surface plasmon resonance (SPR) method.

A serious obstacle for binding measurements is the need for high enzyme amounts, because the intensity of the measuring signal is mostly rather weak; but, as it is directly related to the enzyme concentration, large amounts up to several milligrams increase the accuracy of the data. In comparison, for enzyme assays only catalytic enzyme amounts are required.

Box 4.2: Differences between Enzyme Assays and Binding Measurements

Enzyme assay	Binding measurement
Catalytic enzyme amounts: enzyme $\ll$ substrate	Comparable amounts of enzyme and ligand
Detection sensitivity depends on product formation $\rightarrow$ low enzyme amounts	Detection sensitivity depends on the enzyme amount $\rightarrow$ high enzyme amounts
Substrate and product are chemically different	Free and bound components are chemically identical
Progressing reaction	Rapid reversible equilibrium
Time-dependent measurement	Time-independent measurement

Box 4.3: Binding Methods – Advantages and Disadvantages

Method	Advantages	Disadvantages
Equilibrium dialysis	Applicable for any binding system Robust against disturbances Simple device	Low sensitivity Radioactive-labeled ligands required Long duration
Ultrafiltration	Fast method Simple device	Low accuracy Larger sample amounts
Column chromatography	Gentle method No special device required	Long duration Low accuracy, larger sample amounts
Ultracentrifugation	Relatively accurate No special device (but the presence of ultracentrifuge presupposed)	Laborious method Larger sample amounts
Difference spectroscopy	Accurate method	Detectable absorption difference and sensitive double-beam photometer required
Fluorescence spectroscopy	Accurate method Low sample amounts	Detectable fluorescence change and sensitive spectrofluorimeter required Various disturbances (quenching)

While enzyme assays measure time-dependent reactions, equilibria are time independent and can be observed for a longer time. However, the limited stability of enzymes can be a problem, because equilibria depend strongly on the temperature and must be performed at physiological conditions, for example, 37 °C. Other influences, such as pH, ionic strength, and unspecific interactions, must also be considered. In preliminary binding experiments carried out under the final conditions, the stability of the enzyme must be controlled, testing the constancy of its activity.

Similarly, the stability of the ligand must be controlled. Catalytic conversion by the enzyme must be completely excluded if the ligand is an enzyme substrate. This can be achieved by the removal of essential cofactors or cosubstrates. In the case of single substrate reactions, inactive substrate analogs must be used, although their binding behavior may be modified.

The rules presented in Box 2.6 for substrate concentrations in enzyme assays are valid also for binding measurements. If only binding needs to be detected, high (saturating) ligand amounts are sufficient. For the investigation of the binding mechanism and the determination of the dissociation constant, the range around its value (from about 1/10th to 10-fold) should be covered. However, for binding experiments, also the high enzyme amount must be considered. The binding (or Michaelis–Menten) equation refers to the concentration [A] of the free, unbound ligand, therefore the ligand concentration applied to the assay must be reduced by the amount of ligand bound to the enzyme. As this amount is usually unknown (it should be obtained just by the experiment), it can be roughly estimated from the enzyme concentration, and a correspondingly higher ligand range may be chosen.

As the different binding methods have been described in detail elsewhere (Bisswanger, 2008), only the general principles of the respective methods with some practical assays are presented here.

4.2 Binding Measurements by Size Discrimination

Principally, all methods used in biochemistry for the separation of components according to their size can be modified to serve for binding assays. As already mentioned, the general disadvantages are the long duration and the requirement of high enzyme amounts. On the other hand, these methods are generally applicable, since any binding step proceeds with an increase in the size of the bound complex, while spectral changes observed with the other types of methods occur only by chance. The respective separation method must be modified in a manner that the equilibrium is maintained throughout the experiment. For instance, if one tries to isolate the enzyme–ligand complex by gel filtration on a chromatography column, it will immediately dissociate and only the separated components are eluted.

4.2.1 Equilibrium Dialysis

Using the example of this classical binding method, the principles of binding measurements by size discrimination are discussed (Figure 4.2a). The method is not very accurate and sensitive, but robust and applicable for nearly each binding system, as long as a convenient detection method for the ligand is available. An appropriate device can be constructed in several modifications and chamber volumes (Englund *et al.*, 1969). Even without such devices, equilibrium dialysis can be performed simply with dialysis bags (e.g., Visking®, one layer of such a bag serves also as a semipermeable membrane for the dialysis devices). Enzyme solution is filled in such a bag, which is closed by a knot and immersed in a small tube containing the same volume of a ligand solution. Several tubes with varying ligand concentrations can be used, while the enzyme concentration should be the same in all bags.

A scheme of the equilibrium dialysis apparatus is shown in Figure 4.2b. It consists of two identical cylinders prepared from synthetic material (Teflon, Plexiglas). A central hole is drilled into both cylinders, and they are compressed together in a holder by screws in a manner that both holes face one another, with the

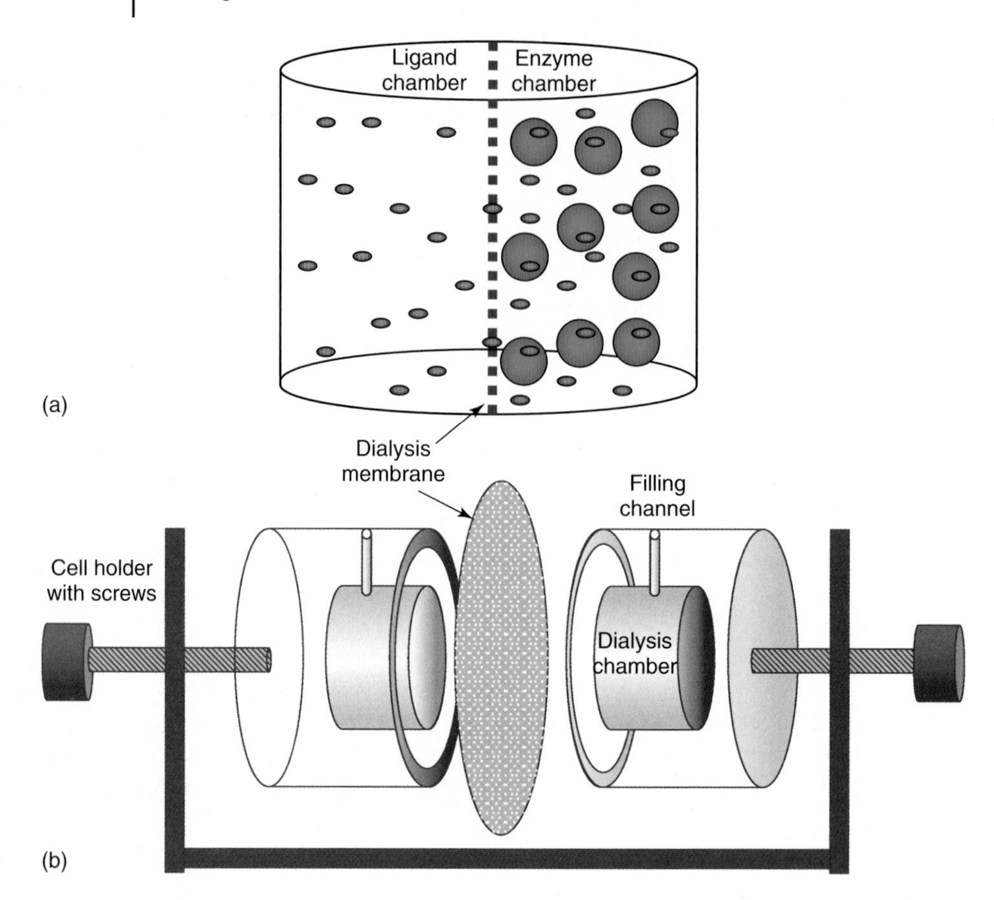

Figure 4.2 Principle of equilibrium dialysis (a) and scheme of an equilibrium dialysis cell (b).

dialysis membrane stretched between them. For density, ringlike protrusions are fitted at the contact sites. Lateral channels to the central hole allow filling and removing of assay solutions. The dimension of the central hole determines the assay volume, for example, 6 mm diameter, 4 mm height for 50–100 μl. The macromolecule is filled in one chamber and the ligand solution in the other, both separated by the semipermeable dialysis membrane, which allows penetration of only the low-molecular ligand (Figure 4.2a). For one experiment, 10–20 dialysis cells should be used, varying the ligand concentration, while the enzyme concentration should be the same in all chambers. The cell holders are integrated in the cylinder, which rotates slowly, driven by an electric motor. The cylinder is placed in an incubator or submersed in a water bath to maintain a constant temperature.

During dialysis, the ligand distributes equally between both chambers, but in the enzyme chamber, the ligand bound to the enzyme adds to the free one. Therefore, in this chamber, an equilibrium between bound and free ligand is formed and the latter one equalizes with the ligand chamber. After equilibration, the ligand concentration in the ligand chamber represents the free ligand in equilibrium [A], while the enzyme chamber contains the same and, in addition, the bound ligand ([A] + $[A]_{bound}$). Subtraction of the value of the ligand chamber from that of the enzyme chamber yields the amount of bound ligand.

Before carrying out a binding experiment, some control experiments should be performed. The dialysis time should be as short as possible, because the experiment must be carried out at physiological temperature (37 °C) where many enzymes are unstable. On the other hand, the dialysis time must be long enough to establish a complete equilibration between both chambers. This time depends on the special apparatus, the dialysis membrane, and the size of the ligand. To determine the minimum dialysis time, a ligand solution is filled in one chamber and the other contains the same volume of water. Several dialysis cells are charged in the same manner. After distinct time intervals, aliquots are removed from both chambers of a cell and the ligand concentration is determined. While the concentration in the ligand chamber decreases, it increases in the buffer chamber, both striving toward the same final value, which should be just the half of the initial ligand concentration. The time required to reach this value should be taken as minimum dialysis time.

During the binding experiment, the enzyme and ligand must remain stable. This should also be established in preliminary assays, testing the activity before and after the experiment. The detection method for the ligand must be as sensitive as possible. Absorption measurements and even colorimetric assays are mostly not sensitive enough. Radioactive labeling of the ligand is a favored procedure, but it is connected with considerably experimental effort.

4.2.1.1 Binding of Indole to Bovine Serum Albumin

This representative experiment should demonstrate the principle of equilibrium dialysis. It can also be adapted to related methods, such as ultrafiltration and ultracentrifugation. Bovine serum albumin (BSA) is used instead of an enzyme. It is available in large amounts and, therefore, sample volumes of 1 ml can be taken. This allows the application of a less sensitive colorimetric detection method, avoiding radioactive labeling. The experiment must be performed carefully to minimize the scatter, which is considerably large with this method.

Serum albumin, the main transport protein in blood, possesses a remarkable structure, consisting of a series of repetitive units, forming three similar domains with binding sites for fatty acids, bile pigments (e.g., billirubin), and tryprophan (Peters, 1985). In the present assay, indole is taken as a tryptophan analog. It can be detected with Ehrlich's reagent, with which it forms a rosindole chromophore (Figure 4.3), absorbing at 572 nm. A standard curve for the chromophore is prepared in the first experiment, followed by the determination of the minimum dialysis time and the preparation of the binding function. Finally, calculation and evaluation of binding experiments are discussed as the example of this experiment.

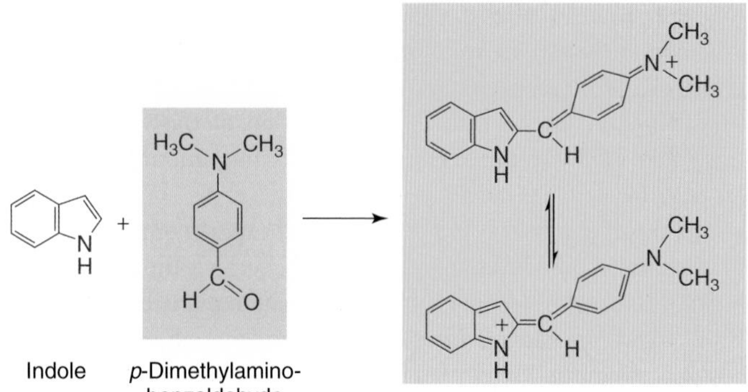

Figure 4.3 Indole reaction with Ehrlich's reagent.

Reagents and solutions

0.1 M potassium phosphate pH 7.0

0.1 mM BSA ($M_r = 68\,000$; 68 mg in 10 ml 0.1 M potassium phosphate pH 7.0)

0.1 M indole ($M_r = 117.2$; 1.17 g in 100 ml isopropanol)

p-Dimethylaminobenzaldeyde (Ehrlich's reagent, $M_r = 149.2$)

5.0 N KOH ($M_r = 56.1$; 28 g in 100 ml water)

Toluene

Ethanol

Ehrlich's reagent:

- Dissolve 9 g *p*-dimethylaminobenzaldeyde under stirring in 205 ml ethanol
- Add 13 ml concentrated sulfuric acid (very slowly, wear protective glasses!)
- Bring to 250 ml with ethanol
- Dilute one part of the reagent with two parts of ethanol immediately before use

The concentrated reagent is stable for several months at room temperature.

Materials and instruments

Dialysis membrane (commercial dialysis tubes, e.g., Visking®, ~3 cm broad)

Dialysis apparatus (Figure 4.2b, with 10–20 dialysis cells, 1 ml void volume of each chamber, ~12 mm ∅, 9 mm deep)

Desktop centrifuge

Photometer (visible range)

Glass cuvettes

Indole assay and standard curve

Prepare 10 samples with increasing indole concentration, use buffer solution for the blank, and pipette in a 1.5 ml reaction tube:

- 0.25 ml indole solution (between 10^{-5} and 10^{-3} M, dilute the stock solution with 0.1 M potassium phosphate pH 7.0 accordingly)
- 10 μl 5 M KOH
- 1 ml toluene

Close tightly and shake thoroughly

Centrifuge for 5 min 5000 rpm

Transfer 0.2 ml from the upper organic phase into a new reaction tube

Add 1.2 ml Ehrlich's reagent (twofold diluted as described above)

20 min at room temperature

Measure the absorption at 572 nm (or 578 nm in a filter photometer)

Preparation of dialysis membranes

Boil a length of the dialysis tube in water for 5 min with some (~1 mM) EDTA to complex heavy metal ions

Slash the double layer laterally and open to get a single layer

Cut into slices large enough to completely cover the area between both cylinders of the dialysis unit (about $3 \times 3\ cm^2$) and press both cylinders together with the aid of the screws, a wet slice stretched between them

Fill the chambers quickly; the slices should not get dry

Determination of the minimum dialysis time

Mount 10 dialysis cells with membrane slices between both cylinders into the holder

Quickly fill one chamber of each cell (e.g., the right one) with 1 ml 0.5 mM indole solution in 0.1 M potassium phosphate pH 7.0, and the other (left) one with 1 ml 0.1 M potassium phosphate pH 7.0

Let the device rotate slowly

After a distinct time interval (e.g., 5 min), remove 0.25 ml aliquots from both chambers of one cell

After another time interval (e.g., 10 min), remove 0.25 ml aliquots from both chambers of the next cell

Continue this procedure for all cells (e.g., 15, 20, 30, 45, 60, 80, 100, 120 min, intervals can be extended later)

Perform the indole assay with all aliquots, including the initial indole solution as reference value, and plot the absorption values against the dialysis time. Two curves are obtained: one for the indole chambers and the other for the buffer chambers. They should meet at the minimum dialysis time

Preparation of the binding curve

Mount 10 dialysis cells with membrane slices between both cylinders in the holder. Fill one chamber of each cell (e.g., the right one) with 1 ml 0.1 mM BSA solution, the other (left) one with indole solutions, the concentrations increasing from 2×10^{-5} to 2×10^{-3} M. The concentration range refers to the expected dissociation constant for indole ($K_d = 8 \times 10^{-5}$ M), considering a twofold dilution owing to the distribution between both chambers and binding to the protein. The dialysis cells should rotate slowly at a constant temperature of 25 °C, for the minimum dialysis time plus 30 min to make sure that the equilibration is complete. Thereafter, aliquots of 0.25 ml per chamber are removed for the indole assay. The data obtained are treated as described in the following section.

References

Englund, P.E., Huberman, J.A., Jovin, T.M., and Kornberg, A. (1969) *J. Biol. Chem.*, **244**, 3038–3044.

Peters, T. (1985) *Adv. Protein Chem.*, **37**, 161–245.

4.2.2 Evaluation of Binding Experiments

As both the binding equation and the Michelis–Menten equation are similar (Box 4.1), the evaluation of the data is also accordingly similar. Only optical titrations need a special treatment and are described in Section 4.3.1.2. Plotting of bound against free ligand yields a hyperbolic curve for normal binding, reaching saturation at $n[E]_0$. The dissociation constant corresponds to the free ligand concentration at half saturation (Figure 4.1). The free ligand concentration [A] must result from the experiment; for equilibrium dialysis, it is the concentration in the ligand chamber. This is in contrast to kinetic experiments according to the Michaelis–Menten equation, where the applied (total) substrate concentration is taken as free: $[A]_0 = [A]$, since the enzyme concentration ("catalytic amounts") and, thus, also the amount of bound substrate are very low and can be neglected. For binding experiments, this treatment is not possible, because of the high enzyme concentrations needed. The concentration of bound ligand is obtained from the difference between both chambers:

$$[\text{indole}]_{\text{bound}} = [\text{indole}]_{\text{enzyme chamber}} - [\text{indole}]_{\text{ligand chamber}}$$

Thus, for the evaluation of the equilibrium dialysis experiments, all data needed are directly measurable. This, however, is not the case for the other binding experiments, such as ultrafiltration of ultracentrifugation. While the free ligand [A] can usually be determined without any difficulty, this does not hold for the bound ligand, since the volume of the enzyme solution changes during the experiment. However, since the bound ligand is the difference between the total ligand at the beginning of the experiment and the free ligand

$$[A]_{\text{bound}} = [A]_0 - [A]$$

it can be calculated. However, this indirect determination can yield wrong results, because unspecific effects can reduce the amount of the total ligand, such as adsorption to the walls of the device or the membrane or by precipitation. With equilibrium dialysis, this is not a problem, because both free and bound ligands are directly measured. Unspecific losses reduce the initial concentrations and shift the ligand range to lower concentrations, but the measured values are reliable; however, only higher initial ligand concentrations should be used.

The disadvantages of the hyperbolic representation of the data have already been discussed in Section 2.2.4.2. The three linear transformations of the Michaelis–Menten equation shown are also possible with the binding equation, but only two of them are in use. They are shown in Box 4.1, where for simplicity $[A]_{\text{bound}}$ is replaced by the fraction of ligand bound to the enzyme $r = [A]_{\text{bound}}/[E]_0$. Saturation is at n. The features of the linearized plots derived from the Michaelis–Menten equation including the determination of constants (Figure 2.15) are similar for the linearized binding curves.

The **double-reciprocal plot** (Figure 2.15a) is convenient, but not very reliable in the evaluation of the constants due to the strong distortion of the error limits, which are compressed to high ligand concentrations near the coordinate origin and extended to low concentrations, allowing no simple linear regression analysis. This impedes the unequivocal determination of the number of binding sites n from the reciprocal ordinate intercept as well as the value of the dissociation constant from the reciprocal abscissa intercept. Therefore, this diagram is seldom used and cannot be recommended. The main advantage of this representation is the separation of both the independent ([A]) and the dependent variable ($[A]_{\text{bound}}$).

The most frequently applied linear representation of binding data is the **Scatchard plot**, which is related to the kinetic Eadie–Hofstee plot (Figure 2.15c), interchanging only the coordinate axes. The abscissa intercept directly yields n, and the dissociation constant is obtained from the slope ($-1/K_d$) or the ordinate intercept as n/K_d.

4.2.3
Ultrafiltration

Ultrafiltration is mainly used for protein concentration (Section 3.4.5.2), but the same devices can serve for measuring of binding equilibria. Various apparatus are commercially available and new devices are currently developed, and so a detailed description cannot be given. Both the reaction volume, which should be small, and the ultrafiltration membrane are crucial. Its exclusion limit should be fairly between the size of the ligand and the enzyme. Further, the material of the membrane is of importance. Synthetic materials are offered, which avoid accumulation and aggregation of the protein at its surface.

Ultrafiltration can be done manually with syringes connected to a small filter attachment. For larger volumes, ultrafiltration cells are used, which direct the solution through the membrane by compressed air. Other devices use the centrifugal force of desktop centrifuges. On frits the solution is sucked through ultrafiltration membrane by a vacuum.

All these devices can be used for binding experiments. A solution, containing the enzyme and the ligand in defined concentration, is filtrated for a short time such that only a small portion (5–10%) passes the filter to avoid larger changes of the macromolecule concentration. The amount of ligand in the filtrate is determined; it corresponds to the fraction of free ligand [A] in the equilibration solution. When subtracted from the amount of the total ligand applied, $[A]_0$, the bound ligand $[A]_{bound}$ is obtained. Although the amount of total ligand added to the experiment is known, it should be measured in the nonfiltered part of the solution to detect losses due to unspecific binding. Very sensitive detection methods, such as radioactive or fluorescence labeling of the ligand or subsequent modification to chromophores, must be used. It should also be considered that the detection method discriminates not between free and bound ligands; this must be achieved by the binding method.

For quantitative binding measurements, a series of experiments with varying ligand at a constant enzyme concentration must be carried out under comparable conditions. Special devices are reported, allowing performance of a series of assays in parallel, such as the ultrafiltration apparatus of Paulus (1969) see Animated Suppl. 4.1.

4.2.4
Gel Filtration

Column chromatography is generally applied in biochemical laboratories; therefore, this type of binding experiments need no further special device. This fact and the advantage of easy handling are opposed by the limited accuracy and the relatively high amounts of reagents needed. Molecular sieve gels consisting of dextran, agarose, or polyacrylamide are available. Since for binding measurements only the macromolecular fraction must be separated from the low-molecular ligand, fast-flow gels such as Sephadex® G-25 are recommended. The principle of gel chromatography – retarded passage through the column of smaller particles because they penetrate into the inner space of the gel particles, while larger molecules does not and migrate faster – need not be explained here.

A binding experiment is carried out in a beaker containing the ligand–enzyme mixture in a gel suspension. After mixing and settling of the gel, the ligand concentration is measured from an aliquot of the supernatant. For calculation, the volume V_o outside the gel particles and the inside volume V_i must be determined in preliminary experiments. V_o is obtained from a similar experiment with only the enzyme, which will distribute only outside the gel particles. A corresponding experiment with only the ligand yields the total volume V_t, from which V_i is obtained by subtraction of V_o ($V_i = V_t - V_o$). In the final experiment, the same ligand solution, including the enzyme, is used. Because part of the ligand binds to the enzyme and does not enter the inner volume, the ligand concentration in the supernatant is larger just by the amount of bound ligand compared with the experiment with ligand alone. This is a simple procedure, but its accuracy is low.

Better results are obtained if the gel material is filled into a chromatographic column. However, for binding measurement, the principle of chromatography, separation of particles according to their size, must be avoided, because it will separate the enzyme and ligand and even the bound ligand will dissociate. Complete separation must be prevented using a short column and a large sample volume. Under these conditions, a ligand–enzyme solution will be eluted in three merging zones: the first zone of enzyme depriving its bound ligand is of minor interest. The second zone contains enzyme with its still bound ligand, which is in equilibrium with the free ligand. During elution, the smaller free ligand [A] is retarded and appears as the third zone. Its concentration is determined and $[A]_{bound}$ can be calculated from the total amount or the middle zone.

More sophisticated and, thus, more precise is a method of Hummel and Dreyer (1962) see Animated Suppl. 4.2. The column filled with the gel (Sephadex® G-25) is equilibrated with a ligand solution of defined concentration. The ligand eluted must have the same concentration as the applied solution. If this condition is attained, a small sample of the enzyme mixed with the ligand solution is applied to the column. A distinct portion of ligand, according to the binding equilibrium, binds to the enzyme, and, consequently, is removed from the free ligand solution. The enzyme together with its bound ligand migrates faster than the free ligand, leaving back a trough of the missing ligands within the otherwise constant free ligand solution. The level of the original ligand solution (measured by absorption) indicates the free ligand in equilibrium. The bound ligand elutes together with the enzyme and adds to the level of free ligand, yielding a peak, which is just equal to the following trough of the missing ligand, removed from the solution due to binding. The amount of bound ligand can be obtained by integrating one of both the areas. Since the peak area also contains the enzyme absorption, the trough area is more reliable.

4.2.5 Ultracentrifugation

Ultracentrifuges belong to the usual category of equipment in biochemical laboratories and besides them no further device is required for binding experiments. Small-volume centrifuges (Beckman TL-100, Airfuge®) with sample volumes as low as 0.1 ml are advantageous. From a solution containing enzyme and ligand, upon high-speed centrifugation, only the enzyme sediments, carrying the bound ligand with it. So, the meniscus region becomes deprived of the bound ligand, while the free ligand remains at its position and can be determined from an aliquot withdrawn from the meniscus. Complete sedimentation of the enzyme to the bottom of the tube must be avoided, because pelleting disturbs the binding equilibrium. This method is easy, but not very precise.

Chanutin, Ludewig, and Masket (1942) described a more accurate technique, which can be performed in normal, but preferentially in small-volume ultracentrifuges see Animated Suppl. 4.3. The procedure is principally the same as described above, but for this method several tubes (according to the number of rotor holes) with varying ligand and constant enzyme amounts are centrifuged in parallel. For this method, the general binding equation (cf. Box 4.1) is rearranged so that it depends on the total amount of ligand:

$$[\mathrm{A}]_0 = [\mathrm{A}]_{\mathrm{bound}} + [\mathrm{A}] = \frac{n[\mathrm{E}]_0[\mathrm{A}]}{K_\mathrm{d} + [\mathrm{A}]} + [\mathrm{A}] \tag{4.2}$$

While the concentrations of the total ligand $[\mathrm{A}]_0$ and the enzyme $[\mathrm{E}]_0$ change owing to sedimentation and become variables, the not sedimenting free ligand [A] can be regarded as constant in Eq. (4.2). Accordingly, $[\mathrm{A}]_0$ and $[\mathrm{E}]_0$ are linearly related, yielding a straight line, if plotted one against the other. This behavior should result from an experiment, where aliquots are taken from different regions of each tube (meniscus, middle region, and bottom) after centrifugation and the total concentrations of ligand and enzyme are determined. For each tube, a straight line should be obtained in a graph $[\mathrm{A}]_0$ versus $[\mathrm{E}]_0$, which extrapolate to [A] as ordinate intercept. The reciprocal slope of each line, plotted against 1/[A], will also yield a linear relationship, extrapolating to $1/n$ at the ordinate intercept and $1/K_\mathrm{d}$ at the abscissa intercept:

$$\frac{1}{\mathrm{slope}} = \frac{K_\mathrm{d}}{n[\mathrm{A}]} + \frac{1}{n} \tag{4.3}$$

Various other ultracentrifugation techniques for binding measurements have been reported. Good results are obtained with the gradient method, where a linear sucrose gradient from the meniscus (5%) to the bottom (20%) of the tube is prepared before centrifugation with a conventional gradient mixer see Animated Suppl. 4.4. The enzyme–ligand solution is applied to the top of the gradient and centrifuged, until the enzyme band reaches about two-third of the gradient. Thereafter, the gradient is fractionated and analyzed for the distribution of the enzyme and ligand. From the relative amount of ligand still bound to the enzyme, the binding constant can be determined (Draper and Hippel, 1979).

References

Chanutin, A., Ludewig, S., and Masket, A.V. (1942) *J. Biol. Chem.*, **143**, 737–751.
Draper, D.E. and Hippel, P.A. (1979) *Biochemistry*, **18**, 733–760.
Hummel, J.P. and Dreyer, W.J. (1962) *Biochim. Biophys. Acta*, **63**, 530–532.
Paulus, H. (1969) *Anal. Biochem.*, **32**, 91–100.

General Reference

Bisswanger, H. (2008) *Enzyme Kinetics, Principles and Methods*, 2nd edn, Wiley-VCH Verlag GmbH, Weinheim.

4.3 Spectroscopic Methods

Spectroscopic techniques avoid serious disadvantages of binding measurements based on size discrimination, such as the requirement of large amounts of enzyme and their relative inaccuracy, but their utility depends on the presence of a detectable signal, which is often difficult to discover. The spectral features of both the enzyme and the ligand in its free and bound state must carefully be observed for any detectable difference. Preferentially UV/Vis and fluorescence spectroscopy are applied, due to the relatively easy performance and interpretation, and also other spectroscopic methods such as circular dichroism (CD) or electron spin resonance (ESR) can be used for binding measurements. If no spectral change can be detected upon binding, the enzyme can be modified, mostly by covalent binding of a chromophore, acting as a probe and signaling the binding of the ligand. Various types of chromophores are available, with respect to both the chromophoric part as well as the reactive group, which mediates the covalent attachment to a functional group of the protein, such as a hydroxyl, amino, carboxyl, or thiol residue. Often, many preliminary experiments are necessary to find out an appropriate label. Usually, the chromophore cannot be conducted to a distinct residue or place on the protein, but will attack all accessible groups, sometimes causing unfavorable modification, partly accompanied by blocking functional residues in active sites and inactivation of the enzyme. Limiting the reaction time with the chromophore can reduce undesired reactions. The presence of substrates or cofactors can protect against the attack of active sites.

In the following sections, difference spectroscopic (UV/Vis) and fluorimetric titrations are described and examples for binding experiments are given. The examples develop strong signals, which cannot be regarded as quite representative, since most systems show only weak signals, which are difficult to visualize and need much experience. However, the described assays, which can be done even with less sensitive instruments, are helpful to study the principle of the methods.

4.3.1
Difference Spectroscopy

Difference spectroscopy is a method suited especially for structural studies on proteins. A sensitive UV/Vis spectrophotometer with a double-beam arrangement (Figure 2.23) and the ability to scan spectra is needed. With computer-controlled instruments equipped with a screen various spectra can be scanned, superimposed, and calculated, for example, sample spectra subtracted from a reference spectrum. A direct scanning diode array photometer is of great advantage for this method. Older photometer types need a recorder, which should be directly connected with the wavelength drive of the photometer.

To interpret spectral changes, the contributions determining the protein spectrum must be analyzed. Proteins are composed of 20 proteinogenic amino acids, and consequently spectra even of very different proteins are considerably similar. Larger differences occur only if characteristic amino acids, such as tryprophan, are missing, or if additional components, such as cofactors or prosthetic groups, are present.

The following general features can be observed with normal protein spectra. While the visible range beyond 320 nm shows no absorption, three regions in the UV range can be discerned:

1) *<200 nm*: Absorption of the peptide bond and aliphatic amino acid, disturbances from buffer ions or oxygen (in the light path or dissolved in aqueous solutions); this far UV region is hardly accessible due to the low intensities of UV lamps
2) *200–230 nm*: Intense absorptions of aromatic amino acids (phenylalanine, 206 nm; tyrosine, 224 nm; and tryptophan, 219 nm) and of histidine (211 nm), cysteine, cystine, arginine, and secondary structure elements (α-helix and β-sheet)
3) *260–300 nm*: Characteristic absorptions of aromatic amino acids (but less intense than in the 200–230 nm range): phenylalanine (257 nm), tyrosine (274.6 nm), tryptophan (280 nm), intensity relationship Phe : Tyr : Trp = 1 : 7.2 : 28.4. The protein peak at 280 nm is mainly determined by tryptophan

A typical protein spectrum is shown in Figure 3.2c. Although the left, short wavelength peak is more intense, the peak at 280 nm is more selective and can directly be assigned to the respective amino acids. It is more suited for protein analysis and protein assays (cf. Section 3.4.1.5). Protein spectra may be calculated from the contributions of the single amino acids; however, various amino acids are shielded in the native protein structure. Therefore, native spectra are usually less intense. This fact can be used to study the denaturation of proteins from the increase in intensity of its UV spectrum. The wavelength region between 200 and 230 nm allows observation of conformational changes of proteins, especially of secondary structural elements, but because of the various intense contributions to this peak, small changes are difficult to detect. For difference spectroscopy, both the short and the long wavelength peak are suitable mainly because of the contributions of aromatic amino acids.

A difference spectroscopy experiment in a double-beam photometer starts with the same enzyme solution in both the reference and the sample cuvette see Animated Suppl. 4.5. When scanning the protein spectrum is recorded; however, the photometer subtracts the reference from the sample signal, so that a zero line results, pretending the absence of any absorbing substance. In the second step, a ligand is added only to the sample cuvette to observe the binding, but the intrinsic ligand spectrum should also be veiled. This can be achieved using tandem cuvettes, which possess a separating window, dividing the cuvette into two equal compartments. The enzyme solution is given only into one compartment of each cuvette, and the other is filled with buffer solution. The ligand is added to the enzyme compartment of the sample cuvette, and the same amount to the buffer compartment of the reference cuvette. Thus, the photometer subtracts the ligand absorption. However, any spectral changes caused by binding in the sample cuvette will be visualized see Animated Suppl. 4.6. Similarly, other influences on the enzyme structure, such as pH, ionic strength, solvent, and temperature, can be observed, whereby only the sample cuvette is subject to the modifying agent.

Difference spectroscopy is a very sensitive method, but it needs much experience to get reliable results. High enzyme concentrations are required to make the small binding signal detectable, but it produces an intense basic absorption. Therefore, even weak influences or small deviations between the sample and the reference cuvette can severely disturb the signal, such as minor variations in the thickness of the cuvettes or small concentration differences of the solutions. Any addition to the sample cuvette changes the volume and thus the concentration. Therefore, a similar volume, for example, of buffer, must always be added to the reference cuvette.

4.3.1.1 Difference Spectroscopic Titration of Ligands Binding to Catalase

As a representative experiment binding of inhibitors to the enzyme catalase is studied and the evaluation of spectroscopic titrations is discussed. Two different ligands are used. Cyanide is a strongly binding inhibitor inducing an exceptional intense difference spectrum, while the weaker binding fluoride develops weaker effects and, thus, demonstrates more realistic conditions. Catalase as an inexpensive enzyme can be used in larger quantities.

Catalase ($M_r = 240\,000$, cf. assay in Section 3.3.1.3.1) is a homotetrameric enzyme, composed of four subunits, each carrying a porphyrin ring. The central iron ion possesses six coordination sites, four occupied by the porphyrin system, the fifth by a structural histidine, while the sixth remains open for the reaction with the substrate H_2O_2. The inhibitors bind to this site and induce the transition of the porphyrin from the high-spin to the low-spin state, the iron moving into the plane of the ring system, accompanied by a strong bathochromic spectral shift, which is observed in this experiment.

Further, the dependency of the dissociation constant on the temperature according to the van't Hoff equation

$$\left(\frac{\mathrm{d}\ln K_\mathrm{d}}{\mathrm{d}T}\right)_P = \frac{\Delta H}{RT^2} \qquad (4.4)$$

is studied, where T is the absolute temperature (kelvin), H the reaction enthalpy, R the gas constant, and P the pressure, which must remain constant. Integration, taking ΔH as temperature independent within a narrow temperature range yields the following equation:

$$\ln K_\mathrm{d} = \frac{\Delta H}{RT} + C \qquad (4.5)$$

where C is an integration constant. Plotting ln K_d against $1/T$ results in a straight line; the slope contains the reaction enthalpy. The temperature dependency of the dissociation constant is in analogy to that of the catalytic constant in enzyme reactions described by the Arrhenius equation (cf. Section 2.2.7.6), where the activation energy is obtained instead of the reaction enthalpy.

Reagents

0.1 M potassium phosphate pH 7.5

Catalase from bovine liver (20 mg ml^{-1}). The enzyme is available as crystal suspension, which must be suspended before removing a sample. Dilute to 0.5 mg ml^{-1} with 0.1 M potassium phosphate pH 7.5, the crystals must completely be dissolved, and remove a faint remaining turbidity by centrifugation. Determine the protein content of the solution at 405 nm, absorption coefficient: $\varepsilon_{405} = 38 \times 10^4\ \mathrm{l\,mol^{-1}} \times \mathrm{cm}^{-1}$

0.1 M KCN (potassium cyanide; $M_\mathrm{r} = 65.1$; 65 mg in 10 ml, prepare freshly, toxic, volatile in solution, store in closed vessels, cover the cuvette during titration)

0.5 M sodium fluoride (NaF, $M_\mathrm{r} = 42$; 215 mg in 10 ml)

Equipment

Recording UV/Vis double-beam photometer

Tandem cuvettes (facultative)

Quartz cuvettes

Microliter syringe

Enzyme spectrum

The spectral features of the protein are studied in a preliminary experiment. A spectrum is scanned in quartz cuvettes with the diluted enzyme solution between 200 and 650 nm against 0.1 M potassium phosphate pH 7.5 as reference.

Thereafter, potassium cyanide is added in three steps to the enzyme solution and the spectrum is scanned after each addition. The cyanide concentrations in the cuvette should be in the lower saturation range (about $K_d/2$, for $K_d \sim 1 \times 10^{-6}$ M) for the first step, in the middle saturation range ($3 \times K_d$) for the second step, and full saturation ($50 \times K_d$) for the last step. Consider that the ligand concentration must be relatively high to compensate for the reduction of free ligand due to binding to the enzyme. All spectra should be superimposed to visualize the spectral shift of the porphyrin absorption. In the final step, the buffer solution in the reference cuvette is substituted by the enzyme solution (without cyanide). Now the true difference spectrum becomes visible.

The first spectrum with catalase alone should show three major peaks, the largest in the short wavelength range (~210 nm), the second at 280 nm, and the third at 405 nm. The first two are essentially protein peaks (with some contributions of the porphyrin system); the visible peak results only from porphyrin. Only this peak exhibits the strongest shift upon addition of cyanide and in this range the characteristic difference spectrum should be observed; the further procedure can be limited to this visible range. The difference spectrum corresponds to the first derivation of the original peak, showing a negative peak, followed by a positive one. Therefore, the baseline of the photometer must be adjusted in the middle of the screen, in particular, the recorder display.

Spectral titration with potassium cyanide

As already mentioned, tandem cuvettes are needed for spectral titrations to correct both for the enzyme and the ligand absorption. However, the ligands, cyanide and fluoride, used for the present experiment develop no absorption in the interesting spectral range; therefore, normal cuvettes (glass cuvettes in the visible range) can be taken. To correct for the enzyme absorption, catalase solution (0.5 mg ml^{-1}) is given to both the sample and the reference cuvette (the same enzyme solution should be used for both cuvettes and should not be diluted directly into the cuvettes). A linear base line should result, deviations indicate concentration differences. If necessary, replace the solutions. Titration is carried out by successive additions of the ligand to the enzyme solution in the sample cuvette. Because each addition dilutes the enzyme solution in the sample cuvette, the same volume of buffer must be given to the reference cuvette to compensate for concentration differences.

About 30 additions are required for the preparation of a binding curve. They cause a considerable increase of the sample volume and dilution of the enzyme solution. Although adequate additions of buffer to the reference compensate for the concentration difference, the dilution effect should be kept as small as possible, to keep the enzyme concentration constant. If 10 μl are added for each addition to 3 ml enzyme solution in the sample cuvette, after 30 additions the volume will be increased by 10%, while additions of 1 μl change the volume only by 1% in maximum. For such small volumes, precise microliter syringes should be used, which are more accurate than automatic adjustable pipettes. A severe problem is mixing in the cuvettes after each addition. Small magnetic stirrer is available, which can be mounted directly below and allow stirring in the cuvette. If such a device is not on hand, stirring can be done manually with small sticks, preferentially made of plastic material, and the aliquot to be added can be placed on its flat tip (Figure 2.9 IV). The problem is cleaning of the stirrer. If this is done after each addition, together 30 times, a considerable part of the enzyme solution will be removed from the cuvette. Therefore, the stirrer should not be cleaned at all and used always for the same cuvette. It is not recommended to immerse the pipette tip directly into the solution for adding the ligand (Figures 2.9 I and 2.10) and during a titration series the cuvette should never be removed from its holder and shaken for mixing. It will hardly be possible to place the cuvette back into exact the previous position.

After each addition, the complete difference spectrum is scanned, the peak area including the decreasing site to the nonabsorbing range (from 340 to 600 nm in our example). This serves as the baseline for the evaluation of the binding curve, instead of the instrumental baseline, which can change during the experiment due to drifts or lamp fluctuations. All difference spectra are superimposed; care must be taken for proper relation to the ligand additions.

Performance of titration

The goal of the titration experiment is to obtain a saturation curve to evaluate the binding process. It must, however, be regarded that the directly obtained function, the **titration curve**, is not identical with the hyperbolic saturation function corresponding to the binding equation (Figure 4.1). The main difference is that the signal is obtained in dependence on the added ligand $[A]_0$, while for the binding equation the dependence on the free ligand [A] is required. Thereon, the signal corresponds to an absorption value, which, although being related, does not directly indicate the absolute amount of bound ligand. Therefore, the titration experiment must be performed from the beginning in a manner to allow proper evaluation of the titration curve. It can be divided into three parts, which must be perfectly worked out by the titration experiment:

1) *Low concentration range*: Low concentration range of the ligand from zero to one-third of the saturation. Within this range, the curve should be nearly linear and a tangent through the coordinate origin must be drawn. About 15 measuring points, especially at low saturation level 50–100 times below the dissociation constant, should cover this range, since incorrect tangent positioning causes a severe distortion of the results. The addition

of successive ligand to the sample cuvette should raise the total ligand concentration in the sample solution in small increments. For the present experiment, considering a K_d of 1×10^{-6} M, the addition of the first ligand should yield a concentration of 2×10^{-8} M in 3 ml catalase solution. This is achieved by the addition of 1 μl of a 6×10^{-5} M ligand solution. The successive addition of 1 μl (five times) raises the total concentration in the cuvette finally to 1×10^{-7} M. Now titration can proceed in larger steps (4×10^{-8} M) using a 1.2×10^{-4} M ligand solution. The addition of 1 μl (10 times) yields a final concentration of 0.5×10^{-6} M within the range of one-third saturation (considering the distribution of free and bound ligands)

2) *Medium concentration range*: In this range, the curve deviates significantly from the linear increase; its bending determines the dissociation constant. About 10 values should mark the curvature. The addition of 1 μl (five times) each from a 3×10^{-4} M ligand solution results in a cuvette concentration of 1×10^{-6} M and another addition (five times) from a 3×10^{-3} M solution yields 5×10^{-6} M, covering the half saturation range

3) *Saturation range*: Saturation cannot simply be extrapolated. Therefore, titration must be continued until complete saturation is reached. This is the case if upon further addition of ligand the superimposed difference spectra show no more increase. At this point, titration should be continued until 4–5 spectra overlap, to be quite sure of saturation. In the present experiment, the addition of the 3×10^{-3} M solution (five times) and further addition of a 6×10^{-2} M ligand solution (five times), 1 μl each, will result in saturation (1.1×10^{-4} M)

For the determination of the reaction enthalpy according to the van't Hoff equation (Eq. (4.4)), the experiment is carried out in a same manner at different temperatures. At least three temperatures (15, 25, and 35 °C) should be chosen, presuming that the values for each dissociation constant are based on a multitude of measuring points. As the affinity decreases with increasing temperature, higher ligand concentrations are needed for 35 °C.

Titration with sodium fluoride (at 25 °C only) can be carried out accordingly. Because of the smaller signal and weaker affinity, the catalase concentration should be increased (2–5 mg ml^{-1}, depending on the sensitivity of the photometer), and the concentration range for the ligand must be 10-fold higher.

4.3.1.2 Evaluation of Spectroscopic Binding Curves

Superposition of the complete spectra is important for controlling the regular progression of titration and for discovering irregularities, possibly caused by artificial influences. For preparing the titration curve, however, only the increase

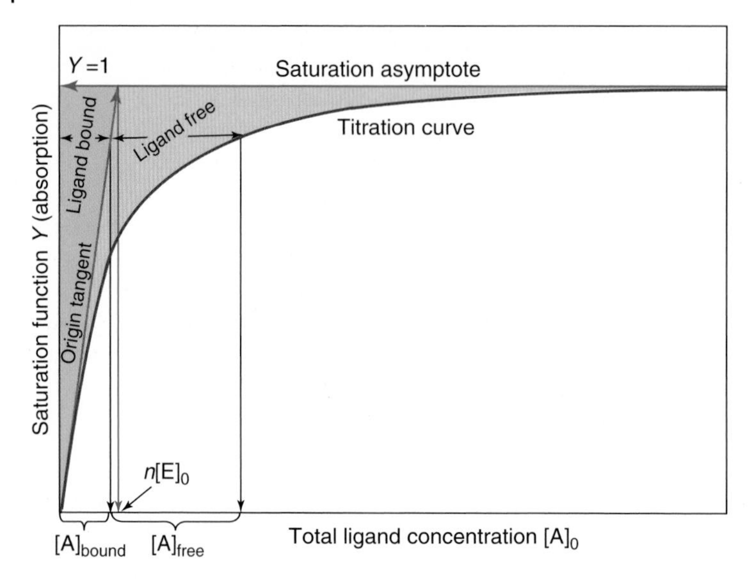

Figure 4.4 Evaluation of a spectroscopic titration curve. For details, see text.

at a distinct wavelength is required, usually the peak maximum. Its deviation from zero is plotted against the total ligand concentration $[A]_0$ in the cuvette. As already mentioned, instead of the baseline of the instrument, a reference value in the nonabsorbing range of the respective spectrum should be taken. It is important that for all spectra of the series always the same maximum and reference wavelength is taken, irrespective of any shifts (which are usually of artificial origin) – do not follow the shift, but keep at the chosen wavelengths.

The resulting titration curve must show a nearly linear increase in the low concentration range, through which a tangent is drawn (Figure 4.4). In this range, the ligand binds almost quantitatively to the enzyme, $[A]_0 \cong [A]_{bound}$, and, correspondingly $[A] \cong 0$. With increasing concentration the ligand binds no longer quantitatively, some free ligand remain in solution. Now $[A]_0$ distributes between $[A]_{bound}$ and [A], and the curve deviates from the tangent (which represents the bound ligand). The extent of deviation indicates the amount of free ligand. Thus, the distance between the tangent and actual titration curve (respectively, the measured value) is [A], while that from the tangent to the ordinate corresponds to $[A]_{bound}$, the distance from ordinate to the titration curve, that is, the sum of both sections, is $[A]_0$. Accordingly, $[A]_{bound}$ and [A] can be obtained from any point of the titration curve, and from these values, plotted in the Scatchard or the double-reciprocal diagram the respective constants are derived. An asymptote aligned to the saturation range of the titration curve meets the origin tangent at the abscissa coordinate $[A]_0 = n[E]_0$. To estimate the number of binding sites n, the enzyme concentration must be determined, for the present experiment from the absorption at 405 nm (see earlier text: "Reagents").

The described treatment of the titration curve also has some problems. Crucial is the origin tangent, wrong positioning, for example, owing to scatter of the measured values or weak affinity distorts the evaluation. The tangent relies on the assumption of complete ligand binding in the low concentration range. This is not quite established in the present experiment, where the catalase concentration is relatively low due to the intense difference spectrum and, therefore, some free ligand appears already at low concentrations. Under these circumstances, the tangent is easily aligned too flat, with the consequence of overestimation of the number of binding sites, but underestimation of the portion of free ligand. These problems can be circumvented by direct linearization of the titration curve, but not without creating new problems. Rearrangement of the binding equation (Box 4.1) according to Stockell (1959)

$$\frac{[A]_0}{[E]_0 Y} = \frac{K_d}{[E]_0(1 - Y)} + n \tag{4.6}$$

yields a linear diagram by plotting $[A]_0/[E]_0 Y$ against $1/([E]_0(1 - Y))$. The saturation function Y is the measured value (absorption), multiplied by a constant factor, which is obtained by setting absorption = 1 at saturation (ordinate value of the saturation asymptote), and $[A]_0$ is the added ligand. The tangent is not needed and free and bound ligands need not be determined. However, the saturation asymptote is still required and incorrect determination of the saturation level produces severe systematic deviations.

4.3.2 Fluorescence Spectroscopy

Fluorescence is often superior to absorption spectroscopy, due to the more pronounced spectral changes, the about 100-fold higher sensitivity, and the lower enzyme requirement. These advantages compensate for the more expensive and complicate instruments and various disturbances. Fluorescence is sensitive against changes in polarity, which occur if a ligand from the polar medium of water binds to an apolar site on the enzyme. However, either the ligand or the enzyme must emit fluorescence light. In proteins, the three aromatic amino acids show fluorescence emission, but only tyrosine and tryptophan with detectable intensity. While the tyrosine fluorescence is often quenched, tryptophan, if present, can always be observed, but this amino acid is seldom directly at the binding site to be influenced by the binding process. Ligands are mostly nonfluorescent and, therefore, chromophores must be introduced into either the protein or the ligand. In the experiment described below, the chromophore 8-anilinonaphthalene sulfonate (ANS) is taken as ligand. In water, it possesses a weak fluorescence, but upon binding to the apolar binding site of bovine serum albumin (BSA) its fluorescence maximum shifts toward lower wavelengths accompanied by a strong intensity increase.

In the first experiment, the spectral features of both the protein and the ligand are studied. Two types of fluorescence spectra exist, the excitation and the emission spectrum. The excitation spectrum corresponds to the absorption spectrum, since a fluorescent substance can only emit light when it is excited. The excitation spectrum is scanned at a constant emission wavelength, and, reverse, at a constant excitation wavelength the emission spectrum is obtained. When compared with excitation, emission is always at higher wavelengths see Animated Suppl. 4.7.

Attention must be paid to several disturbances. Principally, a large peak is obtained when excitation and emission wavelengths overlap. The light from the lamp is scattered in the cuvette and gets into the emission light path. Scatter can be caused by reflections of the cuvette windows and the cuvette housing, by particles (dust) in the solution and by Rayleigh scattering. The scatter peak can vary in its intensity. It can be diminished by reducing scattering sources, but it will not disappear completely. To shorter and to longer wavelengths of excitation, Raman peaks appear especially when measuring in the high sensitive range. They can easily be mistaken for real fluorescence. To identify them, the excitation wavelength should be changed. While real fluorescence peaks remain at their position, Raman peaks migrate in the same direction with the excitation wavelength. This feature can also serve to move a Raman peak out of the interesting region, if it overlaps with a fluorescence peak.

The fluorescence signal is obtained as relative intensity. Unlike the absorption, which can be calibrated between the two limits 0 and 100% transmittance, fluorescence has only a lower limit of zero emission but no upper limit. Thereon, fluorescence intensity depends on the respective instrument, the lamp intensity, the light path, and the photomultiplier sensitivity. Therefore, measurements on different instruments cannot directly be compared, as with absorption photometer. For most purposes, the relative fluorescence obtained by a special instrument is sufficient. For absolute determination, a stable reference substance, such as quinine sulfate, may be used.

Various influences cause severe reduction of the fluorescence signal (*quenching*), such as contamination in the solution, dissolved oxygen, high concentration of the chromophore or contaminating substances, and elevated temperature. Therefore, all substances, including the solvent, must be of utmost purity; oxygen should be removed with nitrogen gas or by applying a vacuum.

4.3.2.1 Binding of ANS to Bovine Serum Albumin

The hypsochromic shift of its fluorescence maximum and the strong intensity increase upon binding of ANS is observed in this experiment. Because of the strong signal, there is no special demand for a high sensitive instrument. Although the experiment is described for an instrument that is able to scan both excitation and emission spectra, instruments with only variable emission wavelengths can be used likewise.

Reagents and solutions

0.1 M potassium phosphate pH 7.5

0.1 mM bovine serum albumin (BSA, $M_r = 68\,000$; 68 mg in 10 ml 0.1 M potassium phosphate pH 7.5)

2×10^{-6} M BSA in 0.1 M potassium phosphate pH 7.5

10 mM ANS (8-anilinonaphthalene sulfonate, ammonium salt, $M_r = 316.4$; 31.6 mg in 10 ml H_2O)

10 mM ANS in 10 ml ethanol

Alcohols as indicated (ε_r = dielectric constants):

	ε_r
Water	80.18
Ethylene glycol	37.7
Methanol	32.6
Ethanol	24.3
n-Propanol	20.1
Isopropanol	18.3
n-Butanol	17.1
Isobutanol	17.7
n-Hexanol	13.9
n-Octanol	10.3

Solutions of 1×10^{-7} M ANS in the alcohols indicated (take the 10 mM ANS solution in water to prepare the solutions in the lower alcohols, and for the higher alcohols (>ethanol) take the 10 mM ANS solution in ethanol)

1×10^{-6} M ANS in water

Equipment

Recording spectrofluorimeter

Fluorescence cuvettes (quartz and glass)

Microliter syringe (for ANS, not for protein)

General considerations

Fluorescence measurements must be performed with highest accuracy; any contamination or soiling (dust, air bubbles) in the cuvette should be avoided. Handle cuvettes only with gloves, do not touch the window sites. Pass all solutions, except protein and alcohols, through a sterile filter to remove dust particles. If during an experiment such as titration, unexpected phenomena occur (spectral shifts, new peaks), mix the solution in the cuvette without changing anything else and repeat the measurement: artificial phenomena will change or disappear. The cuvette should be always placed in the same position into the cell holder and should not be removed during titration. Give additions to the cuvette as shown in Figure 2.9b.

Fluorescence spectra

Scan the excitation and emission spectra of a 2×10^{-6} M BSA solution in 0.1 M potassium phosphate pH 7.5 and of a 1×10^{-7} M ANS solution in ethanol. Adjust the excitation wavelength for the protein to 280 nm (use a quartz cuvette) and scan the emission spectrum to longer wavelengths. Avoid the scatter peak at the excitation wavelength and start scanning 10 nm beyond this wavelength (290 nm). The maximum emission of tryptophan is at 350 nm, but it shifts to shorter wavelengths if integrated into the protein structure. The excitation spectrum is scanned below the constant emission wavelength, which should correspond to the maximum of the emission peak. Scan the whole UV region and stop scanning 10 nm below the emission wavelength to avoid the scatter peak. According to the absorption spectrum, two maxima should appear. The short wavelength peak (210 nm) must be more intense, but because of the weak lamp intensity in the far UV range, it may appear smaller than the peak at 280 nm. Some instruments correct for this effect.

The spectra of the ANS solution in ethanol (excitation at 366 nm and emission at 480 nm) are similarly scanned.

Dependence of ANS fluorescence on polarity

Upon decreasing polarity, the emission maximum of ANS shifts to lower wavelengths and its intensity increases. To demonstrate this effect scan the spectra of ANS in the 1×10^{-7} M solutions of alcohols with increasing chain length as indicated above. At a constant excitation wavelength (366 nm), the emission spectra are scanned from 400 to 600 nm (glass cuvettes can be used). In water, the 1×10^{-6} M ANS solution is used because of the very low intensity. Plot the relative fluorescence intensities and the wavelengths of the emission maxima against the dielectric constants ε_r of the alcohols. Correct for the tenfold concentration of ANS in the aqueous solution.

Standard curve of ANS

A standard curve for the ANS fluorescence in water is prepared with the 10 mM ANS solution in water (avoid any contaminations with ethanol or other alcohols), starting with a concentration of 1×10^{-6} M in the cuvette. Increase the amount of ANS stepwise until 5×10^{-5} M; scan for each step an emission spectrum. Plot the maximum intensities against the ANS concentration. A linear relationship should result, which deviates in the higher concentration range from linearity due to concentration quenching. In this nonlinear range, quantitative measurements are not recommended.

Titration with BSA

Spectroscopic titration is usually performed increasing the ligand concentration at a constant amount of the macromolecule. The following experiment is performed in the reverse manner, varying BSA and keeping the ligand constant. With respect to the binding equilibrium, both components can be regarded as equivalent and the binding equation can likewise be formulated with the macromolecule as variable. Usually, such a procedure is unfavorable, for example, when the macromolecule possesses more than one binding site (which is not the case here). Beyond that because of their large molecular mass, macromolecules cannot be prepared in high molar concentrations and saturation cannot be reached completely. This is also valid for the present experiment, but this procedure has some advantages; therefore, both components will be varied independently. For titration with BSA, a constant amount of ANS (1×10^{-8} M in buffer) must already be present in the cuvette. BSA shows itself no fluorescence in the observed spectral range, so that any spectral change is a signal for the binding process and a normal saturation curve will be obtained. Free ANS, on the other hand, shows already a weak fluorescence, which intensifies upon binding to BSA. Upon stepwise addition of BSA according to the rules of spectroscopic titration (cf. Section 4.3.1.1), the fluorescence intensity increases, accompanied by a blue shift of the maximum. This is observed by scanning the emission spectrum after each addition from 400 to 600 nm. When saturation is reached, the fluorescence maximum will remain constant and no shift and increase should be observed.

To establish the experimental conditions, preliminary tests with only few concentration steps are carried out before performing the complete titration experiment. The constant component, that is, ANS, is the limiting factor in the experiment and determines the saturation level. If it is too high, it will not be possible to add sufficient amounts of the variable component to reach saturation. On the other hand, very low concentrations yield no detectable fluorescence (depending on the sensitivity of the instrument). If the appropriate ANS amount is determined, the complete titration should be carried out as described for difference spectroscopy (Section 4.3.1.1) especially regarding the considerations about concentration ranges of the variable component ($K_d \sim 1 \times 10^{-6}$ M). The titration curve is evaluated as described in Section 4.3.1.2.

It must be considered that during titration the maximum shifts to shorter wavelengths, but to obtain the binding curve a fixed wavelength must be used, namely, the maximum at saturation, where ANS is present only in the bound form. From the fluorescence intensity at saturation compared with the fluorescence intensity of the free ANS at the beginning of the experiment, the extent of fluorescence enhancement due to binding can be estimated and the dielectric constant of the BSA binding site can be determined in comparison to the fluorescence intensities of ANS in different alcohols.

Titration with ANS

This "normal" binding procedure with increasing ANS in the presence of constant BSA has the disadvantage in that two effects contribute to the observed fluorescence enhancement, binding, and the innate ANS fluorescence. In difference spectroscopy, the double-beam arrangement together with tandem cuvettes compensates for such effects (cf. Section 4.3.1). While tandem cuvettes are not useful for fluorescence measurements, the double-beam system is principally possible, but identical light paths for both beams are difficult to realize.

Assuming one binding site for ANS, the same molar amount of constant BSA as determined in the above BSA titration for ANS can be used to save another preliminary test. At low concentrations, the innate ANS fluorescence is negligible compared with the strong binding effect. In the higher concentration range, however, free ANS contributes increasingly to the total fluorescence, masking the saturation level and the end of titration is difficult to recognize (at higher concentrations quenching pretends saturation). When plotting the titration curve saturation can be perceived from the slight linear increase, following the steep increase at the initial range of the curve. The inclination should be the same as that obtained for the standard curve with free ANS, which can be subtracted to obtain a normal saturation curve. It must be considered that for both curves the fluorescence intensities at the wavelength of the maximum of bound ANS, not of free ANS, must be taken.

References

Bisswanger, H. (2008) *Enzyme Kinetics, Principles and Methods*, 2nd edn, Wiley-VCH Verlag GmbH, Weinheim.

Brand, W. (1967) *Meth. Enzymol.*, **11**, 776–856.

Donovan, J.W. (1973) *Meth. Enzymol.*, **27**, 497–525.

Herskowits, T.T. (1967) *Meth. Enzymol.*, **11**, 748–775.

Stockell, A. (1959) *J. Biol. Chem.*, **234**, 1286–1292.

Stryer, L. (1968) *Science*, **162**, 526–533.

Wettlaufer, D.B. (1962) *Adv. Protein Chem.*, **17**, 303–390.

4.4
Other Binding Methods

4.4.1
Radioactive Labeling

The risks of radioactivity become increasingly obvious and the respective methods are replaced more and more by alternative techniques. Radioactive labeling is still applied in connection with ultrafiltration, where a solution with the macromolecule and a radiolabeled ligand is filtered through an ultrafiltration membrane. The free ligand passes the membrane and appears in the filtrate, while the macromolecule together with its bound ligand is retained on the filter. After filtration, the filter is transferred to a tube containing a scintillation fluid and measured in a scintillation counter, indicating the amount of bound ligand. The free ligand can be similarly determined from the filtrate. Although radioactive labeling is a very sensitive method, the filtration technique is less accurate, since the conditions for the macromolecule concentrated on the filter are different from that of the solution. It is a fast and simple method, yielding at least a qualitative estimation of the binding affinity.

A suitable isotope for radiolabeling of ligands is ^{14}C, which has a half-life time of 5730 years and a low specific radioactivity with a short reaching radiation (few centimeters). Practically, each organic substance can be labeled with this isotope. However, by improper handling (release of $^{14}CO_2$), it can be incorporated into the organism. ^{32}P is a favored isotope due to its high specific radioactivity and short half-life (14.3 days), but has strong far-reaching (about 1 m) radiation. Because of its moderate half-life (12.26 years) and short-range radiation, ^{3}H (tritium) is a convenient isotope for various applications. However, it is often displaced from the labeled compound by cold hydrogen.

4.4.2
Surface Plasmon Resonance

This relatively new method, based on the sensor technique, is applicable for routine assays but requires a special instrument, which is available together with convenient software for the evaluation of the data. A thin gold film is attached with its upper site to a glass plate and a prism and to its lower site a carboxymethyldextran matrix is fixed, to which either the macromolecule or the ligand is covalently bound (Figure 4.5).

Monochromatic light penetrates the prism and meets the metal plate, forming an evanescent wave with the conducting electrons (surface plasmon). At a particular wavelength resonance occurs, the light is totally reflected, and the reflection angle (SPR angle) is measured. The evanescent wave associates with the electromagnetic field and interacts with the dextran matrix. Any change of the matrix structure, such as fixation of the macromolecule and binding of the ligand, modifies the reflection angle. The binding process can, therefore, be followed from changes of

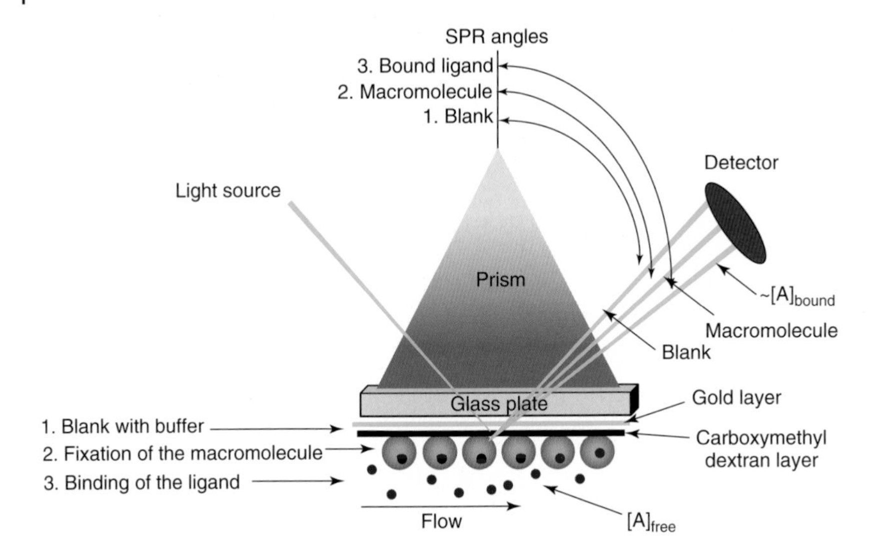

Figure 4.5 Schematic representation of a surface plasmon device.

the SPR angle. In the first step, the matrix is rinsed with the buffer solution and the SPR angle is determined for the blank (Figure 4.5). In the second step, the macromolecule is fixed and the SPR angle is measured again. Finally, the fixed macromolecule is rinsed with the ligand solution; the resulting SPR angle indicates the amount of ligand bound. The free ligand can be determined from the rinsed solution.

References

Fägerstam, L.G., Frostell-Karlsson, A., Karlsson, R., Persson, B., and Rönnberg, I. (1992) *J. Chromatogr.*, **597**, 397–410.

Wilson, W.D. (2002) *Science*, **295**, 2103–2105.

5
Enzymes in Technical Applications

Immobilization is an indispensable prerequisite for the technical application of enzymes. Enzymes as biocatalysts are present in reaction solutions in small quantities, and recovery is not economical; however, such losses cannot be ignored when dealing with expensive enzyme preparations. Fixation of enzymes to solid supports (**matrix**) may solve this problem. It allows complete recovery and repeated utilization of the same enzyme preparation. Although enzyme immobilization seems to be perfectly suited for industrial processes, it is not a broadly applied technique. Among thousands of different enzymes, only a few (approximately 15–20) are, in fact, involved in large-scale processes. This fact reflects the numerous difficulties of immobilizing enzymes, and even a large volume of relevant literature cannot solve these problems. In 1916, adsorption of invertase onto charcoal was the first example of an immobilized enzyme; this was followed by an extensive investigation of this field, starting in the middle of the twentieth century.

Because each enzyme has its own special features, general rules for immobilization cannot be given. Successful immobilization depends on various factors and numerous procedures must be tested until an appropriate method is established. Before starting immobilization, the main points discussed in the following sections must be considered:

- Mode of immobilization
- Type of matrix to which the enzyme should be fixed
- Type of connection between the matrix and the enzyme (**spacer**)

Immobilized enzymes have a broad range of applications in technical processes, enzyme reactors for synthetic procedures, enzyme electrodes for analytic purposes and for process controls, and finally, for medical purposes.

5.1
Modes of Enzyme Immobilization

Various modes of immobilization are described, but no really perfect method exists; advantages and disadvantages of each have to be weighed (Box 5.1). Gentle fixation to take care of the sensitivity of enzymes comes at the cost of stability of the fixation. On the other hand, strong fixation often impairs the enzyme activity. Restraints of

Practical Enzymology, Second Edition. Hans Bisswanger.

free diffusion of the substrate to and product from the immobilized enzyme must be considered and technical implications such as stability (compressibility, resistance against oxidation, and microbial attack) of the material must be considered. The accessible surface per volume unit for immobilization of the enzyme is of particular importance; this should be as large as possible. An example is immobilization to glass beads: a resistant material, but with a limited accessible surface. By employing an expensive technique, the surface can be considerably enlarged to highly porous glass (CPG). Another important aspect is the degree of charging of a distinct matrix unit with enzyme molecules; this essentially depends on the amount of reactive groups on the matrix.

Box 5.1: Immobilization Methods: Advantages and Disadvantages

Method	Advantages	Disadvantages
Adsorption	Noncovalent fixation Easy exchange of the immobilized enzyme	Unstable fixation Unintended release of immobilized enzyme
Entrapment	Noncovalent fixation No direct interaction with the matrix	Elevated temperature for polymerization reaction Diffusion limitation for substrates and products
Encapsulation	Noncovalent fixation No direct interaction with the matrix Coimmobilization of enzymes catalyzing sequential reactions	Diffusion limitation for substrates and products
Cross linking	Matrix-free Easy recovering by filtration or centrifugation	Covalent modification Shielding of active sites within the particles
Covalent immobilization	Stable fixation Disturbing activities (e.g., proteases) are also fixed and prevented from causing damage	Covalent modification

5.1.1 Adsorption

The simplest method of immobilization is noncovalent adsorption to solid supports and this is also used for purification. Various chromatographic methods, such as ion exchange (Figure 5.1), adsorption, hydrophobic, and affinity chromatography, are based on the binding of proteins to solid matrices under the appropriate

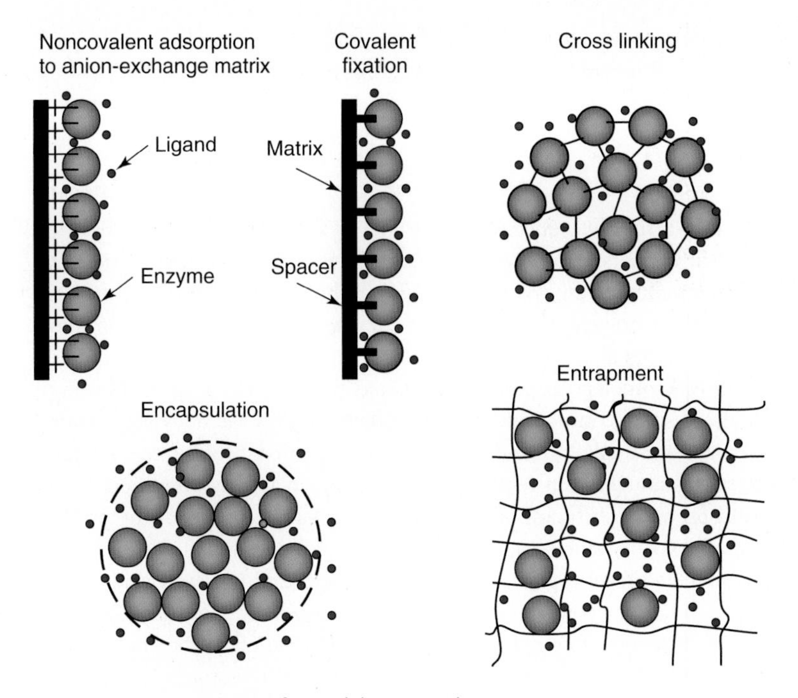

Figure 5.1 Various types of immobilization techniques.

conditions. If the conditions of binding are maintained instead of elution, the enzyme can be regarded as immobilized (Box 5.1).

Conventional **ion-exchange** materials such as DEAE (diethylaminoethyl) or CM-cellulose (carboxymethyl-cellulose)or more gentle materials such as DEAE- and CM-dextrans (DE- and CM-Sepharose®) can be used, or as adsorbing materials such as hydroxylapatite and calcium phosphate. Binding of proteins strongly depends on pH and ionic strength, with lower ionic strength favoring binding, and higher ionic strength favoring elution. Binding conditions for a special protein or an enzyme can easily be found and besides the advantage of noncovalent fixation, the enzyme can be quickly substituted by a fresh preparation, under elution conditions. However, the enzyme may be unintentionally released from the matrix, for example when the ionic strength or the pH changes owing to the reaction conditions. Reaction components and additives such as metal ions or sulfhydryl reagents can also bind to the ion-exchange material and modify its features.

The opposite conditions, binding at high and releasing at low ionic strength are characteristic of **hydrophobic** materials such as phenyl Sepharose. However, detergents or organic components such as ethanol, even in low concentrations, promote elution of the enzyme. Interaction with the nonpolar matrix often destabilizes the native enzyme structures.

Binding of enzymes to affinity material is highly specific and, thus, more stable. Unspecific interactions do not counteract the binding. Elution is specifically achieved not only by the affinity component in its free form but also by high ionic strength or pH variation. However, the affinity component interacts mostly with the active site of the enzyme so that it cannot act as an immobilized catalyst.

5.1.2 Entrapment

Entrapment into a narrow network (Figure 5.1) is a very gentle immobilization technique. Steric barriers retain the enzyme, its native structure is maintained, and no direct interaction with the matrix is required. The mesh width of the network must be narrow to prevent release of the enzyme, but large enough to allow diffusion of substrates and products. Therefore, this method is not applicable to enzymes with macromolecular substrates such as proteases or RNases. However, even diffusion of small compounds can be hindered within the network, forming a zone of substrate depletion and product accumulation around the enzyme. This effect causes an apparent reduction of the enzyme activity due to decreased substrate and product inhibition. Beyond that, complete exchange of the product formed by the new substrate in the reaction batch is not possible.

Materials such as agarose, gelatin or polyacrylamide, and polysaccharides from algae (alginate, κ-carrageenan) are applied for entrapment. The formulation of the network requires heating of the material. In the case of polyacrylamide, the temperature increases due to the polymerization reaction. This can be harmful for thermosensitive enzymes. The entrapment technique is also used for immobilization of cells. This has the advantage of not requiring elaborate enzyme purification, but the problems of diffusion limitation for the substrate and product are even more serious – unfavorable side reactions cannot be excluded and the degree of charging with high enzyme activities is limited.

Polymerization is performed directly in the concentrated enzyme solution under defined conditions. The solidified gel is minced into smaller particles or beads and filled in an enzyme reactor or a column to enlarge the accessible surfaces for the substrate and product exchange.

5.1.3 Encapsulation

This technique is particularly suited for the immobilization of sensitive enzymes and cells, when any interaction with the native enzyme structure should be avoided. The enzyme solution is enclosed into small vesicles by a porous membrane (Figure 5.1). This method also allows coimmobilization of two or more enzymes catalyzing sequential reactions. A certain disadvantage can be the undesired enclosure of disturbing activities, such as proteases or, if proteases themselves are the subject of immobilization, autocatalysis can occur. In addition, there is the problem of diffusion limitation through the vesicle wall.

A simple model for entrapment is a dialysis bag containing the enzyme solution immersed in a substrate solution. However, this is not a very efficient device, because of large diffusion distances for the substrate and product into and out of the bag, respectively, and because of the relatively small surface of the bag involved in maintaining an effective substrate–product exchange. Hollow fibers, available as cartridges, are more suited for entrapment. The enzyme solution is enclosed in the hollow fibers, which are rinsed from the outside by the substrate solution. The product formed is removed by the same procedure. Hollow fibers are rather resistant to mechanical and chemical influences and therefore, well suited for permanent technical usage. Living cells can also be enclosed in the fibers and provided with a fresh medium from the outside. The large surface of the hollow fibers favors the substrate–product exchange, but the enzyme cannot be quantitatively recovered from the fibers.

Microencapsulation at least partly overcomes the problem of large diffusion distances. Frequently, polyamide or nitrocellulose is used as a material for the capsule; these substances are directly formed in the concentrated enzyme solution. Enzymes are also enclosed in liposomes. The walls of the capsule must be permeable for substrates and products, but impermeable for the enzyme. Diffusion limitations of the substrate and product must also be considered. In medicine, encapsulation of enzymes is used for application of drugs.

Reference

Chambers, R.P., Cohen, W., and Baricos, W.H. (1976) *Meth. Enzymol.*, **44**, 291–317.

5.1.4 Cross Linking

The goal of all immobilization techniques is to enlarge the size of the enzymes to enable their recovery. This goal is attained by the cross-linking method directly by increasing the enzyme mass, without any need of a matrix (Figure 5.1). Many enzyme molecules are aggregated to large particles, which can be easily collected by centrifugation or filtration. The covalent binding can influence the native structure and activity of the enzyme. Aggregation is performed in concentrated enzyme solutions. If larger amounts or concentrated solutions of the enzyme are not available, particles can also be formed by coimmobilization with inert proteins, such as BSA, which often have a stabilizing effect. A broad variety of bivalent reagents (Table 5.1) exist; the most common agents for cross linking, glutaraldehyde and dimethyl suberimidate, have already been described (cf. Sections 3.4.4 and 3.5.4.7). Cross-linking reagents differ both with respect to the reactive groups as well as the chain length. Most reagents are directed against amino groups and others react with carboxyl, hydroxyl, or sulfhydryl groups of the protein. The reagents possess either two identical (*homo*) or two different (*hetero*) reactive groups. The chain length of the reagent determines the distance and flexibility of cross linking. Artificial multienzyme complexes can be formed by coimmobilizing different enzyme activities.

Despite these advantages, cross linking is not a frequently applied method. Manipulation with the enzyme aggregates is not quite easy; they are sensitive to

Table 5.1 Cross-linking reagents (according to Pierce Handbook 1989, Pierce Europe).

Compound	Abbreviation	Functionality	Reacts with functional groups	Cleavable by	Spacer length (Å)
N-[4-(*p*-Azidosalicylamido)butyl]-3′-(2′-pyridyldithio) propionamide	APDP	Hetero	$-SH/-SH, -NH_2$	–	21.0
p-Azidophenylglyoxal	APG	Homo	$-NH_2$, arginine	–	9.3
Bis[sulfosuccinimidyl] suberate	BS	Homo	$-NH_2$	–	11.4
Bis-maleimidohexane	BMH	Homo	-SH	–	16.1
Dimethyladipimidate	DMA	Homo	$-NH_2$	–	8.6
Dithiobis(succinimidyl propionate) Lomant's reagent	DSP	Homo	$-NH_2$	Thiols	12.0
Disuccinimidyl tartrate	DST	Homo	$-NH_2$	Oxidizing reagents (periodate)	6.4
1-Ethyl-3-[3-dimethylaminopropyl] carbodiimide	EDC	Hetero	$-NH_2/-COOH$	–	–
Ethyleneglycol bis[succinimidylsuccinate]	EGS	Homo	$-NH_2$	Hydroxylamine	16.1
m-Maleimidobenzoyl-*N*-hydroxysuccinimide ester	MBS	Hetero	$-NH_2/-SH$	–	9.9
3-[2-Pyridyldithio] propionyl hydrazide	PDPH	Hetero	$-SH/ < -COOH$	Thiols	9.2
N-Succinimidyl[4-azidophenyl]-1,3′-dithiopropionate	SADP	Hetero	$-NH_2$ photo-reactive	Thiols	13.9
N-Succinimidyl 3-[2-pyridyldithio]propionate	SPDP	Hetero	$-NH_2/-SH$	Thiols	6.8
Ethyleneglycol bis [sulfosuccinimidylsuccinate]	Sulfo-EGS	Homo	$-NH_2$	Hydroxylamine	16.1
Tris-[2-maleimidoethyl]amine	TMEA	Homo tri-functional	-SH	–	10.3

harsh treatment. Since covalent binding is required, there is no essential advantage over attachment to solid supports, which is superior for technical processes.

5.1.5 Covalent Immobilization to Solid Supports

5.1.5.1 Supports

Covalent fixation of enzymes to solid supports is the most important immobilization method (Figure 5.1). The strong covalent bonds resist harsh treatment, which cannot be completely avoided in technical processes. The materials employed must possess the following features:

- Surface compatible to the enzyme
- Reactive groups for the attachment of the protein
- High binding capacity
- Inertness with respect to the reaction components
- Long-term stability under reaction conditions

For technical applications, compact stable materials are applied: silicon and aluminum oxide, glass, ceramics, and synthetic polymers such as polyamide, polystyrene, polyacrylate, polyacrylamide, polyester, vinyl-, and allopolymers. They endure rough conditions such as high pressure in continuous flow systems and enzyme reactors. However, many enzymes cannot tolerate such supports and suffer considerable activity losses upon immobilization. For laboratory use, more gentle materials are preferred, such as polymeric carbohydrates (agarose, dextrans, chitins, cellulose) and proteins (collagen, gelatin, albumin). Their hydrophilic surfaces are well suited for direct contact with the enzyme structure, and a large number of functional groups, especially hydroxyl residues, are available for covalent fixation of the enzyme. For large-scale application, such materials are too sensitive and expensive. They are susceptible to microbial attack and do not resist conditions of high pressure; the fast flow rates required in enzyme reactors cannot be achieved.

5.1.5.2 Spacer

The mode of binding of the enzyme to the support is of as much importance as its material. Frequently, the enzyme is directly fixed to the surface of the support by a condensing reaction (Figure 5.2a, 1), but direct contact with the matrix is often incompatible for the enzyme due to special features of the surface, its polarity, or charges. Functional regions such as the active site of the enzyme can be shielded and the access of the substrate can be restricted by steric hindrance or electrostatic repulsion. Spacer compounds introduced between the matrix and enzyme avoid direct contact and improve the mobility of the enzyme and the accessibility of the substrate. They confer flexibility onto the enzyme; its active site can reach a larger area within the surrounding medium (Figure 5.2a, 2).

1,6-Hexanediamine (diaminohexane) is one of the most frequently used spacers. It connects the matrix with the enzyme by forming amide bonds with free carboxyl

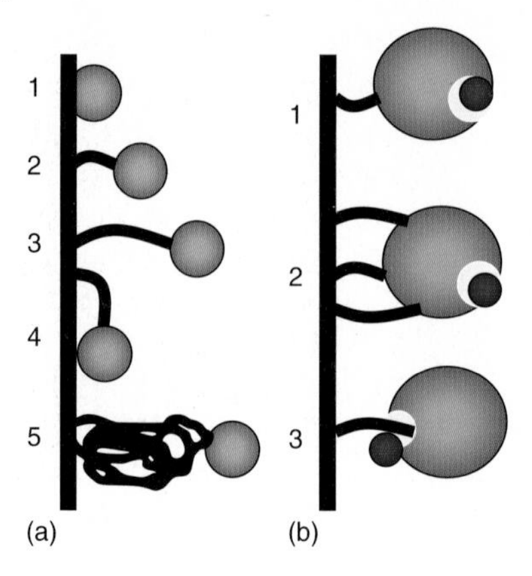

Figure 5.2 Modes of immobilization of enzymes to solid supports. (a) Different spacer types: 1, direct fixation without spacer of the enzyme to the support; 2, short spacer; 3, long spacer; 4, very long spacer bending backward onto the matrix surface; and 5, bulky protein spacer. (b) Different enzyme–spacer interactions: 1, interaction with nonessential parts of the enzyme; 2, multiple interactions; and 3, blocking of the enzyme by direct binding of the spacer to the active site.

groups of both components, creating a distance of about 1 nm between them. Varying the number of methylene groups between the terminal amine groups can modify the distance. Less than six methylene groups between the two reactive amides bring the enzyme closer to the support, thereby increasing the interaction. More methylene groups increase the distance, especially if the contact is unfavorable (Figure 5.2a, 3). However, in any case, long spacers must not separate the enzyme from the matrix; the flexible hydrocarbon chain can bend backward, bringing the enzyme molecule even closer to the surface (Figure 5.2a, 4). Therefore, long spacers must not be of advantage; rather, there exists an optimum length. Larger distances can be realized by limiting the flexibility of the spacer using rigid and bulky molecules such as inert proteins as spacers (Figure 5.2a, 5).

The preference for a special spacer is determined not only by its length. A further important criterion is the mode of attachment. Usually, the spacer is a bifunctional reagent with two terminal reactive groups. It can be either homobifunctional, such as 1,6-hexanediamine, glutardialdehyde, or dimethylsuberimidate, or heterobifunctional with different reactive groups (cf. Table 5.1). This feature decides which residue of the matrix and the protein (amino, hydroxyl, carboxyl, sulfhydryl) can be connected by the spacer. For the reaction with the spacer, several aspects must be considered. Especially for heterobifunctional reagents, the reaction sequence is of importance; that is, whether the spacer should be at first attached to the matrix or to the enzyme. Connecting to the enzyme as the first step has the advantage that possible inactivation due to blocking of essential groups can be controlled in a better manner. Although the spacer should increase the distance to the matrix for diminishing unfavorable interactions, the enzyme comes into direct contact with the spacer and can be influenced by its charges or polarity. Apolar aliphatic compounds, such as the hexamethylene chain, are not compatible with all enzymes.

To connect the enzyme to the matrix, only one spacer molecule is sufficient (Figure 5.2b, 1); however, more than one can react if enough reactive groups are available. This limits the flexibility of the immobilized enzyme (Figure 5.2b, 2). If the reaction of the enzyme with the spacer is not controlled, binding to the catalytic center and inactivation of the enzyme can occur (Figure 5.2b, 3). Such undesired reactions can be prevented by protecting the active site during the reaction by using a substrate and/or a cofactor. On the other hand, the active site occupies only a small part of the whole enzyme surface and the share of enzyme molecules inactivated by binding of the spacer to this region may be rather small. For this, it is favorable to connect the spacer at first with the matrix, so that the now-immobilized spacer attacks the enzyme only from a distinct site and not from all sites such as the free spacer. Finally, it must be considered that all proteins present in the enzyme solution will be coimmobilized. However, this must not, in any case, be disadvantageous; for instance, proteases if immobilized, can no longer attack enzymes fixed to the matrix apart from them.

5.2 Methods for Enzyme Immobilization

In the following section, different immobilization procedures are described in representative experiments. Because of the multitude of enzymes, spacers, and supports, these protocols can give only a brief insight into immobilization techniques.

5.2.1 Microencapsulation in Nylon Beads

Reagents and solutions

Hexanediamine solution (1,6-hexanediamine, 1,6-diaminohexane, $M_r =$ 116.2, 0.44 g; sodium bicarbonate, $M_r = 84.0$, 0.16 g; sodium carbonate, $M_r = 106.0$, 0.66 g; dissolve, one after the other, in 10 ml H_2O)

Enzyme solution (~2 mg ml^{-1})

18 mM sebacoyl dichloride solution ($M_r = 239.1$; 0.4 ml sebacoyl dichloride in 20 ml chloroform and 80 ml cyclohexane, freshly prepared)

0.9% NaCl in 0.05 M potassium phosphate pH 7.0

Procedure

Mix equal volumes of enzyme and 1,6-hexanediamine solutions

Very slowly drop the sebacoyl dichloride solution into it using a microliter syringe (10 μl) so that the nylon beads formed do not touch one another

Decant the organic phase after 5 min

Wash the capsule after complete volatilizing of the organic solvent with 0.9% NaCl in 0.05 M potassium phosphate pH 7.0

Reference

Chang, T.M.S. (1976) *Meth. Enzymol.*, **44**, 201–218.

5.2.2
Entrapment in Polyacrylamide

Polymerization of acrylamide used in polyacrylamide gel electrophoresis (PAGE) is used accordingly for entrapment of enzymes. The long polymer chains are connected by the cross-linker *N,N′*-methylenebisacrylamide (Figure 5.3). The polymerization reaction is initiated by *N,N,N′, N′*-tetramethylethylenediamine (TEMED) and ammonium persulfate. The mesh width of the polymer network must be smaller than the diameter of the enzyme, but larger than the substrates and products to allow their free diffusion. This condition can be controlled by varying the amount of cross linker.

Reagents and solutions

Acrylamide ($M_r = 71.1$)

N, N′-methylenebisacrylamide ($M_r = 154.2$)

N, N, N′N′-tetramethylethylenediamine (TEMED, $M_r = 116.2$)

Ammonium persulfate (ammonium peroxydisulfate, $M_r = 228.2$, 10% solution in water, prepare fresh)

0.1 M Tris/HCl pH 7.2

0.1 M potassium phosphate pH 7.5

Enzyme solution (8 mg in 8 ml 0.1 M Tris/HCl pH 7.2)

Procedure

Add 1.5 g acrylamide and 80 mg *N, N′*-methylenebisacrylamide to 8 ml enzyme solution

Add 10 μl TEMED and 30 μl 10% ammonium persulfate

Keep at 37 °C for 20 min and control the temperature; avoid an increase

Cut the polymerized gel into pieces with a spatula or a knife

Add 10 ml 0.1 M potassium phosphate pH 7.5

Triturate the gel particles with a homogenizer (Waring blender)

Wash three times on a frit with 0.1 M potassium phosphate pH 7.5

The gel particles can be directly used in enzyme assays.

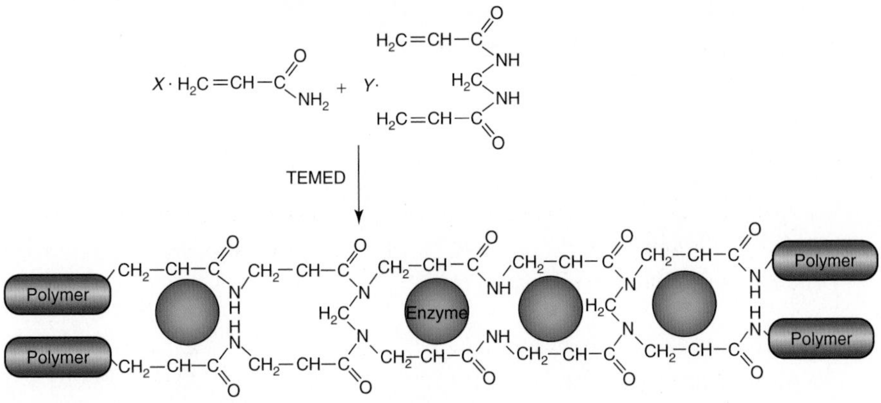

Figure 5.3 Entrapment of enzymes by polymerization of acrylamide with methylenebisacrylamide to polyacrylamide. *X* and *Y* are stoichiometric factors according to the relative concentrations of the respective components.

5.2.3
Covalent Immobilization on Glass Surfaces

For the immobilization of enzymes in technical processes, glass is a valuable material due to its resistance to strong acids, solvents, and high pressure. However, the inert and relatively small accessible glass surface is disadvantageous. Reactive groups are introduced by silanization. The following protocol describes this procedure with the example of normal glass, followed by an immobilization procedure using controlled-pore glass (CPG) with a larger accessible surface.

Amino groups with a silane component are introduced to the surface of glass beads. The amino groups are derivatized with *p*-nitrobenzoylchloride and the resulting nitro groups are reduced to amino groups. With nitrous acid they are converted to diazonium cations, to which enzymes are coupled via a tryrosyl residue (Figure 5.4). Care must be taken during manipulation with the toxic and aggressive reagents. A magnetic stirrer should not be used to avoid damage of the modified glass surface. Use of glass equipment during the silanization procedure should be avoided, as it will also react and the enzyme will get fixed to it.

Reagents and solutions

Glass beads, 1 mm diameter

3-Aminopropyl-triethoxysilane

p-Nitrobenzoylchloride

Dichloromethane

Triethylamine

Sodium nitrite ($NaNO_2$, $M_r = 69.0$)

Hydrochloric acid, concentrated (HCl)

Hydrochloric acid, 2 M

Sodium dithionite ($Na_2S_2O_4$, $M_r = 174.1$)

Nitric acid (5%)

0.1 M potassium phosphate pH 7.5

Enzyme solution (1 mg ml^{-1} in 0.1 M potassium phosphate pH 7.5)

Activation of glass beads

10 g glass beads in 40 ml 5% nitric acid

Heat for 1 h at 80–90 °C

Wash three times with H_2O and transfer into a PE vessel

Add 10 ml H_2O and 3 g 3-aminopropyl-triethoxysilane

Adjust the pH to 3–4, first with concentrated and finally with 2 M HCl; use pH indicator paper (pH electrode can be destroyed!)

Incubate at 65 °C for 12 h

Wash the beads on a suction filter (china, no glass frit) five times with H_2O and three times with acetone (20 ml each)

Dry at 55 °C for 20 min

Diazotization

For the remaining procedure, glass vessels can be used.

Add 25 ml dichloromethane, 1.5 g triethylamine, and 5 g *p*-nitrobenzoylchloride to the activated glass beads

Boil under reflux for 4 h. Moisture must be carefully excluded

Wash the beads three times with 10 ml dichloromethane

Dry for a short period at 55 °C

Transfer the beads into a solution of 1 g $NaNO_2$ in 100 ml of 2 M HCl for 20 min at room temperature for diazotization of free amino groups

Wash thoroughly with H_2O

Reflux for 1 h in 40 ml of a solution of 2 g sodium dithionite in H_2O

Wash the resulting arylamine beads three times with 20 ml H_2O

Transfer in a round-bottomed flask and add 88 ml 2 M HCl

Mount the flask on a rotary evaporator and rotate in an ice bath for cooling to 0 °C

Add 1.12 g $NaNO_2$ slowly in portions within 20 min; between additions the flask rotates without vacuum

Rotate for another 10 min with a weak water jet vacuum (about 20 mbar) to remove nitrous gaseous fumes

Wash the resulting diazo-glass three times with 20 ml ice-cold H_2O

The beads must be kept at 4 °C and should be used immediately for immobilization.

Immobilization of enzyme

Add 10 ml enzyme solution (1 mg ml^{-1} in 0.1 M potassium phosphate pH 7.5) to the glass beads

Rotate the flask slowly at 4 °C over night (12–16 h)

Wash three times with 5 ml 0.1 M potassium phosphate pH 7.5

The supernatant buffer solutions are collected for the determination of nonimmobilized protein and enzyme activity to estimate the efficiency of immobilization.

5.2.4
Covalent Immobilization on Controlled-Pore Glass

Controlled-pore glass (CPG) is a special sintered glass with a large surface due to the microporous structure. For enzyme immobilization, it can be obtained in activated form. Here, CPG is used with 46 nm pore width and glycerol bound via a siloxane bridge to the glass surface. The glycerol groups are oxidized by metaperiodate, resulting in an aldehyde glass (Figure 5.5). A Schiff's base is formed, to which *p*-phenylenediamine binds. This binding is stabilized by reduction with sodium borohydride. The terminal amino group is converted by nitrous acid to a diazonium cation, to which the enzyme is immobilized via a tyrosyl residue.

Reagents and solutions

CPG Glycophase (Bioran®, Pierce Chemical Comp., Rockford, IL, USA; Schott Glas, Mainz, Germany)

Sodium metaperiodate solution ($NaIO_4$, $M_r = 213.9$; 91 mg $NaIO_4$ in 70 ml H_2O, prepare fresh)

$NaBH_4$ ($M_r = 37.8$)

p-Phenylenediamine solution ($M_r = 108.1$; 77 mg in 70 ml H_2O)

Sodium nitrite ($NaNO_2$, $M_r = 69.0$)

Hydrochloric acid (HCl), 2 M

0.1 M potassium phosphate pH 7.5

Enzyme solution (1 mg ml^{-1} in 0.1 M potassium phosphate pH 7.5)

Activation of CPG beads

Add 1 g CPG beads to the 70 ml sodium metaperiodate solution in a 100 ml round-bottomed flask

Rotate at a rotatory evaporator for 1 h at 0.3–0.4 bar

Wash the resulting aldehyde glass several times together with 350 ml H_2O

Add the 70 ml *p*-phenylenediamine solution

Rotate at 0.3–0.4 bar for 20 min

Add 7 mg $NaBH_4$ and rotate for another 20 min

Add 7 mg $NaBH_4$ and rotate for 60 min

Wash thoroughly three times with 20 ml H_2O

Add 88 ml 2 M HCl and rotate in an ice bath to cool to 0 °C

Add 1.12 g $NaNO_2$ slowly in portions within 20 min; between the additions the flask rotates without vacuum

Rotate for another 10 min with a weak water jet vacuum (~20 mbar) to remove nitrous gaseous fumes

Wash the resulting diazo-glass three times with 20 ml ice-cold water

The beads must be kept at 4 °C and used immediately for immobilization.

Immobilization of enzyme

Add 10 ml enzyme solution (1 mg ml^{-1} in 0.1 M potassium phosphate pH 7.5) to the glass beads in a 25 ml flask

Rotate slowly overnight (12–16 h), at 4 °C

Wash the glass beads three times with 5 ml 0.1 M potassium phosphate pH 7.5

The supernatant buffer solutions are collected for the determination of nonimmobilized protein and enzyme activity to estimate the efficiency of immobilization.

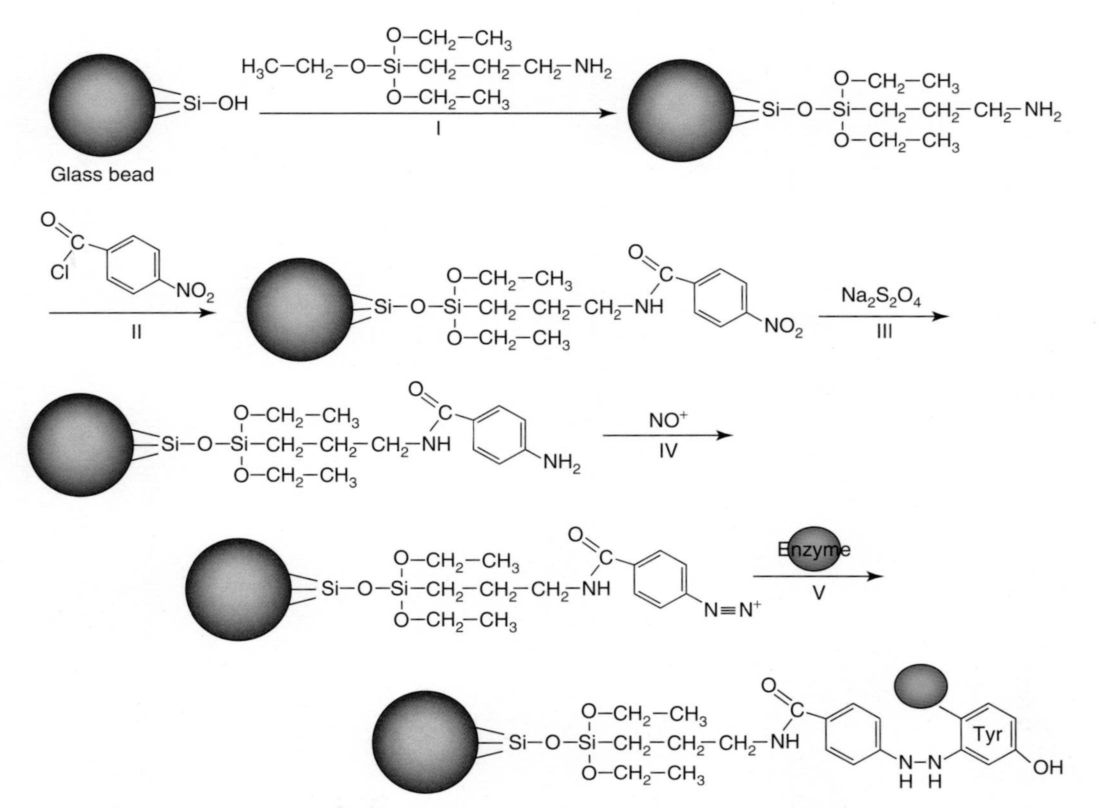

Figure 5.4 Silanization of a glass surface and covalent fixation of an enzyme. I, Silanization of Si–OH groups of the glass surface with 3-aminopropyl triethoxysilane; II, derivatization of the terminal amino group with *p*-nitrobenzoylchloride; III, reduction of the nitro group with sodium dithionite; IV, formation of the diazonium cation with nitrous acid; and V, coupling of the enzyme via a tyrosyl residue.

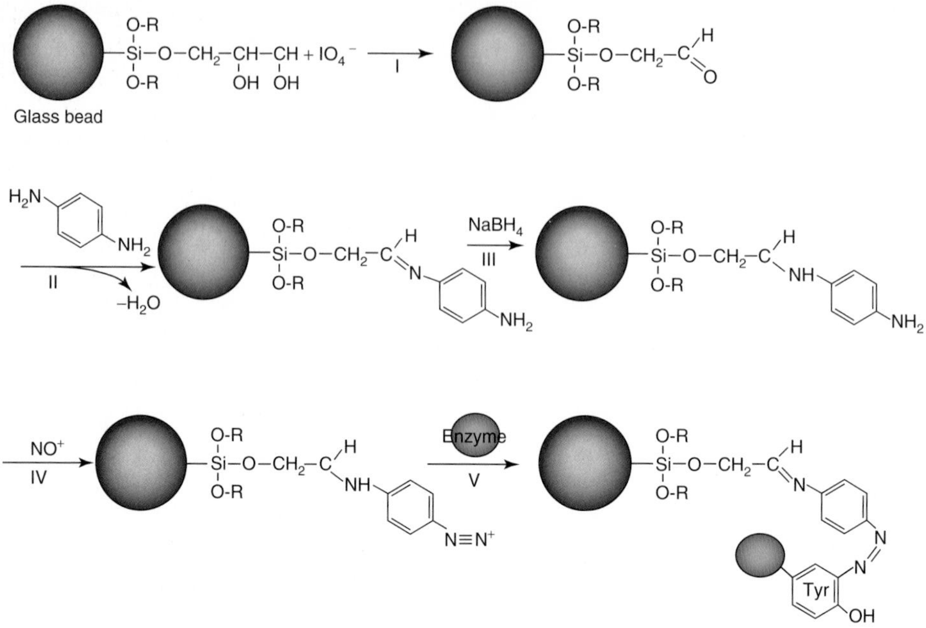

Figure 5.5 Immobilization of enzymes to controlled-pore glass: I, formation of aldehyde glass by periodate oxidation of glycerol, fixed to the glass surface via a siloxan group; II, formation of a Schiff's base with phenylenediamine; III, borohydride reduction of the Schiff's base; IV, formation of the diazonium cation with nitrous acid; and V, coupling of the enzyme via a tyrosyl residue.

References

Gabel, D. and Axen, R. (1976) *Meth. Enzymol.*, **44**, 383–393.
Weetall, H.H. (1976) *Meth. Enzymol.*, **44**, 134–148.

5.2.5
Covalent Immobilization to Polyamide

Synthetic polymers are valuable matrices for the immobilization of enzymes for various technical applications. Polyamides, polyesters, polyethylenes, and polyvinyl alcohols are most frequently used. Polyamides are synthesized by polycondensation of aliphatic dicarboxylic acids with diamines. The most important technical polyamide is Nylon®, a long-chain polymer formed by condensation of equimolar amounts of adipic acid and hexanediamine (polyhexamethylene adipamide, Nylon-6,6,6,6-polyamide; the "6" digits indicate that both the acid and the amine contribute six carbon atoms):

$$n\text{HOOC-(CH}_2)_4\text{-COOH} + n\text{H}_2\text{N-(CH}_2)_6\text{-NH}_2$$
$$\rightarrow [\text{-NH-CO-(CH}_2)_4\text{-CO-NH-(CH}_2)_6\text{-}]_n$$

Nylon is spun directly from the melt and features high tensile strength and elasticity. Advantages for immobilization are its relative hydrophilic character and resistance against microbial attack.

Polycondensation of 6-polyamide (Perlon®, Dederon®) requires only one component, ε-caprolactame. It hydrates to ε-aminocaproic acids and condenses to an alkali-resistant, but acid-labile polymer:

$$\varepsilon\text{-caprolactame} \xrightarrow{+H_2O} nH_2N\text{-}(CH_2)_5\text{-}COOH \rightarrow \text{-}\{NH\text{-}CO\text{-}(CH_2)_5\text{-}\}_n$$

The enzyme is mostly encapsulated or adsorbed to the polymer surface. Covalent fixation is difficult since only the ends of the polymer chains carry reactive groups for direct coupling. To attain a high degree of immobilization of the enzyme to Nylon, two alternative strategies can be used: (i) attachment of reactive groups to the free electron pairs of the oxygen or the nitrogen atoms and (ii) generation of free amino and carboxyl groups for immobilization by partial hydrolysis of amide bonds of the polymer. The conditions for hydrolysis must be strictly controlled to avoid extensive disintegration of the polymer structure.

5.2.5.1 o-Alkylation with Triethyloxonium Tetrafluoroborate

Proteins can be fixed to polyamides via alkylation of the oxygen of the amide bond. The alkylation reagent for this reaction is synthesized from epichlorohydrine (1-chloro-2,3-epoxypropane) and borotrifluoride diethyletherate in diethylether according to the following equation:

$$4(C_2H_5)_2OBF_3 + 2(C_2H_5)_2O + 3ClCH_2\text{-}\overset{O}{\widehat{CH\text{-}CH_2}} \longrightarrow$$

borotrifluoride diethyletherate; diethylether; epichlorohydrine

$$3(C_2H_5)_3O^+BF_4^- + B(\underset{CH_2Cl}{OCH}CH_2OC_2H_5)_3$$

triethyloxonium tetrafluoroborate

The alkylation procedure of polyamide occurs according to the reaction schema shown in Figure 5.6. This is a relatively laborious method because water must be strictly excluded. The alkylation conditions can be modified by varying reaction time and concentration of the alkylation reagent. The toxicity of the substances must be considered; the organic solvents may also resolve components from the synthetic cloth, such as dressing and glues.

Reagents and solutions

Dry diethylether

Epichlorohydrine solution (10%, v/v) in dry diethylether

Borotrifluoride diethyletherate solution (5%, v/v) in diethylether

Dichloromethane

Polyamide cloth

Procedure

Preparation of Triethyloxonium Tetrafluoroborate

Drop 50 ml epichlorohydrine solution from a dropping funnel slowly to 200 ml borotrifluoride diethyletherate solution

Boil 1 h under reflux

Stir for 3 h at room temperature

Wash the precipitating triethyloxonium tetrafluoroborate three times with diethylether

Redissolve in dichloromethane to a concentration of 10% (w/v; triethyloxonium tetrafluoroborate solution)

Alkylation of Polyamide

Incubate a piece of synthetic polyamide cloth in the triethyloxonium tetrafluoroborate solution for 15 min at room temperature to form the imidate salt of the polyamide

Wash on a suction filter (Büchner funnel) with dichloromethane and dioxane, two times each, to remove the exceeding alkylation reagent

The reactive imidate salt of the polyamide must immediately be worked up.

Immobilization of the enzyme can be accomplished using to one of the following strategies:

Direct immobilization by incubation with a concentrated enzyme solution as described in Section 5.2.5.2

To introduce a spacer, incubate the polyamide imidate for 3 h at room temperature in a 10% (w/v) solution of 1,6-hexanediamine in methanol. Wash several times with methanol to remove excess of the reagent

Incubate the polyamide imidate for 3 h at room temperature with a 3% (w/v) solution of adipic acid dihydrazide in formamide to obtain the hydrazide-substituted polyamide. Wash several times with methanol to remove excess of solvent

Activation of hydrazide-substituted polyamide can be carried out with sodium nitrite in hydrochloric acid as shown in Figure 5.6. An azide group is formed to which the enzyme is bound. Alternatively, a Schiff's base is formed with the

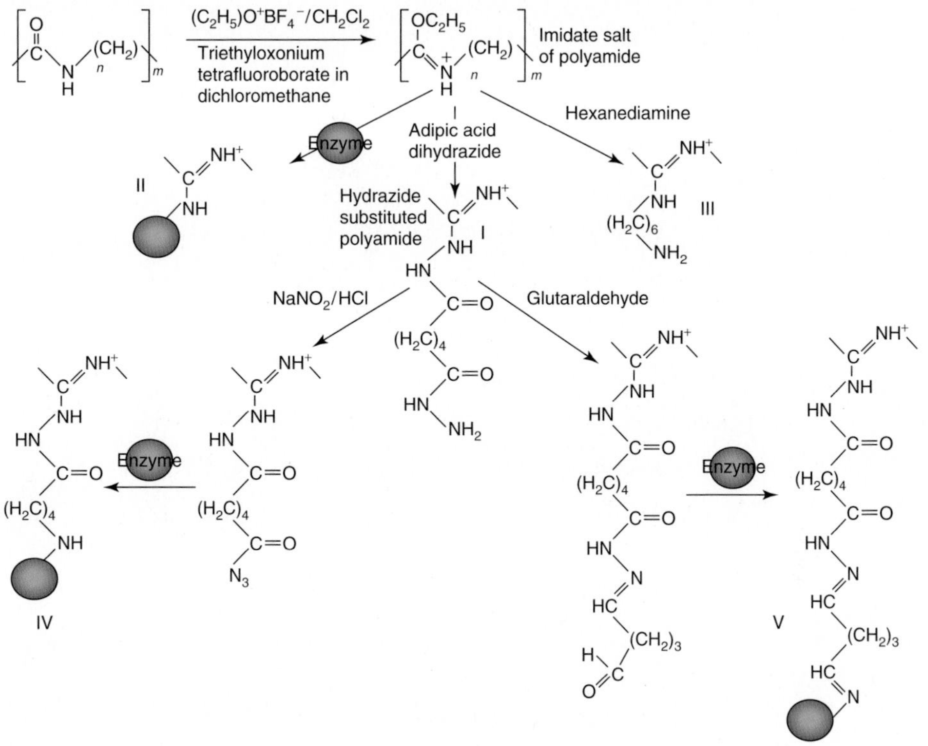

Figure 5.6 Covalent immobilization of enzymes to polyamide. Polyamide is alkylated by treatment with triethyloxonium tetrafluoroborate. The enzyme is fixed to the matrix: I, with adipic acid dihydrazide to form a hydrazide-substituted derivative; II, by direct fixation or 111, with a 1,6-hexanediamine spacer. The hydrazide-substituted polyamide is activated with sodium nitrite in hydrochloric acid (IV) or with glutaraldehyde; the enzyme is subsequently coupled by the Schiff's base formation (V).

terminal amino group of the hydrazide and glutardialdehyde. The enzyme is connected to the second aldehyde group of the reagent via a further Schiff's base (Section 5.2.5.2).

The derivatives are no longer toxic; they can be stored for months at room temperature and reactivated at any time.

5.2.5.2 Immobilization to Amino Groups after Partial Hydrolysis of Polyamide

Partial hydrolysis of polyamide is a simple method for fixation of enzymes, but extensive disintegration of the polymer structure must be avoided. The immobilization reaction proceeds in four steps:

1) Etching of the polymer material to increase the surface and water contact
2) Partial cleavage of amide bonds
3) Activation of free amino or carboxyl groups
4) Coupling of the enzyme to the activated polyamide

Reagents and solutions

Synthetic polyamide cloth, 10 pieces of defined size (i.e., 2 cm^2)

0.2 M borate buffer pH 8.5

Methanol

5% (w/v) glutardialdehyde solution in 0.2 M borate buffer pH 8.5

$CaCl_2\,2H_2O$ ($M_r = 147.0$)

3.65 M hydrochloric acid (HCl)

N, N'-Dimethyl-1,3-propanediamine ($M_r = 102.2$)

0.1 M potassium phosphate pH 7.5

0.5 M NaCl ($M_r = 58.4$) in 0.1 M potassium phosphate pH 7.5

Enzyme solution (0.5 mg ml^{-1} in 0.1 M potassium phosphate pH 7.5)

Etching of Polyamide

Immerse the polyamide pieces in 250 ml of a mixture of 18.6% (w/w) $CaCl_2$ and 18.6% (w/w) water in methanol in a round-bottomed flask

Rotate the flask with a rotatory evaporator (without vacuum) for 20 min at 50 °C

Wash thoroughly with water on a suction filter or a frit

Partial Cleavage of Amide Bonds

Hydrolytic Cleavage

Immerse the polyamide pieces in 3.65 M HCl and rotate for 40 min with a rotatory evaporator (without vacuum) at 50 °C

Wash intensely on a suction filter or frit with H_2O and finally with 0.2 M borate buffer pH 8.5

Nonhydrolytic Cleavage

Dip the polyamide pieces into methanol and dry

Incubate for 12 h at 70 °C in 25 ml N, N'-dimethyl-1,3-propanediamine

Wash intensely with H_2O on a suction filter or a frit

Activation of Free Amino Groups

Rotate (without vacuum) the polyamide pieces after hydrolytic or nonhydrolytic cleavage or after substitution in 25 ml 5% (w/v) glutardialdehyde solution in a round-bottomed flask with a rotatory evaporator, for 15 min at 20 °C

Remove excess of glutardialdehyde by intense washing with 0.1 M potassium phosphate pH 7.5 on a suction filter or a frit

Coupling of the Enzyme to Activated Polyamide

Incubate the polyamide pieces immediately after washing in the enzyme solution (the pieces must be completely submerged) for 3 h at 4 °C

Wash repeatedly on a frit with 0.5 M NaCl in 0.1 M potassium phosphate pH 7.5 to remove the enzyme fraction that has not been immobilized

Wash with 0.1 M potassium phosphate pH 7.5 to reduce the high salt

Store in 0.1 M potassium phosphate pH 7.5 at 4 °C. For long-term storage, wash the pieces with H_2O and dry in air (sensitive enzymes cannot withstand this procedure).

References

Hornby, W.E. and Goldstein, L. (1976) *Meth. Enzymol.*, **64**, 119–134.

Moeschel, K., Nouaimi, M., Steinbrenner, C., and Bisswanger, H. (2003) *Biotechnol. Bioeng.*, **82**, 190–199.

Morris, D.I., Campbell, J., and Hornby, W.E. (1975) *Biochem. J.*, **174**, 593–603.

5.2.5.3 Immobilization to Carboxyl Groups after Partial Hydrolysis of Polyamide

Etching of the polyamide and cleavage of amide bonds occur according to the procedure given earlier (Section 5.2.5.2). Free amino groups are blocked by treatment for 10 min with an ice-cold solution of 1% (w/v) sodium nitrite in 0.5 M hydrochloric acid and further incubated in this solution for 20 min at 40 °C.

Reagents and solutions

0.5 M HCl (hydrochloric acid)

Hexanediamine

Dicyclohexylcarbodiimide

Prepare a solution of hexanediamine and dicyclohexylcarbodiimide, both 1% (w/v) in dichloromethane

0.1 M potassium phosphate pH 7.5

0.5 M NaCl in 0.1 M potassium phosphate pH 7.5

Enzyme solution (0.5 mg ml^{-1} in 0.1 M potassium phosphate pH 7.5)

Activation of free carboxyl groups

Incubate the polyamide pieces for 4 h at 10 °C in the solution of hexanediamine and dicyclohexylcarbodiimide

Wash with dichloromethane, acetone, and ice-cold water, 10 ml of each, to remove free hexanediamine

Incubate for 20 min in 10 ml 1% (w/v) sodium nitrite in 0.5 M HCl at 0 °C

Wash the pieces with 10 ml of ice-cold 1 mM HCl

Coupling of the enzyme to the activated polyamide and storage is carried out as described in Section 5.2.5.2.

5.2.6
Immobilization to Polyester

Polyesters are synthesized by condensation of a dicarboxylic acid with a diol. The frequently used polyethylene terephthalate (PET) (Terylene®, Trevira®, Diolen®) is formed by polycondensation of terephthalate with ethylene glycol, a two-step process with antimony trioxide as catalyst (Figure 5.7). At first, a low molecular weight polymer with terminal hydroxyl groups is formed with an excess of ethylene glycol, which converts into a high molecular polymer by cleavage of ethylene glycol.

Because of their rather nonpolar surface, polyesters are less suited as support for immobilized enzymes. Copolymers from polyester and viscose fibers (Rayon®) are more compatible. The cellulose component increases the hydrophilic character and contributes hydroxyl groups for immobilization. The procedures of immobilization

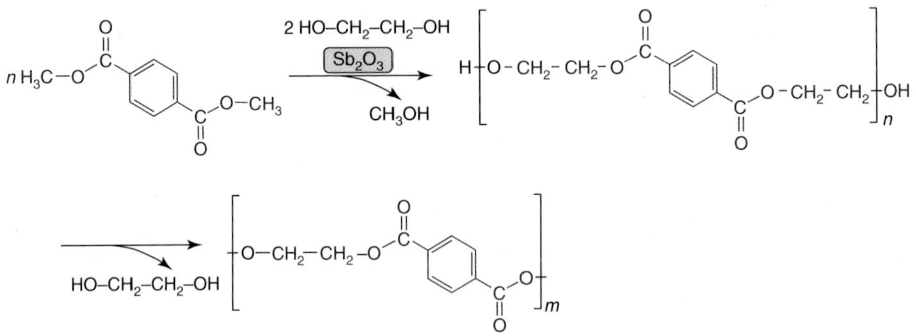

Figure 5.7 The formation of polyester by polycondensation of terephthalate with ethylene glycol. Two molecules of ethyleneglycol react with dimethylterephthalate in the presence of antimony trioxide as a catalyst to a low molecular weight polymer (*n*), which is converted to a high molecular polymer (*m*) by cleavage of ethyleneglycol.

described above for polyamides, such as alkylation of CH-acid atoms or partial hydrolysis, are also applicable for polyesters.

Reagents and solutions

Nonwoven polyester, 10 pieces of a predefined size (i.e., 2 cm^2)

Dichloromethane

Hexanediamine

Dicyclohexylcarbodiimide

Prepare a solution of hexanediamine and dicyclohexylcarbodiimide, both 1% (w/v), in dichloromethane

0.1 M potassium phosphate pH 7.5

0.5 M NaCl in 0.1 M potassium phosphate pH 7.5

Enzyme solution (0.5 mg ml^{-1} in 0.1 M potassium phosphate pH 7.5)

Acid hydrolysis and activation with dicyclohexylcarbodiimide

Etching of the polyester and hydrolysis is carried out as described in Section 5.2.5.2.

For activation of free carboxylic groups, incubate the polyester pieces in 50 ml of the hexanediamine–dicyclohexylcarbodiimide solution for 4 h at 10 °C

Wash with dichloromethane, acetone, and ice-cold water, 10 ml each, on a suction filter or a frit for removing free hexanediamine

Coupling to the activated polyester

Incubate the polyester pieces immediately after washing for 3 h at 4 °C in the enzyme solution (sufficient volume to completely submerge the pieces; e.g., 10 ml)

Wash on a frit with 0.5 M NaCl in 0.1 M potassium phosphate pH 7.5 to remove immobilized enzyme

Wash with 0.1 M potassium phosphate pH 7.5 to reduce the high salt concentration

Store the pieces with immobilized enzyme for a short time (about one week) in 0.1 M potassium phosphate pH 7.5 at 4 °C. For long-term storage, wash the pieces with H_2O to remove the buffer and dry in air (not recommended for very sensitive enzymes).

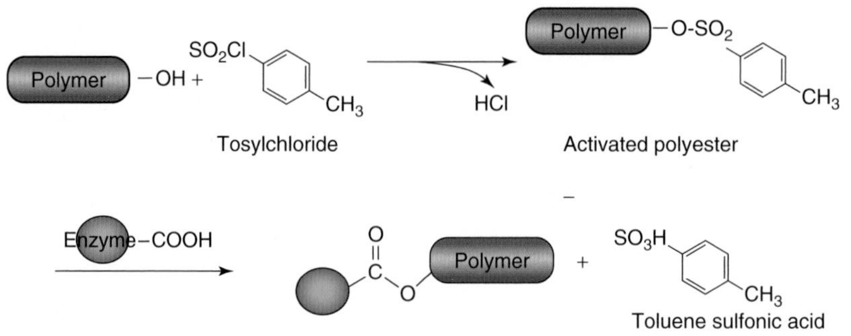

Figure 5.8 Immobilization of enzymes to polyester by tosylchloride activation.

Reference

Daka, N.J. and Laidler, K.J. (1978) *Can. J. Biochem.*, **56**, 774–779.

5.2.7 Immobilization by Alkaline Hydrolysis and Activation with Tosylchloride

With this method, the ester bonds are partially hydrolyzed by alkaline treatment. The activation occurs with tosylchloride in acetone. The enzyme is immobilized to this residue via carboxyl or hydroxyl groups (Figure 5.8).

Reagents and solutions

Polyester pieces (10 pieces of defined size 2 cm^2)

0.1 M potassium phosphate pH 7.5

0.5 M NaCl ($M_r = 58.4$) in 0.1 M potassium phosphate pH 7.5

10% (w/v) NaOH

Acetone

Tosylchloride solution (*p*-toluenesulfonyl chloride, $M_r = 190.7$, 10% (w/v) in acetone)

Enzyme solution (1 mg ml^{-1} in 0.1 M potassium phosphate pH 7.5)

Alkaline hydrolysis and activation by tosylchloride

Incubate the polyester pieces at room temperature for 10 min in 10 ml 10% NaOH

Wash three times with H_2O and five times with acetone, 10 ml each, on a frit

Incubate the pieces immediately after washing, without drying, for 20 min in 10 ml of the tosylchloride solution

Wash three times with acetone, three times with H_2O, and two times with 0.1 M potassium phosphate pH 7.5, 10 ml each, on a frit

Enzyme immobilization

Incubate the polyester pieces in 10 ml enzyme solution (pieces must be completely submerged) directly after washing under gentle shaking overnight at 4 °C

Wash on a suction filter or a frit with 0.5 M NaCl in 0.1 M potassium phosphate pH 7.5 to remove unbound enzyme

Wash with 0.1 M potassium phosphate pH 7.5 for reducing the high salt concentration

Store in this buffer or, for longer time, wash with water and dry in air.

5.2.8 Alkaline Hydrolysis and Activation by Carbonyldiimidazol

This is a more efficient and gentle procedure than the tosylchloride method.

Reagents and solutions

Polyester pieces (10 pieces of defined size 2 cm^2)

0.1 M potassium phosphate pH 7.5

0.5 M NaCl in 0.1 M potassium phosphate pH 7.5

10% (w/v) NaOH

Dioxane

Carbonyldiimidazol solution (CDI, $M_r = 162.2$) 3.5% (w/v) in dioxane

Enzyme solution (1 mg ml^{-1} in 0.1 M potassium phosphate pH 7.5)

Alkaline hydrolysis and activation

Incubate the polyester pieces at room temperature for 10 min in 10 ml 10% NaOH

Wash five times with H_2O, 10 ml each

Immerse for 30 min in dioxane

Wash five times with dioxane to remove residual water

Incubate for 30 min in CDI solution

Wash five times with dioxane and three times with 0.1 M potassium phosphate pH 7.5

Coupling of the enzyme

Submerge the pieces in 10 ml enzyme solution and incubate overnight at 4 °C

Wash on a suction filter or frit with 0.5 M NaCl in 0.1 M potassium phosphate pH 7.5

Wash with 0.1 M potassium phosphate pH 7.5 to reduce the high salt concentration

The pieces can be stored for about one week at 4 °C in this buffer; for long-term storage, wash for a short period in H_2O and dry in air (not recommended for very sensitive enzymes).

References

Boyd, S. and Yamazaki, H. (1993) *Biotechnol. Tech.*, **7**, 277–282.
Howlett, J.R., Armstrong, D.W., and Yamazaki, H. (1991) *Biotechnol. Tech.*, **5**, 395–400.
Nouaimi, M., Möschel, K., and Bisswanger, H. (2001) *Enzyme Microb. Technol.*, **29**, 567–574.

5.3 Analysis of Immobilized Enzymes

5.3.1 General Principles

The analytical methods for immobilized enzymes differ in several aspects from those for soluble enzymes, especially with respect to enzyme assays and protein determination, while some parallels exist with membrane-bound enzymes. Because of the various immobilization principles, general rules cannot be given; the method must be adapted to the special mode of immobilization. Because direct determination of the amount of enzyme or protein immobilized is difficult, this value is often, for simplicity, derived from the difference between the initially applied and the unbound portion after immobilization. This indirect determination is not very reliable since losses due to unspecific binding and adsorption or inactivation caused by the immobilization reaction apparently increase the bound portion of enzyme. Therefore, an accurate analysis must determine both the enzyme activity and the protein content of the immobilized sample. Compared with the respective initial amounts, the efficiency of the immobilization procedure can be estimated.

Most **protein assays**, including the Biuret, Lowry, and Bradford methods, follow the same general procedure: soluble protein is precipitated and the dissolved precipitate is subject to a color reaction, which is determined photometrically. Such a procedure is not possible with immobilized proteins. In some cases, such as adsorption to ion-exchange matrices, the protein can be released from the support. In other cases, the protein can be split off from the support by proteolytic cleavage or by acid or alkaline hydrolysis. However, the portion of immobilized protein will be underestimated if the cleavage is not complete. If, on the other hand, digestion is too efficient, yielding small peptides and free amino acids, these will not be detected by conventional protein assays. Some protein assays, such as BCR, ninhydrin, Coomassie blue, and 2-hydroxy-1-naphthaldehyde, can be modified for immobilized proteins as described in Section 3.4.1.

The same considerations hold for the determination of the **activity** of immobilized enzymes, which can strongly be impaired by the immobilization reaction, the interaction with the support, and the restriction of the accessibility to the substrate. Continuous photometric assays are disturbed by the supports to which the enzymes are immobilized, but the enzyme can be tested in a sequential manner (Figure 5.9). A defined portion of the matrix with the immobilized enzyme is incubated with the assay mixture, under the relevant assay conditions, such as appropriate pH

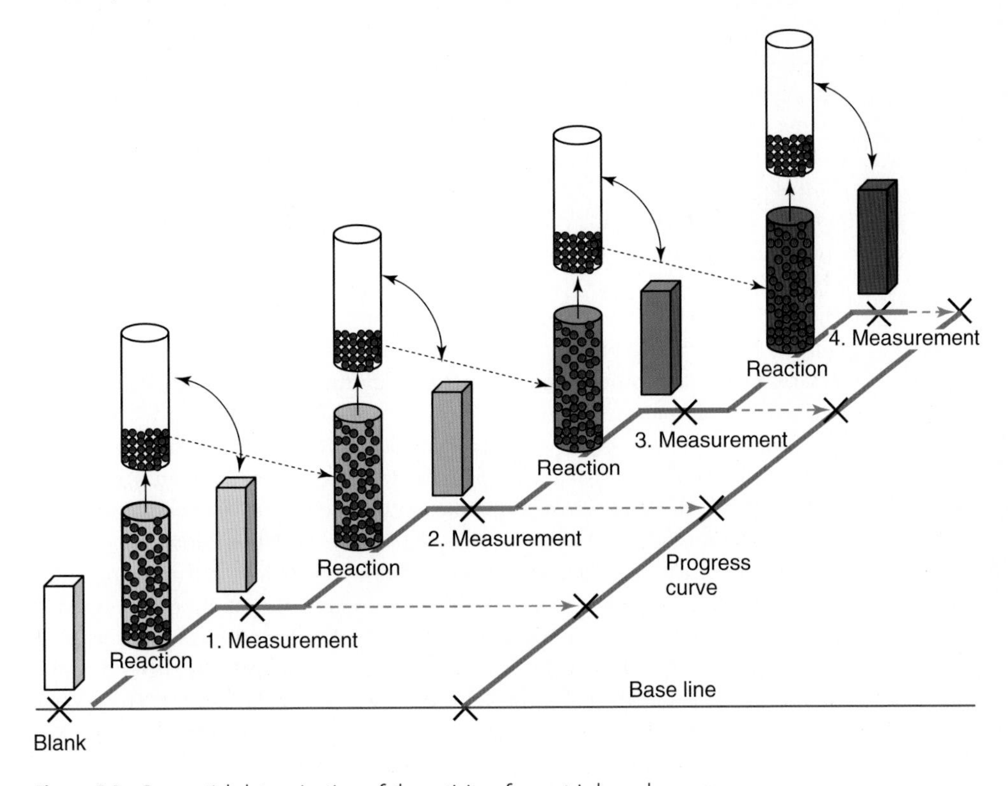

Figure 5.9 Sequential determination of the activity of a matrix-bound enzyme.

and temperature. After mixing or shaking for a certain period, the matrix with the immobilized enzyme is removed (by centrifugation or filtration) and the absorption is measured in the supernatant. This procedure can be repeated several times. The same matrix is added again to the same assay mixture and the reaction continues for a second time period. From several sequential measurements, a progress curve can be derived. The transfer must occur quantitatively with respect to both the assay mixture and the immobilized particles, and the same temperature must be maintained during the reaction and the measuring periods.

Alternatively, the matrix particles can be filled in a column, which is eluted by the assay mixture. The product formed is determined in the eluate and the activity is calculated from the elution time. In the following section, an arrangement for a continuous photometric assay is described.

5.3.2
Continuous Photometric Assays for Immobilized Enzymes

The above-described principle for a sequential photometric assay (Figure 5.9) can be modified into a semicontinuous procedure. The assay mixture is directly filled into the photometric cuvette, together with a sample of the immobilized matrix. The mixture is stirred. After a particular time period, the particles are allowed to settle and absorption is measured in the supernatant. The particles must sediment fast and the light beam must become completely clear. With large particles, several stirring periods, interrupted by short measuring phases, yield a nearly continuous progress curve. The procedure is fast and simple, but the reaction still proceeds in the settled particles. However, the conditions there differ from those in the complete assay mixture. Therefore, the measuring periods should be as short as possible. Even under similar assay conditions, soluble enzymes behave differently when compared to that of immobilized enzymes. Zones of substrate deprivation and product accumulation (product inhibition!) are transitorily formed. Thus, the reaction velocity is influenced by the motion of the particles, and stirring or shaking must be standardized. Further, it must be noted that immobilization to the matrix surfaces need not be quite homogeneous and differing values can be obtained from independent samples.

Continuous measurements can be carried out with an enzyme reactor as shown schematically in Figure 5.10. Matrix particles with the immobilized enzyme are included in a cylinder that is closed from the top and the bottom by a membrane, thus retaining the particles but allowing free penetration of the reaction components. The particles may be either stirred or densely packed like a chromatographic column. In a closed cycle, the assay mixture is continuously pumped through the reactor and a photometric flow cuvette. Valves for filling and removing the assay mixture are arranged in such a manner that the fresh solution first passes through the cuvette for adjusting the blank value before entering the reactor. The system must be completely free from air bubbles, which severely disturb the measurement when entering the cuvette. The reaction volume is determined by the dimensions of the reactor (less the matrix volume), while the volume of the tubes and the cuvette

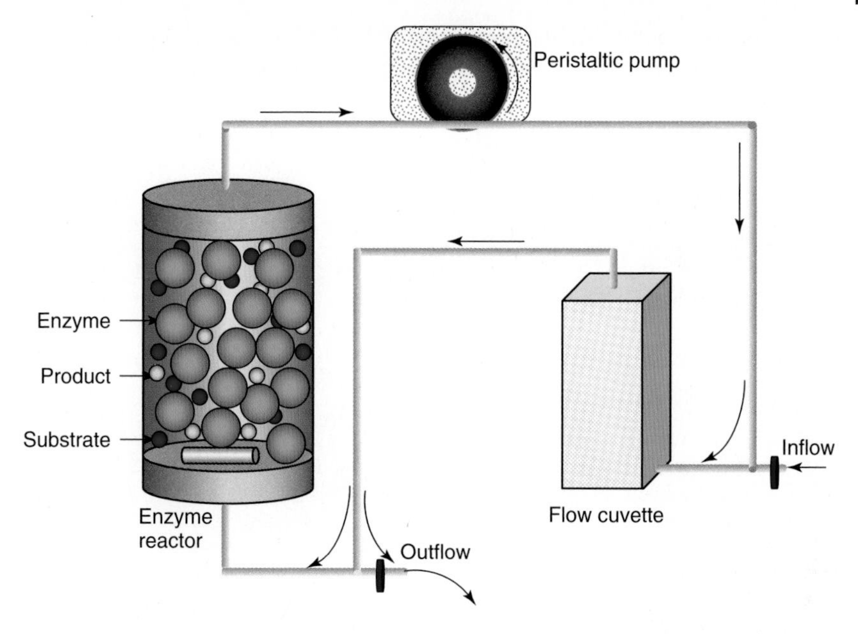

Figure 5.10 Enzyme reactor arrangement for continuous observation of enzyme velocity.

can be treated as dilution of the reacting solution. To determine the total volume of the system, the amount of buffer required to fill the system is measured. The volume of the immobilized enzyme sample depends on the shape and structure of the matrix and may be standardized as dry weight or accessible surface. The reaction velocity obtained by this device depends not only on the amount of immobilized enzyme but also on the stirring and pumping rates. To compare different samples, the system must be standardized; absolute determinations are difficult to accomplish.

5.3.3
Cofactors in Reactions with Immobilized Enzymes

Many enzyme reactions need cofactors. As long as they are covalently or very strongly bound, such as pyridoxal phosphate, porphyrin, or FAD, they are coimmobilized together with the enzyme and must not be considered separately. However, dissociable factors, such as NAD, NADP, ATP, thiamin diphosphate, or metal ions, are lost during the immobilization procedure and must be supplemented in the reaction mixture. For large-scale technical processes, this is a severe problem because they are removed along with the product and regaining these from the mixture is difficult and expensive.Moreover, they contaminate the desired product. A similar problem is with cosubstrates, which are required for the reaction in the same concentration as that of the substrate and yield undesired by-products.

Cofactors such as NAD or ATP can be covalently bound and immobilized to polymers, such as polyethylene glycol or dextran. Regenerating systems are used to recycle the cofactors, which are converted during the reaction. A convenient system is the regeneration of NADH from NAD with formiate dehydrogenase where carbon dioxide is formed. Other regenerating systems are alcohol and glutamate dehydrogenase. ATP can be regained from ADP using the acetate kinase with acetyl phosphate as a substrate.

References

Kragl, U., Kruse, W., Hummel, W., and Wandrey, C. (1996) *Biotechnol. Bioeng.*, **52**, 309–319.
Kula, M.-R. and Wandrey, C. (1987) *Meth. Enzymol.*, **136**, 9–21.

General References about Immobilization

Bickerstaff, G.F. (1997) *Immobilization of Enzymes and Cells*, Humana Press, Totowa.
Chenault, H.K. and Whitesides, G.M. (1987) *Appl. Biochem. Biotechnol.*, **14**, 147–197.
Drauz, K. and Waldmann, H. (1995) *Enzyme Catalysis in Organic Synthesis*, Wiley-VCH Verlag GmbH, Weinheim.
Mosbach, K. (ed.) (1987) *Meth. Enzymol.*, vols. 135–137.
Rehm, H.-J. and Reed, G. (1995) *Enzymes, Biomass, Food and Feed*, vol. 9, Wiley-VCH Verlag GmbH, Weinheim.
Taylor, R.F. (1991) *Protein Immobilization*, Marcel Decker, New York.
Wingard, L.B., Katchalsky-Katzir, E., and Goldstein, L. (1976) *Applied Biochemistry and Bioengineering*, vol. 1, Academic Press, New York.
Woodward, J. (1985) *Immobilised Cells and Enzymes*, IRL Press, Oxford.

5.4 Enzyme Reactors

For large-scale processes with immobilized enzymes, special devices, enzyme reactors or bioreactors have been constructed. While the term *bioreactor* is used more generally for all types of bioconversions, including fermentation of microorganisms or cell cultures, *enzyme reactors* are designed for continuous process control of catalyzed reactions (Figure 5.11). They need to keep the enzyme active under optimum conditions. Substrate solution is permanently supplied and the process should be conducted in a manner that it is quantitatively converted to product (provided the reaction equilibrium allows this). The product formed must be efficiently removed to avoid product inhibition. This depends on the reactor type, the total enzyme activity enclosed in the reactor, and the flow rate. Maintaining constant optimum conditions for the catalyzed reaction and supplementing the system simultaneously with all necessary components requires an exact process control. While the general assay conditions determined by the special enzyme must be equally considered for all reactors, transport processes, such as substrate addition and product removal, flow rate, and stirring depend essentially on the type and dimensions of the respective reactor. Accordingly, various types of enzyme reactors have been designed; three main types are described in the following sections.

5.4.1
Batch Reactor (Stirred-Tank Reactor)

This relatively simple type of enzyme reactor can be regarded as a common reaction vessel for the enzyme assays, scaled up to the desired dimension (Figure 5.11a). It needs no special requisites. A constant reaction temperature and permanent stirring of the reaction mixture must be maintained (*"stirred-tank reactor"*) for optimum interaction between the enzyme and substrate. However, continuous substrate supply and product withdrawal is not possible with this reactor type, and continuous processing can hardly be achieved. The reaction is carried out in a large batch volume and stopped after a particular reaction time, to isolate the products in a separate step. The method has the advantage that optimum conditions can be established at the start of the reaction, and the reaction course can be either permanently controlled by removing samples at distinct time intervals or directly controlled, for example, with biosensors.

5.4.2
Membrane Reactor

The membrane reactor (Figure 5.11b) can be regarded as a further development of the batch reactor. Ultrafiltration membranes at the inflow and outflow openings enclose the enzyme, but allow penetration of substrates and products. The reactor needs no elaborate immobilization procedure with the attendant risks of damaging

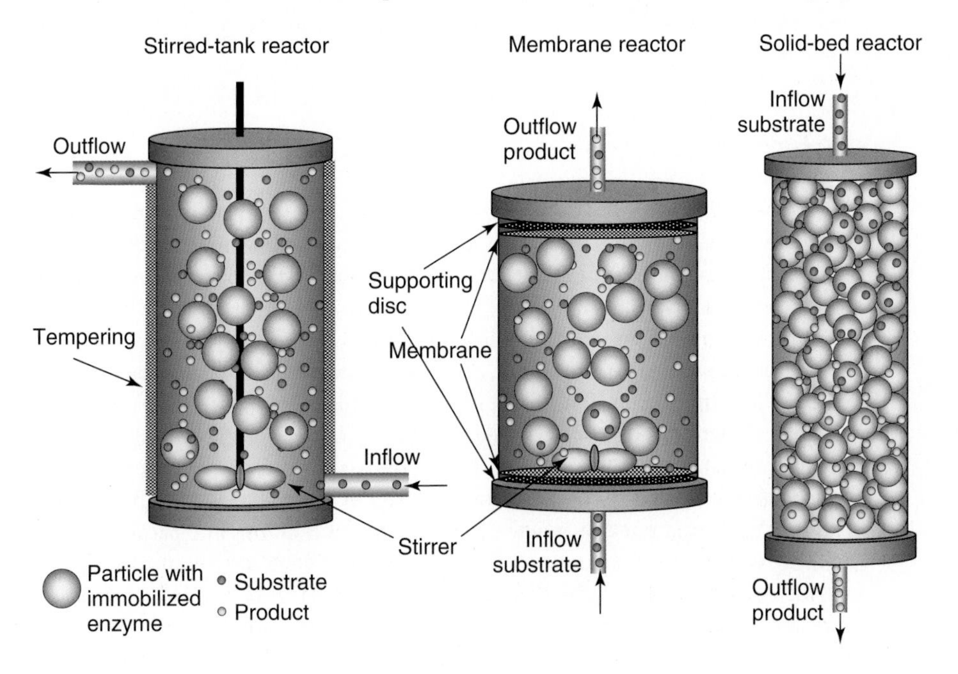

Figure 5.11 Types of enzyme reactors.

the enzyme, and any desired concentration can be used. The reactor can be continuously supplemented from an external source with fresh reaction solution and products can be removed at any time. However, the flow rates through the ultrafiltration membranes are limited and blocking of the membrane may occur during the process, although stirring and permanent flow over the membrane minimizes such effects. Any membrane leakage causes loss of enzyme. Therefore, fixation of the enzyme to particles by covalent immobilization, encapsulation, or cross-linking with inert materials such as BSA simplifies the procedure and the sensitive ultrafiltration membranes can be replaced by coarse filter membranes.

Membrane reactors are especially suitable for multienzyme systems with coupled reactions and cofactor regeneration; for example, coupling of an amino acid dehydrogenase, such as alanine dehydrogenase (Ala-DH, EC 1.4.1.1), with an NAD^+-regenerating enzyme such as formate dehydrogenase (FDH) (EC 1.2.1.2):

$$\text{pyruvate} + NH_4^+ + \text{NADH} \xrightarrow{\text{Ala-DH}} \text{L-alanine} + \text{NAD}^+ + H_2O$$

$$\text{NAD}^+ + \text{formate} \xrightarrow{\text{FDH}} \text{NADH} + H^+ + CO_2$$

To retain NADH enclosed in the reactor, its molecular mass needs to be enhanced by covalent binding to polyethylene glycol.

Enzyme catalysis in two-phase systems can be performed advantageously in membrane reactors, where the membrane serves as a boundary between the aqueous phase at one site and the organic phase at the other.

Use of hollow fibers, instead of membranes, increases the efficiency of membrane reactors. Their large surface intensifies the exchange with the outside medium and the removal of the product.

5.4.3 Solid Bed Reactor

For this modification of the membrane reactor, the enzyme is immobilized on solid particles (Figure 5.11c). The reactor is packed with the particles like a chromatographic column. The enzyme serves as a stationary phase, and the substrate solution is the mobile phase, which is converted during the passage through the column. Most immobilization techniques for enzymes may be used, such as adsorption on ion exchanger, covalent binding to glass, dextran, or cellulose, or cross-linking. Adsorption of the reaction components to the matrix must be avoided. Otherwise, the matrix needs to be regenerated from time to time. In this respect, absorbing the enzyme to the support instead of covalent fixation is advantageous, because both the adsorbed material and the enzyme can be eluted together and substituted by a new material, without changing the packed gel bed in the reactor.

The parameters of the solid bed reactor (reactor dimension, bed volume, elution rate, temperature) can be calculated exactly and adapted to the reaction conditions so that the substrate becomes completely converted to product during its run through the column.

5.4.4
Immobilized Cells

For industrial processes in bioreactors, whole cells instead of immobilized enzymes are strongly preferred. This saves laborious purification procedures and extends the life times of the enzymes. Nonliving cells have the advantage that they need no supplementation of growth medium. On the other hand, viable cells maintain their metabolism and thus, their enzyme activities over the whole life time. If only one distinct enzyme reaction is required, resting cells, maintaining only the reaction of interest, can be used. An example is the conversion of glucose to fructose by glucose isomerase, which can be performed either in a stirred-tank reactor with periodic recharging of fresh biocatalyst or by continuously operating reactors, where new and spent biocatalysts are exchanged constantly at the same rate.

References

Belford, G. (1989) *Biotechnol. Bioeng.*, **33**, 1047–1066.

Bommarius, A.S. (1993) in *Biotechnology*, Bioprocessing, vol. 3 (eds H.-J. Rehm, G. Reed, A. Pühler, P. Stadler, and G. Sephanopoulos), Wiley-VCH Verlag GmbH, Weinheim, pp. 427–466.

Faber, K. (1996) *Biotransformations in Organic Chemistry*, Springer, Berlin.

Karkare, S.B. (1991) in *Protein Immobilization* (ed. R.F. Taylor), Marcel Dekker, New York, pp. 319–337.

Moo-Young (1988) *Bioreactors, Immobilized Enzymes and Cells*, Elsevier, London.

5.5
Biosensors

5.5.1
Enzyme Electrodes

Biosensors combine the specificity of enzymes and other biomolecules with the signal transduction of electrical and optical components. This specificity can refer to substrate, product, or effectors, such as inhibitors and activators, influencing the reaction (an analyte). Similar to the approach for measuring enzyme activities, one component, either (preferably) the product or the substrate, must be detectable by the transducer. Coupling of reactions is also used to make them amenable to this technique.

Electrodes with immobilized enzymes (Figure 5.12) are important applications of the biosensor technique. The enzyme is immobilized on the surface of an electrochemical sensor. The substrate penetrates through a selective membrane into the enzyme layer and is converted to a product. The product formed is detected potentiometrically or amperometrically. *Potentiometric sensors* measure the potential difference between the sample solution in the enzyme layer and an internal filling solution at the electrode. The potential is proportional to the logarithm of the analyte concentration. Potentiometric biosensors are ion-selective electrodes which

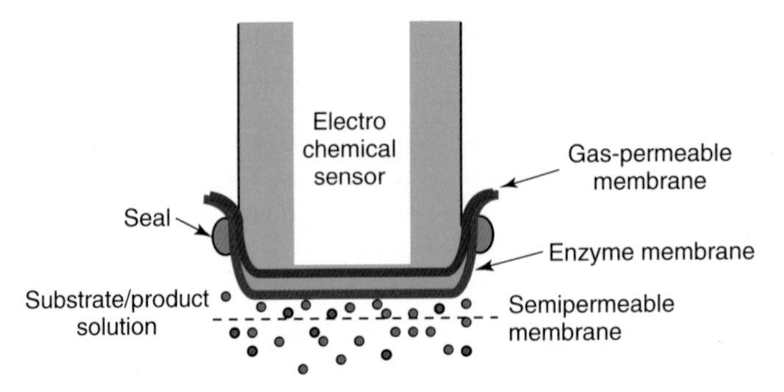

Figure 5.12 Schematic diagram of an enzyme electrode. (Adapted from Römpp Lexikon, *Biotechnology* (1992), (eds. H. Dellwey, R.D. Schmid, and W. Trommer), Thieme Verlag, Stuttgart, p. 254, with permission of Thieme Verlag Stuttgart.)

are combined with a suitable reference electrode for monitoring ions such as NH_4^+, CN^-, or S^{2-} and as the most frequently used ion-selective electrode, the H^+-sensitive electrode is used for pH measurements.

Amperometric sensors determine a current flow between the working electrode and the reference electrode; the response to the substrate concentration is linear. The polarographic oxygen electrode of Clark and Lyons (1962) was one of the first selective electrodes. Either the substrate or the product must be oxidable or reducible to react at the electrode surface. Redox mediators, such as ferricyanide, Meldola Blue, or phenazine methosulfate can be used. The electrode is sensitive in the microvolt and microampere range and time-dependent changes can be monitored.

The immobilization methods described already are also suitable for combining the enzyme with the electrode surface. The enzyme can be entrapped not only in a gel, preferentially polyacrylamide, but also in gelatin or starch; or it is covalently fixed to a matrix. Entrapment is a gentle method, but suitable only for small substrates, which are able to diffuse through the network of the gel. An important example is GOD that is used for the determination of glucose in blood. The enzyme is entrapped on the oxygen electrode because it reacts with oxygen and glucose, forming hydrogen peroxide and gluconic acid. Other examples are acetylcholine esterase and urease, linked to a pH electrode. The enzymes show a considerable lifetime of several months. Thin graphite particles and ferrocene as an electron transfer promoter are incorporated into the nonconductive polyacrylamide gel to facilitate electron transfer (Lange and Chambers, 1985). Alternatively, conductive polymers such as polypyrrole have been applied.

Frequently, covalent immobilization to supports, which in turn are fixed to the sensor surface, is used. Binding of the enzyme to nylon via lysine and a glutardialdehyde spacer is reported for various enzymes like glucose oxidase, urease, ascorbate oxidase, and lactate oxidase. Collagen membranes are also used.

Direct attachment of the enzyme to the electrode surface assures rapid response times. The enzyme can be noncovalently adsorbed with the disadvantage of instability of the binding or can be covalently bound. For covalent binding, the electrode surface must be activated by chemical or electrochemical oxidation to obtain reactive sites, such as carboxyl or phenol groups. Alternatively, the electrode surface can be silanized. The enzyme is coupled by glutardialdehyde or carbodiimide. Electrodes with directly immobilized enzymes remain active for some weeks.

For potentiometric sensors, electrodes selective for distinct ions such as F^-, I^-, S^-, $NH_4{}^+$, and H^+ are used. For example, hydrogen peroxide, formed from glucose and oxygen by glucose oxidase, oxidizes iodide to iodine. Peroxidase, urease, and catalase are also used. Amperometric sensors show high sensitivity. Platinum electrodes are frequently used in enzyme electrodes.

Generally, the biosensor contains two membranes. An outer membrane, permeable to the substrates and products, protects the enzyme layer from the surrounding solution. The direct contact with the electrode mediates the inner *permselective* membrane, which must be permeable only to the components reacting directly with the electrode, such as oxygen, but impermeable to other electrically active substances (e.g., uric acid, ascorbic acid). Therefore, several processes proceed subsequently in the biosensor: the substrate diffuses into the enzyme layer, where it is converted to product, which finally passes the inner membrane to react with the electrode surface.

For the determination of glucose with glucose oxidase (GOD)

$$\text{glucose} + O_2 \xrightarrow{\text{GOD}} \text{gluconolactone} + H_2O_2$$

usually platinum-based electrodes monitoring either cathodic oxygen depletion or anodic hydrogen peroxide production are applied. For the regeneration of $FADH_2$, mediators such as ferricyanide or ferrocene are used instead of oxygen. They react directly with the electrode and prevent inactivation of the enzyme by hydrogen peroxide.

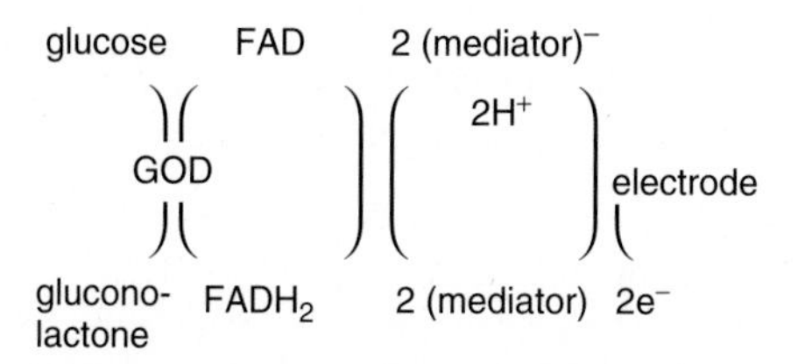

To monitor the urease reaction especially in urine and serum,

$$(NH_2)_2CO + 2H_2O + H^+ \rightarrow HCO_3{}^- + 2NH_4{}^+$$

ammonium-selective electrodes are applied. However, sodium and potassium interfere with this electrode. Therefore, a potentiometric gas-selective electrode detecting the formation of ammonia

$$NH_4{}^+ \rightarrow NH_3 + H^+$$

is superior. This electrode has a sensitivity range between 1 μM and 30 mM. In addition, pH electrodes with a thin layer of immobilized urease are applied.

The alcohol oxidase reaction

$$CH_3CH_2OH + O_2 \rightarrow CH_3CHO + H_2O_2$$

is used for the determination of ethanol and other alcohols, amperometrically pursuing the depletion of oxygen or the formation of hydrogen peroxide. The sensitivity range of alcohol electrodes, for example for determining blood alcohol, is between 0.4 and 50 mg per 100 ml.

Alcohol dehydrogenase (ADH) as well as lactate and malate dehydrogenases are fixed on the surface of a platinum electrode. The enzyme together with the cofactor NAD^+ is attached to the electrode confined by an acetylated dialysis membrane. NAD^+ is reduced by the substrate of the respective dehydrogenase to NADH, which is reoxidized at the anode:

$$\text{ethanol} + NAD^+ \xrightarrow{ADH} \text{acetaldehyde} + NADH + H^+$$

$$NADH \rightarrow NAD^+ + 2e^- + H^+$$

Combination of alcohol dehydrogenase, which forms alcohol from acetaldehyde, and alcohol oxidase (AO), which reoxidizes ethanol under consumption of oxygen, yields a very specific biosensor for the determination of ethanol by measuring the oxygen uptake:

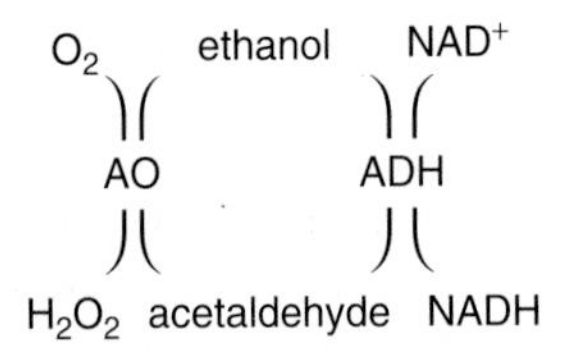

Lactate in serum or foodstuff can be measured in a broad concentration range between 1 μM and 5 mM. NADH formed by the lactate dehydrogenase is detected anodically. With the same system, pyruvate (e.g., in the cerebrospinal fluid) can also be measured. Alternatively, oxygen depletion of hydrogen peroxide production is determined with the lactate oxidase. The cytochrome-dependent LDH reacts with ferricyanide as an electron acceptor, forming ferrocyanide, which can be detected amperometrically or potentiometrically.

Other enzyme electrodes include those for detecting urate by the uricase reaction with a sensitivity range between 10 μM and 100 mM, oxalate in urine with oxalate decarboxylase, or oxalate oxidase to be detected between 0.2 and 10 mM. Creatinine in serum, a degradation product of creatine, is determined with a creatine sensor with immobilized creatinine deiminase:

$$\text{creatinine} + H_2O \rightarrow N\text{-methylhydantoin} + NH_3$$

An ammonia gas sensor with a sensitivity of 1–100 mg per 100 ml is used. For the determination of cholesterol, a pO_2 electrode with immobilized cholesterol oxidase is applied for measuring the oxygen consumption

$$\text{cholesterol} + O_2 \rightarrow \text{4-cholestene-3-one} + H_2O_2$$

in plasma samples in the range of 0.1 μM to 8 mM. Ascorbate (e.g., in fruit juices) within a concentration range of 4 μM to 0.7 mM is determined with ascorbate oxidase immobilized to an oxygen electrode. Various amino acids can be selectively determined with appropriate amino acid oxidases:

$$\text{R-CHNH}_2\text{-COOH} + H_2O + O_2 \rightarrow \text{R-CO-COOH} + NH_4^+ + H_2O_2$$

immobilized to a pO_2 electrode. Ammonia-selective electrodes are also used.

Acetylcholine and choline are determined within a sensitivity range of 1–10 μM, with platinum electrodes, with immobilized acetylcholine esterase and choline oxidase.

References

Clark, L.C. and Lyons, C. (1962) *Ann. N.Y. Acad. Sci.*, **102**, 29–49.
Koncki, R., Leszcyńsky, P., Hulanicki, A., and Głäb , S. (1992) *Anal. Chim. Acta*, **257**, 67–72.
Lange, M.A. and Chambers, J.Q. (1985) *Anal. Chim. Acta*, **175**, 89–97.
Mascini, M., Iannello, M., and Palleschi, G. (1983) *Anal. Chim Acta*, **146**, 135–148.

5.5.2 Immunoelectrodes

This technique makes use of the high affinity of the antigen to the antibody. Antibodies are bound to a membrane at the sensor surface in the membrane enzyme immunosensor. For immobilization, the antibody molecules are orientated, for example, by periodate oxidation of carbohydrate residues at the F_c region, such that only this region binds to the transducer surface, directing the F_{ab} region away from the surface or by coating the transducer surface with protein A, to which, in turn, the F_c region of the antibodies is bound.

To reuse the immunosensor, the antibody–antigen interaction must be disconnected. This is achieved with chaotropic reagents (urea, $MgCl_2$, KCl in high concentrations), but such treatments can cause partial degeneration of the biosensor with respect to response time or sensitivity, especially with indirect immunosensors (one of the immunocomponents is conjugated with an indicator molecule, for example a fluorophore, and binding is detected by fluorescence measurement).

Immunoelectrodes are valuable tools for medical diagnostics; target antigens are viruses, bacteria, and toxins. For example, immunoelectrode-specific antibodies against human chorionic gonadotropin (HCG) are bound via cyanogen bromide onto an electrode. Upon binding of the HCG antigen, the electrode potential shifts in the positive direction.

5.5.3 Other Biosensors

Besides electrodes, other types of sensors are in practical use. An important principle is based on the determination of the **reaction heat**. A semiconductor with a temperature-dependent resistance, a thermistor, detects temperature differences of 0.01 °C. Thermocouples are also applied. Often unspecific effects induce temperature changes and so the strict dependency of the signal on the reaction under study must be assured.

Optical biosensors based on the optical fiber technique are used for different applications, for example for monitoring pO_2, pCO_2, or pH changes by observing a pH indicator reaction.

The principle of **field-effect transistors** is to monitor changes in the metal isolator semiconductor interface. Two sensor regions of the *n*-type, designated as *source* and *drain* (Figure 5.13), are separated by a *p*-type metal oxide gate. Between the source and drain a voltage is applied. In the enzyme field-effect transistor (EnzFET), the metal oxide gate is replaced by an ion-sensitive layer, to which the enzyme is fixed. Ions produced or consumed in the course of the enzyme reaction induce a current in the gate and modulate the voltage between the source and drain.

Bioaffinity sensors detect direct binding. The binding effect changes the readout of the sensor and gives a direct signal of binding. For binding, antibodies can be used, as well as other binding proteins, such as avidin that reacts with biotin, concanavalin A binding to oligosaccharides, protein A reacting with IgG via the F_c region, or various receptor molecules, reacting with their respective ligand. The binding proteins are fixed to platinum- or ion-selective electrodes. Changes in the potential induced by ligand binding are monitored. The sensors can also be based on field-effect transducers, piezoelectric crystals, and optical fibers. Membrane-bound receptors, for example the acetylcholine receptor, when

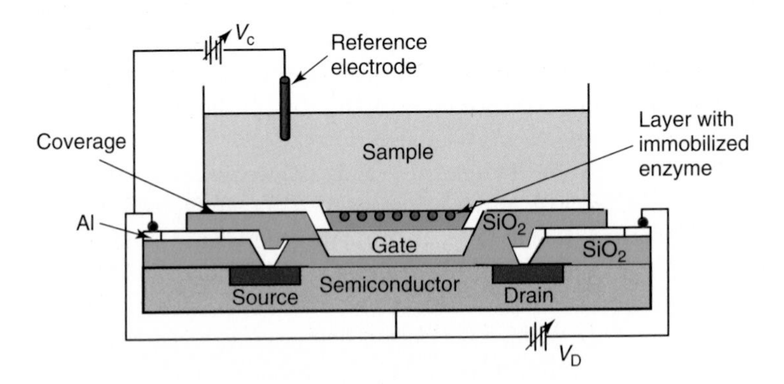

Figure 5.13 Schematic diagram of an enzyme field-effect transistor. (Adapted from Buchholz, K. and Kasche, V. (1997) *Biokatalysatoren und Enzymtechnologie*, Wiley-VCH Verlag GmbH, Weinheim.)

immobilized to the transducer, must be stabilized by integrating them into artificial membranes such as lipid films or vesicles.

References

Buchholz, K. and Kasche, V. (1997) *Biokatalysatoren und Enzymtechnologie*, Wiley-VCH Verlag GmbH, Weinheim.

Cruz, H.J. *et al.* (2002) *Parasitol. Res.*, **88**, S4–S7.

Eggins, B. (1996) *Biosensors: An Introduction*, John Wiley & Sons, Ltd, Chichester.

Guilbault, G.G., Kauffmann, J.-M., and Partriarche, G.J. (1991) in *Protein Immobilization* (ed. R.F. Taylor), Marcel Dekker, New York, pp. 209–262.

Mattiasson, B., Rehm, H.-J., and Reed, G. (eds) (1991) *Biotechnology*, 2nd edn, vol. 4, Wiley-VCH Verlag GmbH, Weinheim.

Taylor, R.F. (1991) in *Protein Immobilization* (ed. R.F. Taylor), Marcel Dekker, New York, pp. 263–303.

5.6
Immobilized Enzymes in Therapy

Immobilized enzymes are gaining increasing importance in therapy. Hereditary enzyme deficiencies are treated by supplying with microencapsulated enzymes. One of the first applications was the enzyme replacement by microencapsulated catalase in catalase-deficient mice. The application of microencapsulated enzyme has the further advantage that immunological reactions are induced. Encapsulation can be achieved with liposomes or red blood cells after hemolysis and resealing. This method is used for the treatment of Gaucher's disease with ß-glucosidase. Phenylalanine ammonia lyase in artificial cells is used for phenylketonuria in rats. Asparaginase serves for the degradation of asparagine, an amino acid essential for the growth of tumor cells. Micorencapsulated enzymes help to circumvent problems of immunogenicity and toxicity. Modification, such as deamination, acetylation, or conjugation with polyamino acids, prolongs the half life of the enzyme and decreases its immunogenicity. Artificial cells containing urease degrade urea. However, the released ammonia must be removed. This can be achieved by applying ammonia adsorbents or by incorporation into amino acids by using multienzyme systems enclosed in artificial cells comprising urease, glutamate dehydrogenase, glucose dehydrogenase (for cofactor regeneration), and transaminase. The cofactors (NADH and NADPH) are retained in the artificial cells either by an artificial lipid polymer cell membrane or by covalent cross linking to dextran. Microencapsulated islet cells secreting insulin to maintain normal glucose levels are used for treatment in *diabetes mellitus*. Artificial cells with tyrosinase contribute to reduce the tyrosine level in hepatic defects, for example liver cirrhosis.

References

Chang, T.M.S. (1977) *Biomedical Applications of Immobilized Enzymes and Proteins*, Plenum Press, New York.

Chang, T.M.S. (1991) in *Protein Immobilization* (ed. R.F. Taylor), Marcel Dekker, New York, pp. 305–318.

Chang, T.M.S. and Poznansky, M.J. (1968) *Nature*, **218**, 242–245.

Holcenberg, J.S. and Roberts, J. (1981) *Enzymes as Drugs*, John Wiley & Sons, Inc., New York.

Appendix

List of enzymes according to the EC numbers

1. **Oxidoreductases**
   - 1.1 Acting on the CH–OH group of donors
     - 1.1.1 With NAD^+ or $NADP^+$ as acceptor
       - 1.1.1.1 Alcohol dehydrogenase, alcohol:NAD^+ oxidoreductase
       - 1.1.1.27 L-Lactate dehydrogenase, (*S*)-lactate:NAD^+ oxidoreductase
       - 1.1.1.28 D-Lactate dehydrogenase, (*R*)-lactate:NAD^+ oxidoreductase
       - 1.1.1.29 Glycerate dehydrogenase, (*R*)-glycerate:NAD^+ oxidoreductase
       - 1.1.1.37 Malate dehydrogenase, (*S*)-malate:NAD^+ oxidoreductase
       - 1.1.1.38 Malate dehydrogenase (oxaloacetate-decarboxylating) "*malic enzyme,*" (*S*)-malate:NAD^+ oxidoreductase
       - 1.1.1.41 Isocitrate dehydrogenase (NAD^+), isocitrate:NAD^+ oxidoreductase
       - 1.1.1.42 Isocitrate dehydrogenase ($NADP^+$), isocitrate: $NADP^+$ oxidoreductase (decarboxylating)
       - 1.1.1.44 6-phosphogluconate dehydrogenase, 6-phospho-D-gluconate:$NADP^+$ 2-oxidoreducatse
       - 1.1.1.47 Glucose 1-dehydrogenase, β-D-glucose:$NAD(P)^+$ 1-oxidoreductase
       - 1.1.1.49 Glucose-6-phosphate 1-dehydrogenase, D-glucose-6-phosphate:$NADP^+$ 1-oxidoreductase, *Zwischenferment*
     - 1.1.2 With a cytochrome as acceptor
       - 1.1.2.3 L-Lactate dehydrogenase(cytochrome), (*S*)-lactate: ferricytochrome-*c* 2-oxidoreductase

Practical Enzymology, Second Edition. Hans Bisswanger.

1.1.3 With oxygen as acceptor
- 1.1.3.4 Glucose oxidase, β-D-glucose:oxygen 1-oxidoreductase
- 1.1.3.5 Hexose oxidase, D-hexose:oxygen 1-oxidoreductase
- 1.1.3.22 Xanthine oxidase, xanthine:oxygen oxidoreductase

1.1.4 With a disulfide as acceptor

1.1.5 With a quinone or similar compound as acceptor

1.1.99 With other acceptors

1.2 Acting on the aldehyde or oxo group of donors

1.2.1 With NAD^+ or $NADP^+$ as acceptor
- 1.2.1.2 Formate dehydrogenase, formate:NAD^+ oxidoreductase
- 1.2.1.3 Aldehyde dehydrogenase (NAD^+), aldehyde:NAD^+ oxidoreductase
- 1.2.1.12 Glyceraldehyde-3-phosphate dehydrogenase (phosphorylating), D-glyceraldehyde-3-phosphate:NAD^+ oxidoreductase

1.2.2 With a cytochrome as acceptor
- 1.2.2.2 Pyruvate dehydrogenase (cytochrome), pyruvate: ferricytochrome-b_1 oxidoreductase

1.2.3 With oxygen as acceptor
- 1.2.3.3 Pyruvate oxidase, pyruvate:oxygen 2-oxidoreductase (phosphorylating)
- 1.2.3.4 Oxalate oxidase, oxalate:oxygen oxidoreductase

1.2.4 With a disulfide as acceptor
- 1.2.4.1 Pyruvate dehydrogenase (lipoamide), pyruvate: lipoamide 2-oxidoreductase (decarboxylating and acceptor-acetylating)
- 1.2.4.2 Oxoglutarate dehydrogenase (lipoamide), 2-oxoglutarate lipoamide 2-oxidoreductase (decarboxylating and acceptor-succinylating)
- 1.2.4.4 3-Methyl-2-oxobutanoate dehydrogenase(lipoamide), branched chain α-keto acid dehydrogenase, 3-methyl-2-oxobutanoate:lipoamide 2-oxidoreductase (decarboxylating and acceptor-2-methylpropanylating)

1.2.7 With an iron–sulfur protein as acceptor
- 1.2.7.1 Pyruvate synthase, pyruvate:ferredoxin 2-oxidoreductase (CoA-acetylating)

1.2.99 With other acceptors
- 1.2.99.2 Carbon monoxide dehydrogenase, carbon monoxide:(acceptor) oxidoreductase

1.3 Acting on the CH–CH group of donors

1.3.1 With NAD^+ or $NADP^+$ as acceptor

1.3.1.6 Fumarate reductase (NADH), succinate:NAD^+ oxidoreductase
1.3.1.8 Acyl-CoA dehydrogenase ($NADP^+$), acyl-CoA: $NADP^+$ 2-oxidoreducatse
1.3.2 With a cytochrome as acceptor
1.3.3 With oxygen as acceptor
1.3.5 With a quinone or related compound as acceptor
1.3.5.1 Succinate dehydrogenase (ubiquinone), succinate:ubiquinone oxidoreductase
1.3.7 With an iron–sulfur protein as acceptor
1.3.99 With other acceptors
1.3.99.3 Acyl-CoA dehydrogenase, acyl-CoA:acceptor 2,3-oxidoreductase
1.4 Acting on the CH–NH_2 group of donors
1.4.1 With NAD^+ or $NADP^+$ as acceptor
1.4.1.1 Alanine dehydrogenase, L-alanine:NAD^+ oxidoreductase (deaminating)
1.4.1.2 Glutamate dehydrogenase, L-glutamate:NAD^+ oxidoreductase (deaminating)
1.4.1.14 Glutamate synthase (NADH), L-glutamate:NAD^+ oxidoreductase (transaminating)
1.4.2 With a cytochrome as acceptor
1.4.3 With oxygen as acceptor
1.4.3.2 L-Amino acid oxidase, L-amino acid:oxygen oxidoreductase (deaminating)
1.4.3.4 Amine oxidase (flavin-containing), amine:oxygen oxidoreductase (deaminating) (flavin-containing)
1.4.3.6 Amine oxidase (copper-containing), amine:oxygen oxidoreductase (deaminating) (copper-containing)
1.4.4 With a disulfide as acceptor
1.4.4.2 Glycine dehydrogenase, glycine:lipoylprotein oxidoreductase (decarboxylating and acceptor-aminomethylating)
1.4.7 With an iron–sulfur protein as acceptor
1.4.99 With other acceptors
1.5 Acting on the CH–NH group of donors
1.5.1 With NAD^+ or $NADP^+$ as acceptor
1.5.1.3 Dihydrofolate reductase, 5,6,7,8-tetrahydrofolate: $NADP^+$ oxidoreductase
1.5.3 With oxygen as acceptor
1.5.3.13 Polyamine oxidase, *N1*-acetylspermidine:oxygen oxidoreductase
1.5.4 With a disulfide as acceptor
1.5.5 With a quinone or similar compound as acceptor
1.5.99 With other acceptors

- 1.6 Acting on NADH or NADPH
  - 1.6.1 With NAD^+ or $NADP^+$ as acceptor
  - 1.6.2 With a heme protein as acceptor
    - 1.6.2.2 Cytochrome-b_5 reductase, NADH-ferricytochrome-b_5 oxidoreductase
  - 1.6.4 With a disulfide as acceptor
    - 1.6.4.2 Glutathion reductase (NADPH), NADPH:oxidized-glutathion oxidoreductase
    - 1.6.4.5 Thioredoxin reductase (NADPH), NADPH:oxidized thioredoxin oxidoreductase
  - 1.6.5 With a quinone or similar compound as acceptor
    - 1.6.5.3 NADH dehydrogenase (ubiquinone), NADH: ubiquinone oxidoreductase
  - 1.6.6 With a nitrogenous group as acceptor
    - 1.6.6.1 Nitrate reductase (NADH), NADH:nitrate oxidoreductase
  - 1.6.8 With a flavin as acceptor
  - 1.6.99 With other acceptors
    - 1.6.99.3 NADH dehydrogenase, NADH:(acceptor) oxidoreductase
- 1.7 Acting on other nitrogenous compounds as donors
  - 1.7.2 With a cytochrome as acceptor
  - 1.7.3 With oxygen as acceptor
  - 1.7.7 With an iron–sulfur protein as acceptor
  - 1.7.5 With a quinone or similar compound as acceptor
  - 1.7.99 With other acceptors
    - 1.7.99.3 Nitrite reductase, nitric oxide:(acceptor) oxidoreductase
    - 1.7.99.4 Nitrate reductase, nitrite:(acceptor) oxidoreductase
- 1.8 Acting on a sulfur group of donors
  - 1.8.1 With NAD^+ or $NADP^+$ as acceptor
    - 1.8.1.4 Dihydrolipoamide dehydrogenase, diaphorase, dihydrolipoamide:NAD^+ oxidoreductase
    - 1.8.1.7 Glutathione disulfide reductase, glutathione:$NADP^+$ oxidoreductase
    - 1.8.1.9 Thioredoxin disulfide reductase, thioredoxin: $NADP^+$ oxidoreductase
  - 1.8.2 With a cytochrome as acceptor
    - 1.8.2.1 Sulfite reductase, sulfite:ferricytochrome-*c* oxidoreductase
  - 1.8.3 With oxygen as acceptor
    - 1.8.3.3 Glutathione oxidase, glutathione:oxygen oxidoreductase
  - 1.8.4 With a disulfide as acceptor

1.8.4.2 Protein disulfide reductase (glutathione), glutathione:protein disulfide oxidoreductase
1.8.5 With a quinone or similar compound as acceptor
1.8.7 With an iron–sulfur protein as acceptor
1.8.99 With other acceptors
1.9 Acting on a heme group of donors
1.9.3 With oxygen as acceptor
1.9.6 With a nitrogenous group as acceptor
1.9.99 With other acceptors
1.10 Acting on diphenols and related substances as donors
1.10.1 With NAD^+ or $NADP^+$ as acceptor
1.10.2 With a heme protein as acceptor
1.10.3 With oxygen as acceptor
1.10.3.2 Laccase, benzenediol:oxygen oxidoreductase
1.10.3.3 L-Ascorbate oxidase, L-ascorbate:oxygen oxidoreductase
1.10.99 With other acceptors
1.11 Acting on a peroxide as acceptor
1.11.1.1 NADH peroxidase, NADH:hydrogen peroxide oxidoreductase
1.11.1.6 Catalase, hydrogen peroxide:hydrogen peroxide oxidoreductase
1.11.1.7 Peroxidase, donor:hydrogen peroxide oxidoreductase
1.11.1.9 Glutathione peroxidase, glutathione:hydrogen peroxide oxidoreductase
1.12 Acting on hydrogen as donor
1.12.1 With NAD^+ or $NADP^+$ as acceptor
1.12.2 With a cytochrome as acceptor
1.12.99 With other acceptors
1.13 Acting on single donors with incorporation of molecular oxygen (oxygenases)
1.13.11 With incorporation of two atoms of oxygen
1.13.11.1 Catechol 1,2-dioxygenase, catechol:oxygen 1,2-oxidoreductase
1.13.11.12 Lipoxygenase, linoleate:oxygen 13-oxidoreductase
1.13.12 With incorporation of one atom of oxygen (interal monooxygenases or internal mixed function oxygenases)
1.13.12.4 Lactate 2-monooxygenase, (*S*)-lactate:oxygen 2-oxidoreductase
1.13.99 Miscellaneous
1.14 Acting on paired donors with incorporation of molecular oxygen
1.14.11 With 2-oxoglutarate as one donor, and incorporation of one atom each of oxygen into both donors

1.14.12 With NADH or NADPH as one donor, and incorporation of two atoms of oxygen into one donor

1.14.12.1 Anthranilate 1,2-dioxygenase (deaminating, decaroxylating) anthranilate, NAD(P)H:oxygen oxidoreductase (1,2-hydroxylating, deaminating, decarboxylating)

1.14.13 With NADH or NADPH as one donor, and incorporation of one atom of oxygen

1.1.4.13.17 Cholesterol 7α-monooxygenase, cholesterol, NADPH:oxygen oxidoreductase (7α-hydroxylating)

1.14.14 With reduced flavin or flavoprotein as one donor, and incorporation of one atom of oxygen

1.14.14.1 Unspecific monooxygenase, microsomal P-450, substrate, reduced-flavoprotein:oxygen oxidoreductase (RH-hydroxylating or -epoxidizing)

1.14.15 With a reduced iron–sulfur protein as one donor, and incorporation of one atom of oxygen

1.14.15.4 Steroid 11β-monooxygenase, steriod, reduced-adrenal-ferredoxin:oxygen oxidoreductase (11β-hydroxylating)

1.14.16 With reduced pteridine as one donor, and incorporation of one atom of oxygen

1.14.16.4 Tryptophan 5-monoooxygenase, L-tryptophan, tetrahydro-biopterin:oxygen oxidoreductase (5-hydroxylating)

1.14.17 With ascorbate as one donor, and incorporation of one atom of oxygen

1.14.17.1 Dopamine 11β-monooxygenase, 3,4-dihydroxy-phenethylamine, ascorbate:oxygen oxidoreductase (β-hydroxylating)

1.14.99 Miscellaneous

1.14.99.5 Stearoyl-CoA desaturase, stearoyl-CoA, hydrogenase-donor:oxygen oxidoreductase

1.15 Acting on superoxide radicals as acceptor

1.15.1.1 Superoxide dismutase, superoxide:superoxide oxidoreductase

1.16 Oxidizing metal ions

1.16.1 With NAD^+ or $NADP^+$ as acceptor

1.16.1.1 Mercury(II) reductase, Hg:$NADP^+$ oxidoreductase

1.16.3 With oxygen as acceptor

1.16.3.1 Ferroxidase, Fe(II):oxygen oxidoreductase

1.17 Acting on CH_2 groups

1.17.1 With NAD^+ or $NADP^+$ as acceptor

1.17.3 With oxygen as acceptor

- 1.17.4 With a disulfide as acceptor
- 1.17.99 With other acceptors
- 1.18 Acting on reduced ferredoxin as donor
  - 1.18.1 With NAD^+ or $NADP^+$ as acceptor
    - 1.18.1.2 Ferredoxin-$NADP^+$ reductase, ferredoxin-$NADP^+$ oxidoreductase
  - 1.18.6 With dinitrogen as acceptor
    - 1.18.6.1 Nitrogenase, reduced ferredoxin:dinitrogen oxidoreductase (ATP-hydrolyzing)
  - 1.18.99 With H^+ as acceptor
    - 1.18.99.1 Hydrogenase, ferredoxin:H^+ oxidoreductase
- 1.19 Acting on reduced flavodoxin as donor
  - 1.19.6 With dinitrogen as acceptor
    - 1.19.6.1 Nitrogenase (flavodoxin), reduced flavodoxin: dinitrogen oxidoreductase (ATP-hydrolyzing)
- 1.97 Other oxidoreductases
    - 1.97.1.3 Sulfur reductase, (donor):sulfur oxidoreductase

2. Transferases
- 2.1 Transferring one-carbon groups
  - 2.1.1 Methyltransferases
    - 2.1.1.23 Protein-arginine *N*-methyltransferase, protein methylase I, *S*-adenosyl-L-methionine:protein-L-arginine *N*-methyltransferase
    - 2.1.1.29 *t*-RNA (cytosine-5-)-methyltransferase, *S*-adenosyl-L-methionine:tRNA (cytosine-5-)-methyltransferase
  - 2.1.2 Hydroxymethyl-, formyl-, and related transferases
    - 2.1.2.1 Glycine hydroxymethyltransferase, 5,10-methylenetetrahydro-folate:glycine hydroxymethyltransferase
  - 2.1.3 Carboxyl- and carbamoyltransferases
    - 2.1.3.2 Aspartate carbamoyltransferase, carbamoylphosphate:L-aspartate carbamoyltransferase
  - 2.1.4 Amindinotransferases
    - 2.1.4.1 Glycine amindinotransferase, L-arginine:glycine amidinotransferase
- 2.2 Transferring aldehyde and ketone residues
    - 2.2.1.1 Transketolase, seduheptulose-7-phosphate:D-glyceraldehyde-3-phosphate glycoaldehydetransferase
    - 2.2.1.2 Transaldolase, seduheptulose-7-phosphate:D-glyceraldehyde-3-phosphate glyceronetransferase
- 2.3 Acyltransferases
  - 2.3.1 Acyltransferases
    - 2.3.1.8 Phosphate acetyltransferase (phosphotransacetylase), acetyl-CoA:orthophosphate acetyltransferase

2.3.1.12 Dihydrolipoamide *S*-acetyltransferase, acetyl-CoA: dihydrolipoamide *S*-acetyltransferase
2.3.1.61 Dihydrolipoamide *S*-succinyltransferase, succinyl-CoA:dihydrolipoamide *S*-succinyltransferase
2.3.1.85 Fatty acid synthase, acyl-CoA:malonyl-CoA *C*-acyltransferase (decarboxylating, oxoacyl- and enoyl-reducing and thioester-hydrolyzing)
2.3.1.85 Fatty acyl-CoA synthase (yeast fatty acid synthase), acyl-CoA:malonyl-CoA *C*-acyltransferase (decarboxylating, oxoacyl- and enoyl-reducing)
2.3.2 Aminoacyltransferases
2.3.2.6 Leucyltransferase, L-leucyl-tRNA:protein leucyltransferase
2.4 Glycosyltransferases
2.4.1 Hexosyltransferases
2.4.1.1 Phosphorylase, 1,4-α-D-glucan:orthophosphate α-D-glucosyltransferase
2.4.1.11 Glycogen(starch) synthase, UDPglucose:glycogen 4-α-D-glucosyltransferase
2.4.1.18 1,4-α-Glucan branching enzyme, 1,4-α-D-glucan:1,4-α-D-glucan 6-α-D-(1,4-α-D-glucano)-transferase
2.4.2 Pentosyltransferases
2.4.2.1 Purine nucleoside phosphorylase, pyrimidine nucleoside:orthophosphate α-D-ribosyltransferase
2.4.2.2 Pyrimidine nucleoside phosphorylase, pyrimidine nucleoside:orthophosphate α-D-ribosyltransferase
2.4.99 Transferring other glycosyl groups
2.5 Transferring alkyl or aryl groups, other than methyl groups
2.5.1.6 Methionine adenosyltransferase, ATP:L-methionine *S*-adenosyl-transferase
2.6 Transferring nitrogenous groups
2.6.1 Transaminases
2.6.1.2 Alanine transaminase, L-alanine:2-oxoglutarate aminotransferase
2.6.3 Oximinotransferases
2.6.99 Transferring other nitrogenous groups
2.7 Transferring phosphorus-containing groups
2.7.1 Phosphotransferases with an alcohol group as acceptor
2.7.1.1 Hexokinase, ATP:D-hexose 6-phosphotransferase
2.7.1.2 Glucokinase, ATP:D-glucose 6-phosphotransferase
2.7.1.11 6-Phosphofructokinase, ATP:D-fructose-6-phosphate 1-phosphotransferase
2.7.1.37 Protein kinase, ATP:protein phosphotransferase
2.7.1.38 Phosphorylase kinase, ATP:D-phosphorylase-6 phosphotransferase

2.7.1.40 Pyruvate kinase, ATP:pyruvate 2-*O*-phosphotransferase

2.7.2 Phosphotransferases with a carboxyl group as acceptor

2.7.2.1 Acetate kinase, ATP:acetate phosphotransferase

2.7.3 Phosphotransferases with a nitrogenous group as acceptor

2.7.3.2 Creatine kinase, ATP:creatine *N*-phosphotransferase

2.7.4 Phosphotransferases with a phosphate group as acceptor

2.7.4.3 Adenylate kinase (myokinase), ATP:AMP phosphotransferase

2.7.4.6 Nucleoside diphosphate kinase, ATP:nucleoside diphosphate phosphotransferase

2.7.6 Diphosphotransferases

2.7.6.1 Ribose phosphate pyrophosphokinase, ATP:D-ribose-5-phosphate pyrophosphotransferase

2.7.7 Nucleotydyltransferases

2.7.7.2 FMN adenylyltransferase, ATP:FMN adenylyltransferase

2.7.7.48 RNA-directed RNA polymerase, nucleoside triphosphate:RNA nucleotidyltransferase (RNA-directed)

2.7.7.49 RNA-directed DNA polymerase, deoxynucleoside triphosphate:DNA deoxynucleotidyltransferase (RNA-directed)

2.7.8 Transferases for other substituted phosphate groups

2.7.9 Phosphotransferases with paired acceptors

2.8 Transferring sulfur-containing groups

2.8.1 Sulfurtransferases

2.8.2 Sulfotransferases

2.8.3 CoA transferases

2.8.3.3 Malonate CoA transferase, acetyl-CoA:malonate CoA transferase

3. Hydrolases

3.1 Acting on ester bonds

3.1.1 Carboxylic ester hydrolases

3.1.1.3 Triacylglycerol lipase (lipase), triacylglycerol acylhydrolase

3.1.1.4 Phospholipase A_2, phosphatidylcholine 2-acylhydrolase

3.1.1.7 Acetylcholine esterase, acetylcholine acetylhydrolase

3.1.1.8 Choline esterase, acylcholine acylhydrolase

3.1.2 Thiolester hydrolases

3.1.2.2 Palmitoyl-CoA hydrolase

3.1.3 Phosphoric monoester hydrolases

3.1.3.1 Alkaline phosphatase, orthophosphoric-monoester phosphohydrolase (alkaline optimum)

3.1.3.2 Acid phosphatase, orthophosphoric-monoester phosphohydrolase (acid optimum)

3.1.4 Phosphoric diester hydrolases

3.1.4.1 Phosphodiesterase I, oligonucleate 5′-nucleotidohydrolase

3.1.4.3 Phospholipase C, phosphatidylcholine cholinephosphohydrolase

3.1.5 Triphosphoric monoester hydrolases

3.1.6 Sulfuric ester hydrolases

3.1.6.1 Arylsulfatase, aryl sulfate sulfohydrolase

3.1.7 Diphosphoric monoester hydrolases

3.1.8 Phosphoric triester hydrolases

3.1.11 Exodeoxyribonucleases producing 5′-phosphomonoesters

3.1.11.1 Exodeoxyribonuclease I

3.1.13 Exoribonucleases producing 5′-phosphomonoesters

3.1.11.1 Exoribonuclease II

3.1.14 Exoribonucleases producing other than 5′-phosphomonoesters

3.1.15 Exonucleases active with either ribo- or deoxyribonucleic acids and producing 5′-phosphomonoesters

3.1.16 Exonucleases active with either ribo- or deoxyribonucleic acids and producing other than 5′-phosphomonoesters

3.1.21 Endodeoxyribonucleases producing 5′-phosphomonoesters

3.1.21.1 Deoxyribonuclease I

3.1.22 Endodeoxyribonucleases producing other than 5′-phosphomonoesters

3.1.22.1 Deoxyribonuclease II

3.1.25 Site-specific endodeoxyribonucleases specific for altered bases

3.1.26 Endoribonucleases producing 5′-phosphomonoesters

3.1.27 Endoribonucleases producing other than 5′-phosphomonoesters

3.1.30 Endonucleases active with either ribo- or deoxyribonucleic acids and producing 5′-phosphomonoesters

3.1.31 Endonucleases active with either ribo- or deoxyribonucleic acids and producing other than 5′-phosphomonoesters

3.2 Glycosidases

3.2.1 Hydrolyzing *O*-glycosyl compounds

3.2.1.1 α-Amylase, 1,4-α-D-glucan glucanohydrolase

3.2.1.2 β-Amylase, 1,4-α-D-glucan maltohydrolase

3.2.1.4 Cellulase, 1,4-(1,3;1,4)-β-D-glucan 4-glucanohydrolase

3.2.1.17 Lysozyme, peptidoglucan *N*-acetylmuramoylhydrolase

3.2.1.20 α-Glucosidase, α-D-glucoside glucohydrolase

3.2.1.23 β-Galactosidase, β-D-galactoside galactohydrolase

3.2.1.26 β-D-Fructosefuranoside fructohydrolase, β-fructofuranosidase, invertase, invertin, saccharase, sucrase
3.2.1.108 Lactase, lactose galactohydrolase
3.2.2 Hydrolyzing *N*-glycosyl compounds
3.2.2.1 Purine nucleosidase, *N*-D-ribosylpurine ribohydrolase
3.2.3 Hydrolyzing *S*-glycosyl compounds
3.2.3.1 Thioglucosidase, thioglucoside glucohydrolase
3.3 Acting on ether bonds
3.3.1 Thioether hydrolases
3.3.2 Ether hydrolases
3.3.2.3 Epoxide hydrolase
3.4 Acting on peptide bonds (peptidases)
3.4.11 Aminopeptidases
3.4.11.1 Leucyl aminopeptidase
3.4.13 Dipeptidases
3.4.13.6 Cys-Gly dipeptidase
3.4.14 Dipeptidyl peptidases and tripeptidyl peptidases
3.4.15 Peptidyl dipeptidases
3.4.16 Serine-type carboxypeptidases
3.4.16.1 Serine-type carboxypeptidase
3.4.17 Metallocarboxypeptidases
3.4.17.1 Carboxypeptidase A
3.4.18 Cysteine-type carboxypeptidases
3.4.18.1 Cysteine-type carboxypeptidase
3.4.19 Omegapeptidases
3.4.19.1 Acylaminoacyl peptidase
3.4.21 Serine endopeptidases
3.4.21.1 Chymotrypsin
3.4.21.4 Trypsin
3.4.21.5 Thrombin
3.4.21.62 Subtilisin
3.4.22 Cysteine endopeptidases
3.4.22.1 Cathepsin B
3.4.22.2 Papain
3.4.22.17 Calpain
3.4.23 Aspartic endopeptidases
3.4.23.1 Pepsin A
3.4.22.15 Renin
3.4.24 Metalloendopeptidases
3.4.24.3 Microbial collagenase
3.4.24.27 Thermolysin
3.4.99 Endopeptidases of unknown catalytic mechanism
3.4.99.46 Multicatalytic endopeptidase complex (proteasome)

- 3.5 Acting on carbon–nitrogen bonds, other than peptide bonds
  - 3.5.1 In linear amide
    - 3.5.1.1 Asparaginase, L-asparagine amidohydrolase
    - 3.5.1.4 Amidase, acylase; acylamine amidohydrolase
    - 3.5.1.5 Urease, urea amidohydrolase
    - 3.5.1.35 D-Glutaminase, D-glutamine amidohydrolase
  - 3.5.2 In cyclic amides
    - 3.5.2.3 Dihydroorotase, carbamoylaspartic dehydrase
  - 3.5.3 In linear amidines
    - 3.5.3.1 Arginase, L-arginine amidinohydrolase
  - 3.5.4 In cyclic amidines
    - 3.5.4.1 Cytosine desaminase, cytosine aminohydrolase
  - 3.5.5 In nitriles
    - 3.5.5.1 Nitrilase, nitrile aminohydrolase
  - 3.5.99 In other compounds
    - 3.5.99.1 Riboflavinase, riboflavin hydrolase
- 3.6 Acting on acid anhydrides
  - 3.6.1 In phosphorus-containing anhydrides
    - 3.6.1.5 Apyrase, ATP diphosphohydrolase
  - 3.6.2 In sulfonyl-containing anhydrides
    - 3.6.2.1 Adenylylsufatase, adenylylsulfate sulfohydrolase
- 3.7 Acting on carbon–carbon bonds
  - 3.7.1 In ketonic substances
    - 3.7.1.1 Oxalacetase, oxalacetate acetylhydrolase
- 3.8 Acting on halide bonds
  - 3.8.1 In C-halide compounds
    - 3.8.1.1 Alkylhalidase, alkyl halide halidohydrolase
- 3.9 Acting on phosphorus–nitrogen bonds
    - 3.9.1.1 Phosphoamidase, phosphamide hydrolase
- 3.10 Acting on sulfur–nitrogen bonds
    - 3.10.1.2 Cyclamate sulfohydrolase, cyclohexylsulfamate sulfohydrolase
- 3.11 Acting on carbon–phosphorus bonds
    - 3.11.1.1 Phosphonoacetaldehyde hydrolase, 2-oxoethylphosphonate phosphonohydrolase
- 3.12 Acting on sulfur–sulfur bonds
    - 3.12.1.1 Trithionate hydrolase, trithionate thiosulfohydrolase

4. Lyases
  - 4.1 Carbon–carbon lyases
    - 4.1.1 Carboxy-lyases
      - 4.1.1.1 Pyruvate decarboxylase, 2-oxo-acid carboxy-lyase
      - 4.1.1.17 Ornithine decarboxylase, L-ornithine carboxy-lyase
    - 4.1.2 Aldehyde lyases

4.1.2.5 Threonine aldolase, L-threonine acetaldehyde-lyase
4.1.2.9 Phosphoketolase, D-xylulose-5-phosphate D-glyceraldehyde-3-phosphate-lyase (phosphate-acetylating)

4.1.3 Oxo-acid lyases
4.1.3.1 Isocitrate lyase, isocitrate glyoxylate-lyase
4.1.3.2 Malate synthase, L-malate glyoxylate-lyase
4.1.3.7 Citrate (*si*) synthase (condensing enzyme), citrate oxaloacetate-lyase
4.1.3.27 Anthranilate synthase, chorismate pyruvate-lyase (amino-accepting)

4.1.99 Other carbon–carbon-lyases
4.1.99.1 Tryptophanase, L-tryptophan indole-lyase

4.2 Carbon–oxygen lyases

4.2.1 Hydrolyases
4.2.1.1 Carbonate dehydratase, carbonic anhydrase, carbonate hydrolyase
4.2.1.2 Fumarate hydratase, fumarase, (*S*)-malate hydrolyase
4.2.1.3 Aconitate hydratase, aconitase, citrate(isocitrate) hydrolyase
4.2.1.20 Tryptophan synthase, L-serine hydrolyase (adding indole glycerol phosphate)

4.2.2 Acting on polysaccharides
4.2.2.1 Hyaluronate lyase, hyaluronidase, hyaluronate lyase

4.2.99 Other carbon–oxygen lyases
4.2.99.2 Threonine synthase, *O*-phospho-L-homoserine phospho-lyase (adding water)

4.3 Carbon–nitrogen lyases

4.3.1 Ammonia-lyases
4.3.1.1 Aspartate ammonia-lyase, aspartase, L-aspartate ammonia-lyase

4.3.2 Amindine-lyases
4.3.2.1 Argininosuccinate lyase, argininosuccinase, *N*-(L-argininosuccinate) arginine-lyase

4.3.3 Amide lyases

4.3.99 Other carbon–nitrogen lyases
4.3.99.1 Cyanate lyase, cyanate C–N lyase

4.4 Carbon–sulfur lyases
4.4.1.2 Homocysteine desulfhydrase, L-homocysteine hydrogen-lyase (deaminating)

4.5 Carbon–halide lyases
4.5.1.1 DTT-dehydrochlorinase, 1,1,1-dichloro-2,2-bis (4-chlorophenyl) ethane chloride-lyase

4.6 Phosphorus–oxygen lyases
4.6.1.1 Adenylate cyclase, ASTP-pyrophosphate-lyase

4.99 Other lyases

4.99.1.1 Ferrochelatase, protoheme ferro-lyase

5. Isomerases

5.1 Racemases and epimerases

5.1.1 Acting on amino acids and derivatives

5.1.1.1 Alanine racemase

5.1.2 Acting on hydroxy acids and derivatives

5.1.2.1 Lactate racemase

5.1.3 Acting on carbohydrates and derivatives

5.1.3.1 Ribulose-phosphate 3-epimerase, D-ribulose-5-phosphate 3-epimerase

5.1.99 Acting on other compounds

5.1.99.1 Methylmalonyl-CoA epimerase, 2-methyl-3-oxopropanoyl-CoA 2-epimerase

5.2 *cis-trans*-Isomerases

5.2.1.3 Retinal isomerase, *all-trans*-retinal 11-*cis-trans*-isomerase

5.3 Intramolecular oxidoreductases

5.3.1 Interconverting aldoses and ketoses

5.3.1.1 Triosephosphate isomerase, D-glyceraldehyde-3-phosphate ketol isomerase

5.3.1.5 Xylose isomerase, D-xylose ketol isomerase

5.3.2 Interconverting keto- and enol groups

5.3.2.1 Phenylpyruvate tautomerase, phenylpyruvate *keto-enol* isomerase

5.3.3 Transposing C=C bonds

5.3.3.1 Steroid Δ-isomerase, 3-oxosteroid Δ5-Δ4-isomerase

5.3.4 Transposing S–S bonds

5.3.4.1 Protein disulfide isomerase

5.3.99 Other intramolecular oxidoreductases

5.3.99.7 Styrene oxide isomerase (epoxide-cleaving)

5.4 Intramolecular transferases (mutases)

5.4.1 Transferring acyl groups

5.4.1.1 Lysolecithin acylmutase

5.4.2 Phosphotransferases (phosphomutases)

5.4.2.2 Phosphoglucomutase, α-D-glucose 1,6-phosphomutase

5.4.3 Transferring amino groups

5.4.3.2 Lysine 2,3-aminomutase, L-lysine 2,3-aminomutase

5.4.99 Transferring other groups

5.4.99.5 Chorismate mutase, chorismate pyruvate mutase

5.5 Intramolecular lyases

5.5.1.1 Muconate cycloisomerase, 2,5-dihydro-5-oxofuran-2-acetate lyase (decyclizing)

5.99 Other isomerases
- 5.99.1.2 DNA topoisomerase

6. Ligases
   - 6.1 Forming carbon–oxygen bonds
     - 6.1.1 Ligases forming aminoacyl-tRNA and related compounds
       - 6.1.1.1 Tyrosine-tRNA ligase, tyrosyl-tRNA synthetase
   - 6.2 Forming carbon–sulfur bonds
     - 6.2.1 Acid-thiol ligases
       - 6.2.1.1 Acetate-CoA ligase, acetate:CoA ligase (AMP-forming)
   - 6.3 Forming carbon–nitrogen bonds
     - 6.3.1 Acid-ammonia (or amine) ligases (amide synthases)
       - 6.3.1.2 Glutamate-ammonia ligase, glutamine synthetase, L-glutamate:ammonia ligase (ADP-forming)
     - 6.3.2 Acid-amino acid ligases (peptide synthases)
       - 6.3.2.3 Glutathion synthase, γ-L-glutamyl-L-cysteine:glycine ligase (ADP-forming)
     - 6.3.3 Cyclo ligases
     - 6.3.4 Other carbon–nitrogen ligases
       - 6.3.4.1 GMP synthase, xanthosine-5′-phosphate:ammonia ligase (AMP-forming)
     - 6.3.5 Carbon–nitrogen ligases with glutamine as amido-*N*-donor
       - 6.3.5.2 GMP synthase (glutamine-hydrolyzing)
   - 6.4 Forming carbon–carbon bonds
       - 6.4.1.1 Pyruvate carboxylase, pyruvate:carbon dioxide ligase (ADP-forming)
       - 6.4.1.2 Acetyl-CoA carboxylase, acetyl-CoA:carbon dioxide ligase (ADP-forming)
   - 6.5 Forming phosphoric ester bonds
       - 6.5.1.1 DNA ligase (ATP), poly(deoxyribonucleotide):poly (deoxyribonucleotide) ligase (AMP-forming)

Index

Practical Enzymology, Second Edition. Hans Bisswanger.